Flexible and Stretchable Electronics

Flexible and Stretchable Electronics

Materials, Designs, and Devices

edited by

Run-Wei Li

Gang Liu

JENNY STANFORD
PUBLISHING

Published by

Jenny Stanford Publishing Pte. Ltd.
Level 34, Centennial Tower
3 Temasek Avenue
Singapore 039190

Email: editorial@jennystanford.com
Web: www.jennystanford.com

British Library Cataloguing-in-Publication Data
A catalogue record for this book is available from the British Library.

Flexible and Stretchable Electronics:
Materials, Designs, and Devices

ISBN 978-981-4800-46-4 (Hardcover)
ISBN 978-0-429-05890-5 (eBook)

Contents

Preface

The last decade has witnessed evolutional development of flexible electronics and they are reshaping the way we live, work, and communicate with the world. Setting themselves apart from the traditional silicon-based semiconductor devices, the unique capability of bending and stretching, together with the merits of light-weight and low-cost fabrications, endows flexible electronics imaginative versatility for soft circuits in portable energy and industrial equipments, wearable consumer digital gadgets, implantable health-monitoring electronic skins, medicine robots and prosthetics, and even the fictional human (brain)-machine docking that appears only in sci-fi movies. As an important technological advance and one of the most promising information technologies, flexible electronics have received broad and intensive attention from both the academic and industrial community ever since their birth twenty years ago.

The fabrication of flexible electronics requires multiscale manufacture on soft substrates, spanning from the nanofeatures, microstructures, to large-area macro-integration. What determines their performance characteristics and, in principle, enables a host of potential applications is the accurate control of multiple functional materials at their interfaces, such as organics, metals, and ceramics, which show completely different mechanical and electrical properties, and multidisciplinary investigations in the area of chemistry, physics, materials science, biotechnology, and electronic engineering. In this book, we provide a timely and systematic overview of the operating mechanisms, materials selection, device structure and circuit design, fabrication methods, and range of applications of flexible electronics, with the aim to deliver a comprehensive summary and present the prospects of the specific field. The materials used for flexible electronics range from those that carry intrinsic molecular and structural flexibility, namely, organic semiconductors, organic–inorganic hybrids, two-dimensional graphene derivatives and analogs, and layered metal oxides, to room-temperature liquid metals that demonstrate the fluidity of common liquid and electrical conductivity of metals. Fundamental circuit elements of resistors, capacitors, and transistors are then produced with these materials to construct sensors, memory and energy storage devices, and circuits.

This book is a result of the great efforts of our contributing authors, who have covered major areas of flexible electronics. In Chapter 1, Drs. Jung-Yao Chen and Cheng-Liang Liu discuss the materials, charge-transport model, deposition methods, and integration of organic field-effect transistor (OFET) for photo detecting. In Chapter 2, Dr. Bing Cao mainly presents an overview on the device structure, operating principle, and the latest advancement of new materials for high-performance organic solar cells (OSCs). Dr. Yimao Cai and coworkers have compiled Chapter 3, which focuses on the use of CMOS-compatible poly(para-xylylene) material in metal–oxide semiconductor field-effect transistors (MOSFETs), resistive random access memories (RRAMs), and sensor applications. In Chapter 4, Dr. Xiaohui Yi and coauthors offer a deep understanding of the designing strategy for flexible RRAM devices, followed by particular emphasis on organic and inorganic–organic hybrid resistive switching materials and devices. Chapters 5 and 6, contributed by Drs. Zhuang's and Shen's research groups, respectively, review the device structure, working principle, materials selection, fabrication approach, and integration of in-plane microcapacitor as on-chip power supply in stretchable electronic skin systems. Drs. Tie Li and Ting Zhang and their coworkers identify the special characters and requirements for flexible and stretchable sensors and summarize the strategies of structural design and functional materials for detecting human information in Chapter 7. Beyond the electroactive materials employed in the above-mentioned device applications, Drs. Xuelin Wang and Jing Liu present the use of liquid metal as soft conductor and electrode in flexible and stretchable electronics in Chapter 8. After an overview of the performance benchmarks for the different materials and devices of flexible electronics in Chapter 9, Drs. Wei Yuan and Zheng Cui introduce the manufacturing processes, inks, post-treatments, and important applications of the high-throughput, cost-effective, and eco-friendly printing technology for producing wearable electronics, portable energy harvesting and storage devices, flexible display panels, and human-skin sensors. Finally, the mechanical design concept, theory, simulation, and applications of smart flexible structures for optimizing the overall performance of stretchable electronics are discussed by Drs. Guoyong Mao and Shaoxing Qu in Chapter 10

As we enter the 21st century Big Data and Artificial Intelligence era, information technology has changed rapidly over the past few years and is mainly being driven by the new practical requirements and application scenarios that we had never expected before. Looking back into the 1980s, none of the prophets could have imagined that a palm-sized telephone will accomplish complex tasks that were being done separately by clumpy desktop computers, brick-like walkie-talkies, mechanical film cameras, etc. Thanks to the fourth technological revolution, the "mission impossible" is becoming "impossible is nothing" and flexible electronics show great promise for the future as they will closely interconnect every human being, every object, and every circumstance. As this is a newborn area, the theoretical framework of flexible electronics is still incomplete and global efforts are needed to facilitate their transition from research and laboratory prototypes to commercial products. This book examines the research progress achieved in this field around the world, and through it we hope to inspire interests to advance flexible electronics technology, especially colleagues and university students.

Run-Wei Li

Gang Liu

Autumn 2019

Chapter 1

Organic Field-Effect Transistors for Flexible Electronics Application

Jung-Yao Chen[a] and Cheng-Liang Liu[b]
[a]*Department of Chemical Engineering, National Chung Cheng University, Chiayi 62102, Taiwan*
[b]*Department of Chemical and Materials Engineering, National Central University, Taoyuan 32001, Taiwan*
chmjyc@ccu.edu.tw; clliu@ncu.edu.tw

During the past decades, with the advent of novel organic semiconductors (such as conjugated polymers and small molecules) and low-cost process requirement, significant interest and progress in the field of organic electronics have been reported. In particular, organic field-effect transistors (OFETs) have become the backbone of the recent developments in the large-area flexible electronics and found usage in numerous applications such as the digital and analog circuits, memories, photovoltaic, and sensors. This chapter gives the present advancements in OFET devices based on an overview of the individual layers of OFETs, charge transport model, deposition methods, and their integration of flexible substrate. Besides, their applications in organic phototransistors based on the OFET architecture are briefly addressed at the end of the chapter.

Flexible and Stretchable Electronics: Materials, Designs, and Devices
Edited by Run-Wei Li and Gang Liu

ISBN 978-981-4800-46-4 (Hardcover), 978-0-429-05890-5 (eBook)
www.jennystanford.com

1.1 Introduction

Organic electronics have been a field of broad academic and industrial interest for the past few decades [1–3]. A wide range of organic semiconductors are currently being explored steadily for organic electronics primarily due to the growing need for substituting Si-based technology with some organic-based materials offering relatively simple and commercial products and applications. Organic devices demonstrate numerous advantages such as light weight, low cost, mechanical and molecular flexibility, and low-temperature processing allows cost-effective production, which eventually leads to huge benefits on various fronts, such as fabrication with flexible substrates.

The transistor is a fundamental element for all modern electronics. Here, transistors based on organic semiconductors as active layer are referred to as OFETs. The performance of OFETs is comparable to but still lower than that of conventional inorganic transistors; however, the production cost and flexibility of Si-based devices are obvious constraints. Consistent advancements in fabrication techniques provide the ideal solution to the realization of flexible and large-area electronic circuits. The charge transport properties can be undoubtedly enhanced with the optimization of fabrication methodology and the synthesis of novel organic-based materials.

Excellent review articles are available in the recent years, which introduce the overall picture of OFETs or deal with each layer in detail [4–14]. This chapter aims to give the readers a brief overview of the building blocks in the chemical structures of organic semiconductors and fabrication and characterization in OFET devices and to provide their flexible electronic applications. This chapter is organized into eight sections, including the current introduction section. Section 1.2 covers the study of the major developments in OFET structures and operation principles. The characteristic parameters of OFETs for analyzing device performance are discussed in Section 1.3. Several first-generation p-type and n-type molecular and polymeric semiconductors used in the active layer are reported in Section 1.4. Major processing techniques involved in the fabrication are illustrated in Section 1.5. Furthermore, Section 1.6 enumerates several representative studies on flexible OFETs. Section 1.7

demonstrates the application of OFETs on phototransistors. Finally, concluding remarks related to OFETs are summarized in Section 1.8.

1.2 Device Structures and Operation Principle

Typical OFETs consist of four main components: active organic semiconducting layer, gate dielectric (insulator), three terminals (gate, source, and drain contact), and supporting substrate. In principle, three-terminal organic electronic devices include intrinsic π-conjugated organic molecules and polymers as active channel materials; inorganic oxide and organic insulator as dielectric; conducting polymers, metals, or carbon materials as electrodes; and Si, flexible transparent polymer, and stretchable polymers as substrate. The active organic semiconductor layer is generally located in the channel between the source and drains, isolated from the electrode by the dielectric. OFETs can act as a modified capacitor, where one plate is from the gate electrode and the second from the source/semiconductor/drain circuit. The application of voltage can modify the resistance of the organic semiconductor; therefore, a typical application of a transistor is acting as an amplifier of the electrical signal or switch.

According to the gate electrode and relative position of organic semiconductors and source/drain electrode, the representative arrangement may vary into top-gate bottom-contact (TGBC), top-gate top-contact (TGTC), bottom-gate bottom-contact (BGBC), and bottom-gate top-contact (BGTC) configuration, as shown in Fig. 1.1. There are also some special novel structures such as dual-gate, side-gate, and vertical structure. The performance of OFETs strongly depends on their structures, which is mainly due to different charge injection paths and areas between source and drain. For TC OFETs, the source and drain electrodes are placed on the top of semiconductors. On the other hand, organic semiconductors are deposited onto prefabricated source and drain electrodes for BC OFETs. Considering BG devices as an example, the application of a gate voltage induces a current channel at the interface between semiconductor and dielectric layer. A BGTC exhibits a higher current due to the relatively large injection area for charge carriers, resulting in a lower contact resistance. In contrast, the BGBC structure demonstrates an inferior performance due to high metal/semiconductor contact resistance, which is an effect of small contact area as well as a non-uniform deposition of the semiconductors with distinct morphologies around the prefabricated source and drain, even in the case of identical materials, process dimensions, and processing skills. Although the device performance is low, the BGBC OFETs allow simper and cost-effective processing/patterning schemes over BGTC OFETs since deposition of organic semiconductors constitutes the last fabrication step with greater flexibility.

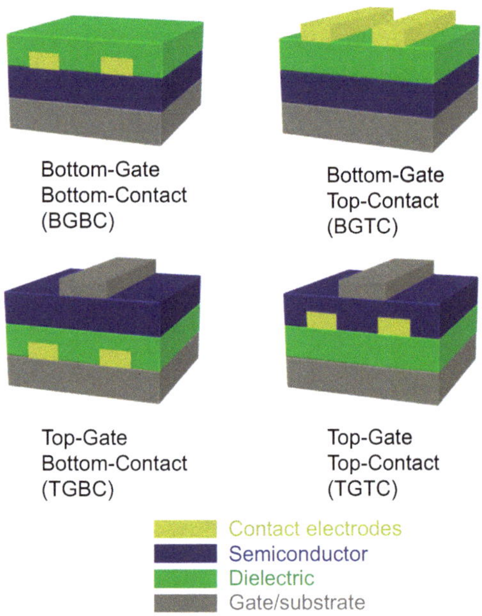

Figure 1.1 Structures and materials of four OFET architectures.

The operation principle of OFETs is similar to the traditional metal-oxide-semiconductor field-effect transistor (MOSFET) used in silicon integrated circuits. OFETs operate as voltage-controlled current source on applying the gate bias (V_g), and mobile charge carriers are accumulated near the semiconductor/dielectric interface that allows the current channel through the active semiconductor layer on applying drain voltage (V_d). It should be noted that the source serves as the grounded electrode. Hole and electron in the semiconductor act as charge carriers mostly depending on the relative energy level relationship between the semiconductor and source/drain contact, which can be divided into three operation modes, including p-type, n-type, and ambipolar channel. Current–voltage (I–V) characteristics of typical OFETs can be expressed from the linear (Eq. 1.1) and saturation region (Eq. 1.2).

$$(I_{\rm d})_{\rm lin} = \left(\frac{W}{L}\right)\mu C_i \left(\frac{V_{\rm g} - V_{\rm th} - V_{\rm d}}{2}\right) V_{\rm d} \tag{1.1}$$

$$(I_{\rm d})_{\rm sat} = \left(\frac{W}{2L}\right)\mu C_i (V_{\rm g} - V_{\rm d})^2 \tag{1.2}$$

where $I_{\rm d}$ is the drain current, W and L are the width and length of the channel, μ is the field-effect mobility, C_i is the capacitance per unit area of the gate dielectric, and $V_{\rm th}$ is the threshold voltage. At low $V_{\rm d}$, $I_{\rm d}$ builds up linearly due to the presence of carriers all along the channel, following Ohm's law. As $V_{\rm d}$ further increases and approaches $V_{\rm g}$, pinch-off channel occurs and $I_{\rm d}$ becomes independent of $V_{\rm d}$. Thus, any further increase in $V_{\rm d}$ does not contribute more toward the increase in I_d. The region is called the saturation operation.

1.3 Important Device Parameters

The performance of OFETs is characterized in terms of the important parameters, including the field-effect mobility, threshold voltage, ON/OFF current ratio, and subthreshold swing. Basically all the parameters can be evaluated by the output and transfer plot, where the former provides a plot of $I_{\rm d}$ versus $V_{\rm d}$ at constant $V_{\rm g}$ and the latter provides a plot of $I_{\rm d}$ versus $V_{\rm g}$ at constant $V_{\rm d}$. The transition from linear to saturation region is observed in the output curves. The operation of performance of OFETs is generally governed by two key mechanisms: (1) charge transport within the organic semiconductor layer and (2) charge injection and extraction at the contact. Four critical parameters—field-effect mobility (μ), current ON/OFF ratio ($I_{\rm ON}/I_{\rm OFF}$), threshold voltage ($V_{\rm th}$), and subthreshold swing (SS)—are measured.

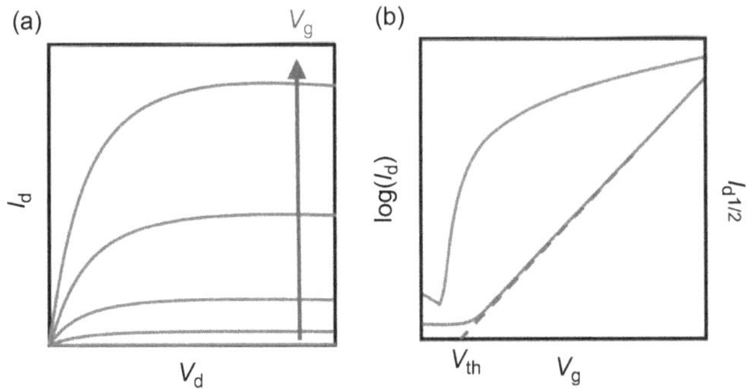

Figure 1.2 (a) Typical transfer and (b) output characteristics of OFET device operation.

1.3.1 Field-Effect Mobility

Mobility is defined as the charge-carrier drift velocity per unit electrical field, which is commonly used as a figure of merit for reporting OFET performance of various organic semiconductor materials by the research community. The field-effect mobility can be extracted from the slope of the linear fit in the saturation region on the plot of $I_d^{1/2}$ versus V_g. Important factors that strongly affect mobility is the crystallinity and grain size of the active semiconducting layer, which depends on how the layer is deposited. Significant improvements in the mobility of organic semiconductors are obtained by designing novel high-performance materials and optimizing the fabrication methods, which may fulfill the possibility of organic electronic devices for real applications.

1.3.2 Current ON/OFF Ratio

In addition to charge-carrier mobility, the other important device parameter for OFETs is the current ON/OFF ratio, which is defined as the ratio of the maximum I_d (accumulation current) in the ON state of the transistor to the minimum I_d (depletion current) in the OFF state of the transistor. During the OFF state, negligible I_d flows when V_g is small and independent of the bias applied between the source and drain contacts (V_d). The OFETs turn ON when V_g is applied, which induces charge carriers in the semiconductors at the interface with dielectric layer. This ratio is strongly influenced by the channel conductance, mobility, thickness, and dielectric constant of the dielectric layer. A high ON/OFF ratio is a desirable quality since it is essentially the signal-to-noise ratio for a transistor.

1.3.3 Threshold Voltage

The threshold voltage V_{th} is the minimum V_g required for accumulating charge carriers at the semiconductor/dielectric interface forming the active channel between the source and drain. By plotting $I_d^{1/2}$ versus V_g in the saturation region and performing the linear curve fitting, V_{th} can be extracted from the intercept of linearly fitted curve. V_{th} shows strong dependence on the dielectric constant of insulator, channel length, and the thicknesses of active and dielectric layer. Typically, a decrease in the thickness and an increase in the dielectric constant of the dielectric layer result in a significant reduction in V_{th} due to the high gate capacitance.

1.3.4 Subthreshold Swing

The subthreshold swing (SS) is the ratio of the change in the gate biasing to the change in I_d in the logarithmic scale, which can be expressed as

$$SS = \frac{\partial V_G}{\partial \log I_D} \tag{1.3}$$

The subthreshold operation of OFETs is closely related to the mobility enhancement for carrier hopping, which determines an efficient usage of the transistor as a switch. The active thin film equality achieved during the fabrication process may affect this value. A discontinuous semiconducting layer leads to the accumulation of defects and higher trap density, which results in less steeper slope (high SS).

1.4 Materials

Tremendous efforts have been put toward the synthesis of new organic materials for enhancing the performance of OFETs. Material engineering is a necessary component for the success of organic electronics and development of molecular devices. The applicability of OFETs has increased rapidly with the progress of high-performance novel organic materials. The performance of OFETs significantly depends on the material constituents for the active layer. However, the selection of the dielectric and electrode layers also has a strong influence. Besides, an appropriate selection of substrate material makes the device flexible at low cost. This section provides insight into the materials of different layers in OFETs.

1.4.1 Organic Semiconductors

Semiconducting materials are the most significant in OFET devices. Organic semiconductor materials are generally classified as small molecules and polymers made from small fundamental groups of conjugated monomer units. Small-molecule semiconductors have been widely studied because they are easy to purify and they form crystalline films for the fabrication of the desired high-performance device. Polymers have good solution processability on a range of desirable substrates, which renders them the most promising materials for large-area flexible electronics. A variety of approaches have given valuable insight into the relationship between functionality and electrical properties. Undoubtedly, the semiconducting π-conjugated system is one of the most important factors determining the ultimate performance of a device. For good semiconducting materials, it should give an extended electron-delocalized π-conjugation, high intermolecular interaction, and appropriate energy level structures in favor of charge injection. This section will focus on the categorization of p-type and n-type organic semiconducting materials, compiling structural and performance data as reported in the literature.

1.4.1.1 p-Type

The molecular design and synthesis of p-type organic semiconductors are hot topics of chemical research. The common strategy for obtaining good hole-transporting semiconductors for OFETs is via a delicate balance of highest occupied molecular orbital (HOMO) energy level, which should lie somewhere around −5 to −5.5 eV. Holes are less affected in ambient condition, and thus p-type

OFETs have preferable performance. Most extensively used p-type conjugated polymers are poly(3-hexylthiophene) (P3HT) [15], poly(3,3″-dialkylquarterthiophene) (PQT-12) [16], and poly(9,9′-dioctylfluorene-co-bithiophene) (F8T2) [17].

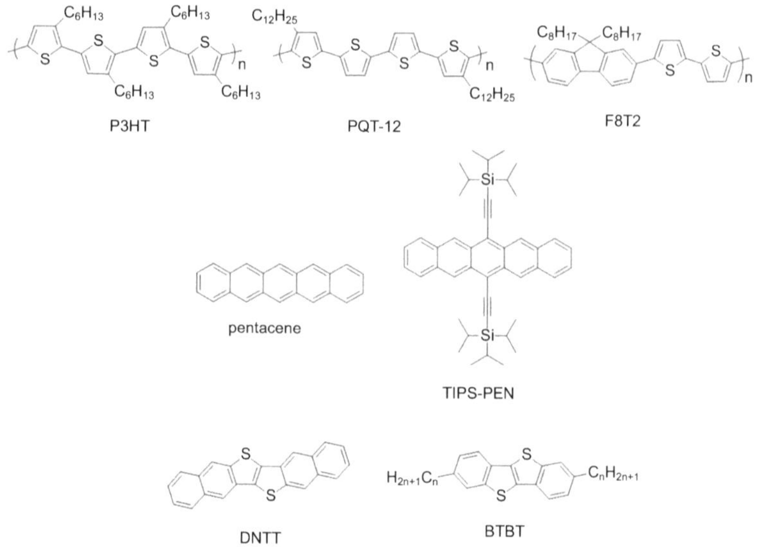

Figure 1.3 Chemical structures of p-type organic semiconductors.

Polycyclic aromatic hydrocarbons are currently among the most widely studied organic π-functional materials. For example, pentacene is a member of the acene series with reported OFET characteristics due to the effect of adequate orbital overlapping among the molecules in the crystal lattice. Thermal-evaporated pentacene films have the mobility of more than 1 cm^2/Vs [18] and moderate stability to oxygen and light. Despite better performance and popularity, pentacene is not soluble to an appreciable degree and can only be deposited using vacuum process, which is not a good candidate for low-cost printing technique. To make the high-performance pentacene-based material compatible with solution-processing fabrication, a functionalized pentacene derivative, 6,13-bis(triisopropylsilylethynyl)pentacene (TIPS-PEN), with a sufficient solubility in comment organic solvents has been introduced by Anthony et al. [19]. Takimiya et al. have also studied thienoacences core with benzene end-capped rings, which was found to result in higher hole mobility compared to the earlier heteroacene-based organic semiconductors. For example, benzothieno[3,2-b]benzothiophene (BTBT) [20] and dinaphtho[2,3-b:2′,3′-f]thieno[3,2-b]thiophene (DNTT) [21] derivatives were founded to be air-stable, high-performance organic semiconductor materials.

Poly(3-alkylthiophene)s are one of the most studied polymeric families for semiconducting applications thanks to their good processability. The length of alkyl chain plays an important role in thin-film charge transport, and the hexyl chain length is reported to be optimized. Regioregular head-to-tail P3HT shows outstanding solubility in a range of organic solvents and mobility of 0.1 cm^2/Vs [22]. However, exposure of a P3HT film to air easily causes polymer doping degradation of the ON/OFF ratio of OFETs. Ong et al. have reported PQT to afford excellent OFET performance under ambient condition. Another popular category of polymeric semiconductor materials is polyfluorene derivatives. It was found that F8T2 with nematic liquid crystalline properties can be preferentially oriented by rubbing polyimide, with a mobility of 0.02 cm^2/Vs [17].

1.4.1.2 n-Type

n-Type organic semiconductors are an important class of materials due to their necessity in the fabrication of complementary logic circuits. The search for high-performance and environmentally stable n-type materials is a quite challenging task. An ideal n-type organic semiconductor requires that its lowest unoccupied molecular orbital (LUMO) energy level is close to the workfunction of source/drain electrode, where the electrons can easily inject from the electrode to the semiconductor layer and the n-channel forms efficiently. Therefore, introduction of an electron-withdrawing group such as halogen, cyano, carbonyl, etc. into the semiconductors is an efficient way to lower the LUMO levels [23]. Additionally, the electron-withdrawing group can increase the electron affinity of the materials [24] as well as the air stability of the anions. It has been reported that n-type semiconductors require electron affinities greater than 4 eV [25].

Figure 1.4 shows the chemical structures of some n-type semiconductors. Fullerene derivatives, metallo-substituted phthalocyanine (Pc) naphthalene diimide (NDI) derivatives, perylene diimide (PDI) derivatives, thiopene-based n-type oligomers, and quinoidal system are typical n-type small molecular semiconductors. Reports on n-type polymers are much rarer than those on p-type polymers. The first report of n-type active semiconducting polymer was ladder-like poly(benzobisimidazobenzophenanthroline) (BBL) with an electron mobility as high as 0.1 cm^2/Vs, but the film can only be prepared from Lewis acid/nitromethane solutions [26, 27]. Recently, Facchetti's group developed naphthalene dicarboximide-bithophene based copolymers with an electron mobility of 0.06 cm^2/Vs and good stability after 14 weeks in air [28].

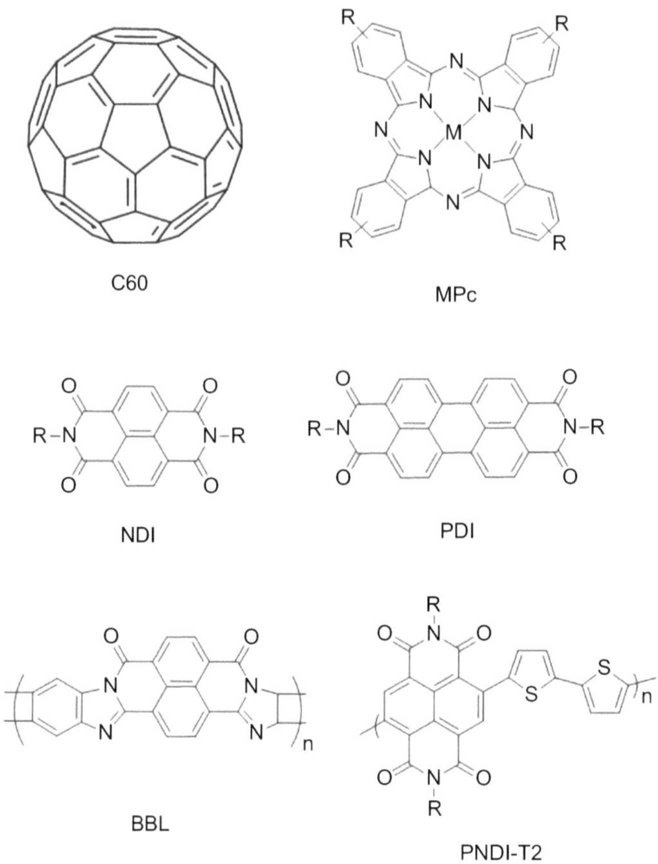

Figure 1.4 Chemical structures of n-type organic semiconductors.

1.4.2 Gate Dielectric Materials

Gate dielectric materials play an important role in the performance of OFETs, especially in terms of lowering the device operating voltage and minimizing gate leakage current. The accumulation of induced charge carriers at the semiconductor/dielectric interface strongly depends on the dielectric constant and the thickness of the insulator layer. The dielectrics are typically classified into inorganic and organic materials. Inorganic-based dielectrics, mainly including SiO_2 and various metal oxides such as AlOx, ZrOx, HfOx, and TiOx, possess high areal capacitance, high dielectric strength, and low-term stability. These excellent dielectric properties are guaranteed only after good film formation, which typically requires expensive vacuum apparatus or high annealing temperatures. Therefore, it is not easy to fabricate inorganic dielectric films on the top of organic semiconductors and plastic flexible substrates. Organic dielectric materials mainly from a variety of polymers, such as poly (methyl methacrylate) (PMMA), polystyrene (PS), and polyimide (PI), demonstrate versatile processability in solution and compatibility with organic semiconductors for flexible electronics. Most of the solution-processed polymers need high enough thickness for use as a dielectric since a thin layer of polymeric dielectric contains pinholes, which may pass substantial amount of gate leakage. Generally, a high operation voltage is required for the organic dielectric-based OFETs due to their low dielectric constant. The balance of dielectric materials between the working voltage and flexible substrate should be considered for use in specific organic electronic devices.

1.4.3 Electrode Materials

One of the critical operation junctions in OFET devices is the source and drain electrodes to the channel. Any barrier between the contacts and the channel will appear in series and impede the flow of charge through the device. Therefore, the selection of electrode materials is important. The contact metal should be selected in such a way that it should not produce high contact resistance. It implies that the barrier difference between workfunction of electrode and relative energy level of organic semiconductor layer should be low for efficient carrier injection. Gold is a popular choice for p-type

OFETs since its high workfunction provides reasonable access to the HOMO level of many semiconducting materials. The addition of a thin layer of nickel, titanium, or chromium with gold improves their adhesion. On the other hand, using low workfunction metal contacts for n-type organic semiconductors should be favorable.

Some alternatives are under consideration for better energy level matching and contact injection. A series of doped conducting polymers such as poly(3,4-ethylenedioxythiophene):styrene sulfonic (PEDOT:PSS) and polyaniline doped with camphorsulfonic acid (PANI:CSA) offer a reasonable energy level match and can potentially locally dope the semiconductor with contact dopant and offer a better surface for oligomer organization than metals.

1.4.4 Substrate Materials

The selection of substrate materials for the device mainly relies on the different kinds of applications. Silicon substrates are used in OFETs not only because of the intrinsic stable conducting properties but also for the ability to produce an oxide layer by thermal oxidation process. Solid glass substrate is also necessary for fabricating transparent devices. On the other hand, plastic substrates have attractive features such as physical durability, high flexibility (which makes them good for roll-to-roll processing), light weight, and potentially low cost when produced in large volumes. Numerous printed organic transistors were made on flexible polymer substrates such as polyethylene naphthalate (PEN), polyethylene therephthalate (PET), and polyimide (PI) for flexible electronics. Furthermore, to realize a human-friendly interface system such as wearable and stretchable electronics, stretchability of the device is critical. Stretchable substrate based on rubber materials can be integrated in OFET devices.

1.5 Overview of Processing Techniques

The processing methods used to deposit organic semiconductors play a crucial rule when determining the performance of OFETs, even though the same material is used. In general, organic semiconductors are deposited either from vapor or solution deposition depending on their solubility. Small-molecule semiconductors such as pentacene or oligothiophene are usually deposited through the vacuum deposition method due to their limited solubility in the organic solvents. On the other hand, conjugated polymers such as P3HT are solution in solvents, which can be deposited by the solution-processing method. These two techniques are described as follows.

1.5.1 Vacuum Deposition

Thermal evaporation in vacuum allows for the deposition of the small molecule and oligomer semiconductors. The material is placed in a vacuum (evacuated to a minimum pressure of 10^{-5} Torr) and heated by Joule heats (delivered through a high melting point metal). When the vapor pressure of the heated material exceeds the background pressure of the material in the chamber, the vapor state stable material evaporates and condenses on cooler surfaces. High molecular weight organic/polymeric semiconductors may decompose at higher temperature, and thus this method is not suitable for them. The main advantages of vacuum thermal deposition are the precise control of the thickness and enhanced purity of the deposited film. By monitoring the deposition rate and substrate temperature, a highly ordered film can be realized. Besides, the patterned films can be fabricated by evaporation through the shadow mask. The primary drawback is the requirement of sophisticated vacuum-based apparatus, which is in contrast to the simplicity of spin-coating or other solution process techniques.

1.5.2 Solution-Processed Deposition

Solution-processed deposition of organic semiconductors may offer a significant reduction in manufacturing cost by eliminating the vacuum deposition process and replacing them with the potentially cheaper solution deposition method. Many strategies have been developed for the solution-coating of organic semiconductors, which are beneficial in realizing large-area electronics at considerably low temperature and lower cost [29]. The common deposition methods include spin coating, drop casting, dip coating, spray coating, blade coating (e.g., doctor blading), and inkjet printing.

One of the most simple and effective ways of realizing a nice organic semiconductor (especially polymer) film is by spin-coating. Controlled parameters of the spin-coating process include spin speed, choice of solvent and concentration in the utilized solution, drying temperature/time, and so on. When a small amount of the coating solution is paced on the substrate with a high speed rotation, the organic semiconductor is uniformly spread over the substrate due to the centrifugal force. The optimized thin-film thickness and uniformity can be achieved. The dip-coating process is generally used to fabricate a uniform film efficiently and economically over cylindrical and flat substrates, which involves the vertical withdrawal of the substrate dipped in a bath of organic semiconductor solution. Key parameters such as pooling velocity, substrate/solution temperature influence the development of concentration gradients and fluid flow within the meniscus. Wet films of varying thicknesses are achieved to produce aligned crystalline domain in the dried film after solvent evaporation. Alternative solution-based processing methods also demonstrate potential for organic electronic fabrication.

1.6 Flexible Organic Transistor Device

Since organic electronics are envisioned to be flexible and light weight, the ability to fabricate organic electronics on flexible substrate is of considerable interest for devices. The mechanical flexibility of organic materials makes them particularly attractive. The criteria of selection of flexible substrate used in OFETs include thermal, chemical, and mechanical stability, surface roughness, and the cost of films. Several representative studies about flexible OFET devices are demonstrated here.

Bao's group did comprehensive studies on OFETs and their flexible electronic applications. The organic electronic devices are easily degraded in both ambient and aqueous environments, which may limit the development of OFETs. Their group overcame the barrier using the isoindigo-based polymer semiconductors with siloxane-terminated solubilizing group, which demonstrates its good solubility and high mobility [30]. Flexible OFET sensors were fabricated on polyimide substrate, with a benzocyclobutene (BCB) layer deposited prior to the poly(4-vinylphenol) (PVP) dielectric layer to decrease the roughness of the polyimide, as shown in Fig. 1.5 [31]. The OFETs exhibited good electronic performance with an absence of hysteresis, threshold voltage of −1.62 ± 1.25 V, mobility of 0.035 ± 0.02 cm^2/Vs, and ON/OFF ratio >130. The flexible OFETs exhibited stable electrical performance not only in ambient condition but also in marine environment (DI water and seawater), which are further capable of selectively sensing heavy metal ions. The low cost and high reproducibility indicated a promising method for large-area integrated flexible electronics.

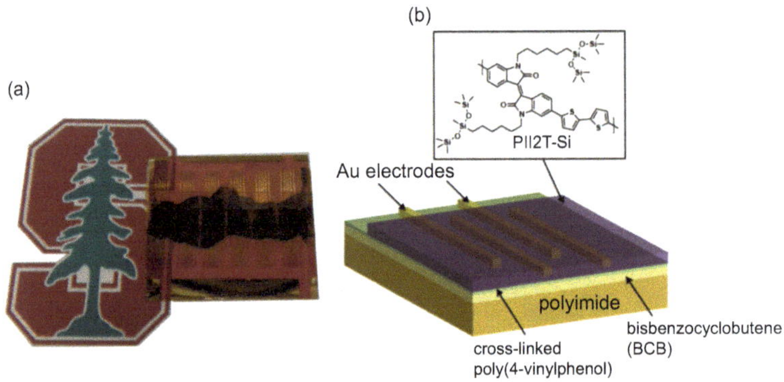

Figure 1.5 (a) Optical image and (b) schematic of fabricated flexible OFET sensor. Reproduced with permission from Ref. [31], Copyright 2014, Springer Nature.

Wang et al. presented a pentacene made on a flexible poly(ethylene terephthalate) (PET) plastic substrate with a very high mobility value of 23.2 cm^2/Vs in the saturation region and a low operating voltage of −3 V [32]. A natural biopolymer silk fibroin was chosen as the gate dielectric and fabricated in thin-film form by low-cost solution process at low temperature. For typical pentacene-based OFETs, a large portion of pentacene in an amorphous form mixed with the pentancene orthorhombic phase of SiO_2 was used as dielectric. The primary role of the silk fibroin dielectric is to increase the pentacene orthorhombic phase and to reduce the amorphous phase in pentacene during thermal evaporation. Thus, a gate dielectric film of ~420 nm thick silk fibroin is spin-coated, and the OFET exhibited output characteristics with pinchoff and current saturation, whereas transfer characteristics were measured at a low frequency. It gave a great potential for a wide variety of applications, such as e-paper, RFID tags, and biosensors.

Lee et al. reported flexible OFETs with controlled nanomorphology of semiconducting polymers on chemically and mechanically stable nanogrooved polymer substrate [33]. The oleophatic nanogrooved polymer dielectric layer was able to achieve high polymer alignment, which yields the mobilities of 19.3 and 10.5 cm^2/Vs on rigid and plastic substrates, respectively. Figure 1.6 shows the schematic device architecture of flexible OFETs and mobility variations as a function of the bending distance. The mobilities are almost constant under tensile and compressive bending with a bending distance of up to 5 mm. This result demonstrates that directed self-assembly can be achieved by soft nanostructures, and flexible plastic OFETs are promising for the realization of high-performance flexible electronics.

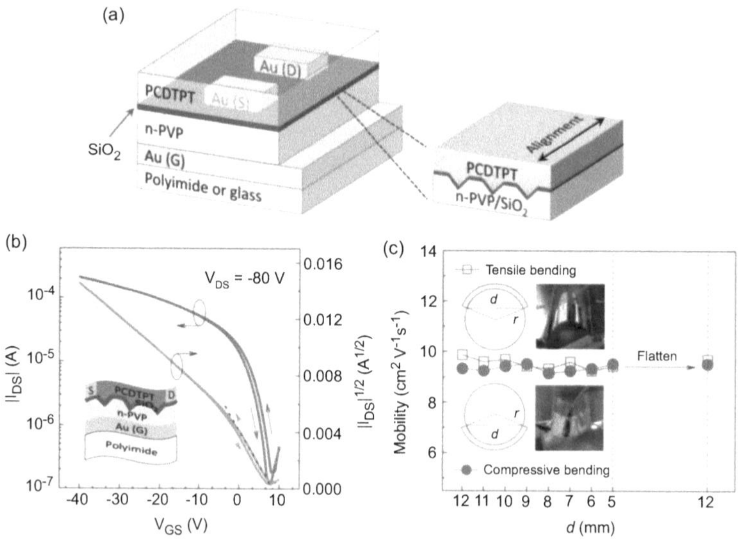

Figure 1.6 (a) Schematic flexible OFET architectures. (b) Transfer curves of flexible device fabricated on transparent polyimide substrates. (c) Mobility plot as a function of bending distance under tensile and compressive bending radius. Reproduced with permission from Ref. [33], Copyright 2016, American Chemical Society.

1.7 Flexible Organic Phototransistor

1.7.1 Introduction

The high demand for sensing and imaging applications such as optical communication, light fidelity health monitoring, and thermal imaging has allowed the rapid progress in photosensing devices, including photodiodes [34] (vertical device configurations) and photoresistors [35, 36]/phototransistors [37–39] (lateral device architectures). Phototransistors typically exhibit higher responsivity than photodiodes because the current generated by photons is amplified by gate voltage modulation without increasing noise [40]. In contrast to two-terminal photosensors, the photo-excited current in phototransistor can be further increased via lateral bias (V_d) to assist the dissociation of light-induced excitons [41]. Besides, phototransistors that combine light detection and signal amplification together in a single device can be easily integrated into conventional electronic circuits due to the unique structure of field-effect transistors (FETs) such as light switchers, light memories, and light detectors [42].

The first phototransistor with the prototype of n-p-n junction where the hole injection by the emitter junction was simply replaced by hole electron pair generation by light was induced to optoelectronics in 1951 by William Shockley [43]. Field-effect transistor gradually replaced the bipolar junction transistor due to the superior energy consumption, which also prompted the invention of phototransistors based on field-effect transistors. Besides, among the various stimulations for human beings, light is a distinct external stimulus with wavelength expanded from ultraviolet region to infrared light radiating from sunlight or even X-ray or γ-radiation from medical equipment, which is usually invisible to our bare eyes. Therefore, exploring stretchable, wavelength-accessible light detectors with high sensitivity within a broad spectrum has been of longstanding interest owing to their potential for use in invisible or detrimental radiative detector during daily life. Furthermore, non-contact and media-less transportation characteristics of electromagnetic wave communication also show promising application in electronic skin labeling for entertainment purpose or human health monitor.

Organic photosensitive material is essential in realizing a flexible photosensitive device. Therefore, the organic photovoltaic material is an appropriate candidate for the charge-transporting material in organic phototransistors. A bulk heterojunction composed of a donor–acceptor system not only provides an effective charge-carrier pathway but also strengthens the photosensitivity through exciton (electron–hole) separation between the junction of donor and acceptor compared to a single component. Photochromic material is a compound that can undergo reversible structural transformations under the influence of light stimuli; this induces a color change, as well as changes in their physical and chemical properties [44]. Through the introduction of photochromic material in the active layer or dielectric layer in an organic phototransistor, the electrical properties can be further verified under light illumination. Besides, the monomer that shows reversible photon-induced polymerization also serves as the photosensitive material in organic phototransistors.

In the following, we will introduce the main parameters for organic phototransistors and categorize the flexible organic phototransistor based on their photon-responsive mechanism as donor–acceptor system, photochromism, and photopolymerization.

1.7.2 Important Device Parameters of Organic Phototransistor

For organic phototransistors, in addition to the normal OFET parameters, photoconductive gain, photocurrent/dark current ratio, photoresponsivity, and external quantum efficiency are the main factors that determine the efficacy of the device.

1.7.2.1 Photoconductive gain (*G*)

The photoconductive gain is given by the following equation:

$$G = \frac{\tau_{\text{lifetime}}}{\tau_{\text{drift}}} \tag{1.4}$$

where τ_{lifetime} is carrier lifetime, $\tau_{\text{drift}} = L^2\mu^{-1}V_{\text{d}}^{-1}$ is the drift time of charge carrier along the channel length (L) with the charge-carrier mobility (μ) and lateral voltage (V_{d}) [45]. When the organic phototransistors are operated under a high bias, the carrier lifetime will be larger than the drift time, which is needed for carriers to travel from one electrode to the opposite electrode. So the gain can well exceed 1 (>100%).

1.7.2.2 Photocurrent/dark current ratio (*P*)

The photocurrent/dark current ratio is given by the following formula:

$$P = \frac{\text{Signal}}{\text{Noise}} = \frac{I_{\text{photon}} - I_{\text{dark}}}{I_{\text{dark}}} \sim \frac{I_{\text{photon}}}{I_{\text{dark}}} \tag{1.5}$$

where I_{photon} and I_{dark} correspond to I_{d} under illumination and dark. P strongly depends on the illuminated intensity and also on the operating bias (V_{g}).

1.7.2.3 Photosensitivity (*R*)

The value of R reveals to what extent the optical power (P_{opt}) is converted into an electrical current with the unit of A/W.

$$R = \frac{I_{\text{photon}} - I_{\text{dark}}}{P_{\text{opt}}} \tag{1.6}$$

1.7.2.4 Quantum efficiency (*η*)

The quantum efficiency, known more specifically as the external quantum efficiency, is the number of electron–hole pairs generated for each incident photon (unit of %):

$$\eta = \left(\frac{I_{\text{photon}}}{q}\right)\left(\frac{P_{\text{opt}}}{hv}\right)^{-1}$$

where I_{photon} is the photogenerated current from the absorption of incident optical power P_{opt} at a wavelength λ (corresponding to a photon energy hv), and q is the electronic charge [46].

1.7.2.5 Photodetectivity (*D**)

The photodetectivity is given by the following formula:

$$D^* = \frac{(A\Delta f)^{1/2} R}{i_{\text{n}}} \tag{1.7}$$

where A is the effective area of the detector, Δf is the electrical bandwidth, R is the photoresponsivity, and i_{n} is the noise current.

1.7.3 Examples of Flexible Organic Phototransistors

The widely available materials for the fabrication of high-performance flexible organic phototransistors are acutely required in order to further the practical applications of organic optoelectronic devices. The possible routes to enhance photoresponsivity include fabrication of the semiconducting layer with the donor–acceptor interface [38, 40, 41, 47] and the introduction of photosensitive materials such as photochromic molecules [48, 49] or photopolymerization molecules [50] in the semiconducting material layer. This section will focus on the mechanism of the above flexible organic phototransistors and their specific examples.

1.7.3.1 Donor–acceptor system

Organic phototransistors based on a donor–acceptor system are a kind of optoelectronic devices that involve the mechanism of photovoltaic and field-effect transistor at the same time. Similar to photovoltaic, the high performances of the hybrid system in organic phototransistors have been attributed to the efficient charge transfer between the heterojunction interface of the donor and the acceptor, which effectively inhibits the recombination of photo-excited excitons in a single component. At the same time, the lateral bias (V_d) and the vertical bias (V_g) can further manipulate the lateral current (I_d) without increasing noise.

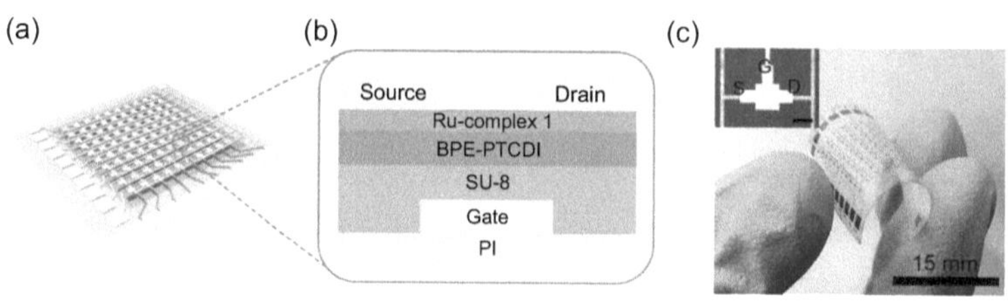

Figure 1.7 (a) Schematic illustration of 10 × 10 Ru-complex 1/BPE-PTCDI/SU-8 organic phototransistor device fabricated on a flexible, transparent PI. (b) Detailed schematic structure of a single organic phototransistor. (c) Photograph of the 10 × 10 organic phototransistor array prepared on a PI substrate. (Inset: optical image of a single device layout; S, D, and G indicate source, drain, and gate, respectively. Scale bar, 200 μm). Reproduced with permission from Ref. [40], Copyright 2016, American Chemical Society.

Rigid rod conjugated polymers have been used as a single component of active layer in flexible organic phototransistor. Chou et al. demonstrated flexible organic phototransistors by using the cross-linked poly(4-vinylphenol) (C-PVP) as a polymer dielectric layer, and *N,N′*-ditridecyl-3,4,9,10-perylenetetracarboxylic diimide (PTCDI-$C_{13}H_{27}$) as an n-type active layer on a transparent polyether sulfone (PES) substrate with R = 0.41 A/W, a tunable photocurrent/dark current ratio through lateral and vertical bias. With the merit of energy offset between the donor and acceptor, organic phototransistors based on a bulk heterojunction system in active channel can enhance the dissociation of photo-excited excitons and further provide the possible routes to improve photoresponsivity. An air-stable n-type organic semiconductor, *N,N′*-bis(2phenylethyl)-perylene-3,4:9,10-tetracarboxylic diimide (BPE-PTCDI)-based organic phototransistors functionalized with ruthenium (Ru)-complex 1 exhibited ~5000 times higher η than that of pristine BPE-PTCDI organic phototransistors and high R of 7230 A/W at V_g = 80 V, owing to the metal–ligand charge transfer (MLCT) from Ru-complex 1 to the active component of the device [40]. In addition, a large 10 × 10 organic phototransistor array (2.5 × 2.5 cm^2) has been prepared on a transparent PI substrate, showing distinct light mapping (Fig. 1.7). One-dimensional poly(3-hexylthiophene-2,5-diyl) (P3HT): phenyl-C_{61}-butyric acid methyl ester (PCBM) electrospun fibers based organic phototransistors also have demonstrated the maximum P and R values of ~40 and 0.2 A/W, respectively [47].

Recently, organic–inorganic hybrid perovskite materials have emerged as effective photoactive components, which possess exceptional semiconducting properties such as intense light-harvesting capability, small exciton binding energy (~20–50 meV) [51], and long charge-carrier lifetime (>100 ns) and diffusion length (>1 μm) [52–54]. These characteristics have allowed its rapid progress in photosensing devices. Therefore, apart from using organic/organic mixtures in flexible organic phototransistors, a light-sensitive transistor based on a composite of conjugated polymer and hybrid organic/inorganic perovskite has also been developed. Xie and Shi et al. have successfully utilized methyl ammonium lead iodide ($CH_3NH_3PbI_3$) and poly(diketopyrrolopyrrole-terthiophene)

(PDPP3T) as complementary light absorbers, which overcome the low exciton diffusion length (~10 nm) of conjugated polymer and broaden the spectral sensitivity [38]. The organic phototransistor demonstrated sensitive photo-response in the near-infrared region but also significantly improved photoresponsivity and photodetectivity in the UV–visible region, compared with those of pure $CH_3NH_3PbI_3$ organic phototransistor even at low driving voltages. For the $CH_3NH_3PbI_3$/PDPP3T composite photodetector, R was calculated to be 10.7 mA/W at 365 nm (UV light), 25.5 mA/W at 650 nm (visible light), or 5.5 mA/W at 937 nm (near-infrared radiation) at a low applied voltage of 1 V, demonstrating its broadband photo-detection characteristics [38].

1.7.3.2 Photochromism

Photochromism involves the reversible transformation of a chemical species between two isomeric forms triggered by light stimuli, which results in the change in absorption spectra. Most studies on photochromic molecules in organic phototransistors have focused on three types of photochromic system: azobenzenes, spiropyrans, and diarylethenes. The mechanisms include changes in the cis–trans isomerizations, conjugation length and ring structure, which can be translated into corresponding adaptations of the optoelectronic properties, such as dipole moment, refractive index, and charge-carrier mobility. Importantly, these dynamic transformations can generate coincident changes in the optical, chemical, electrical, and bulk properties of the system that incorporates them. Photochromic molecules, therefore, play a pivotal role within photo-responsive systems, being able to convert an optical signal via their isomerization to a useful property change [49].

Leydecker et al. demonstrated a flexible organic phototransistor via an active layer consisting of a blend of diarylethene (DAE) moiety, namely DAE-Me, and semiconducting polymer, poly(3-hexylthiophene) (P3HT), serving as the electroactive nanomaterial (Fig. 1.8) [48]. When the organic phototransistor is illuminated with monochromatic light at λ = 546 nm, the DAE-Me is in its open form (DAE-Me-o) with the corresponding HOMO level lying outside the bandgap of P3HT, which results in a scattering center for charge carriers. On the other hand, under 313 nm illuminating light, the DAE-Me turns to the closed isomer (DAE-Me-c), which features a HOMO level that is situated within the bandgap of the polymer matrix and can, therefore, reduce the drain current through accepting holes. Interestingly, the high stability of both open and closed isomers further provides the memory effect with retention capabilities of 500 days.

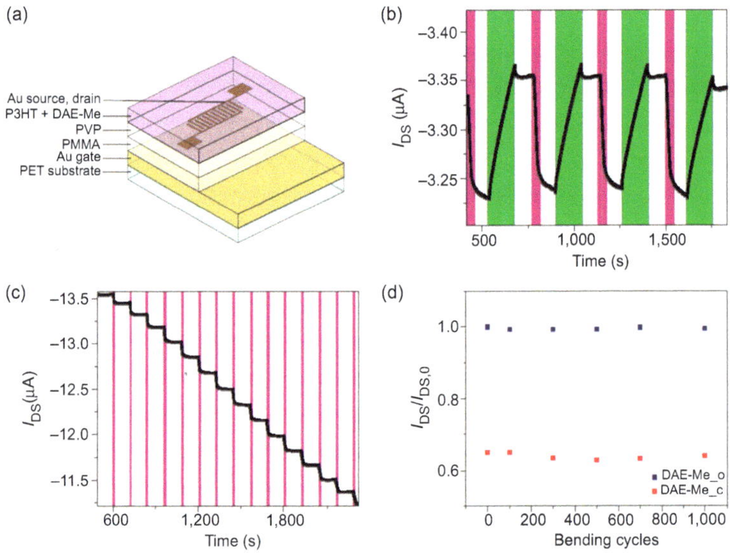

Figure 1.8 Integration of the memory unit onto a flexible polyethylene terephthalate (PET) substrate. (a) Structure of the devices on PET. (b) Experimental dynamic I_d–time curves, corrected for bias stress, with write–erase cycles (313 nm for 20 s/546 nm for 140 s) at $V_G = -60$ V and $V_d = -10$ V. (c) I_d–time curve corrected for bias stress showing small irradiation steps at 313 nm, underlining the ability to behave as a multilevel memory (5 to 20 s incremental irradiation steps). (d) Drain current as a function of bending cycles involving devices based on either DAEs in the closed form (after 10 min irradiation at 313 nm) or in the open form. Each cycle consisted of 5 s of bending at a bending radius of 8.0 mm, followed by 5 s of resting time. Reproduced with permission from Ref. [48], Copyright 2016, Springer Nature.

1.7.3.3 Photopolymerization

"Click reaction," coined by Kolb et al. in 2001, is a chemical transformation with such advantageous characteristics as high

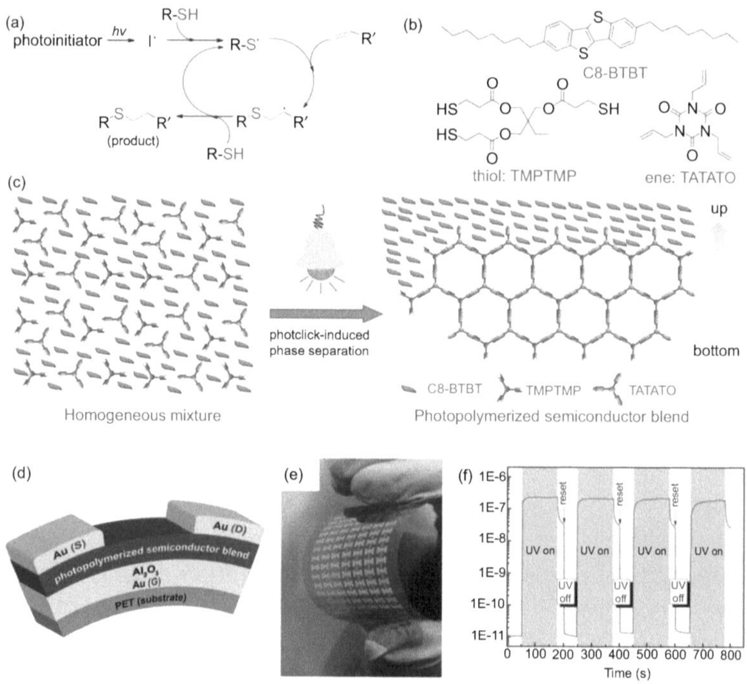

Figure 1.9 Photopolymerized semiconductor blend formed via thiol-ene photoclick reaction-induced phase separation: (a) reaction mechanism of the photo-mediated thiol-ene click reaction; (b) chemical structures of C8-BTBT, TMPTMP, and TATATO; and (c) photopolymerized semiconductor blend formed via thiol-ene photoclick reaction-induced phase separation using the homogeneous mixture of C8-BTBT, TMPTMP, and TATATO. (d) Schematic illustration of the structure of the flexible organic phototransistor. (e) Optical picture of the photopolymerized organic phototransistor arrays on a flexible and transparent PET substrate. (f) Light response of the photopolymerized organic phototransistor upon exposure to 4 mW/cm^2 of 350 nm light at V_g = 0 V and V_d = −10 V. A pulse of gate bias (V_G = −2 V and V_D = 0 V) was used to reset the device to the low-conductivity state in the dark. Reproduced with permission from Ref. [50], Copyright 2018, American Chemical Society.

efficiency, regioselectivity, and functionality tolerance as well as mild reaction conditions, fast reaction rates, and simple product isolation procedures [55, 56]. Peng et al. took the advantage of photo-initiated click reaction to display a novel protocol of organic phototransistor where the heterojunction semiconductor interfaces are generated in situ via a photo-mediated thiol-ene click reaction with trimethylolpropane tris(3-mercaptopropionate) (TMPTMP) and 1,3,5-triallyl-1,3,5-triazine-2,4,6(1H,3H,5H)-trione (TATATO) as monomers, diphenyl(2,4,6-trimethylbenzoyl)phosphine oxide (TPO) as the photoinitiator, and the p-type semiconductor, 2,7-dioctyl[1]benzothieno[3,2-b][1]benzothiophene (C8-BTBT) [50]. Under 350 nm light exposure, the in-plan phase separation of thiol-ene monomers network (bottom layer) and C8-BTBT (top layer) leads to an increase in hole density derived from electron–hole pair formation and thus the positive shift in threshold voltage, as shown in Fig. 1.9. The photopolymerized organic phototransistors show no deterioration even after 50 ON/OFF cycles with R = 2.5 A/W.

1.8 Conclusion

This chapter presents an overview of OFETs, particularly aiming at flexible electronic applications. The fundamental properties of organic semiconductors and the basic operation of OFET devices were reviewed. A procedure was explained for the extraction of important OFET parameters to provide a basic evaluation of OFET performance. Organic semiconductors and device architecture considered for specific processing skills in OFET fabrication were also discussed. Flexible OFETs and organic memories were integrated to realize functional flexible electronic applications. OFETs boast of a bright future with a wide range of potential applications. It still faces a number of challenges and needs interdisciplinary research to develop a commercially viable OFET technology.

Acknowledgments

We appreciate the financial support by the Ministry of Science and Technology (MOST) of Taiwan. Jung-Yao Chen appreciates the financial support from the Young Scholar Fellowship Program by the Ministry of Science and Technology (MOST) of Taiwan, under Grant MOST107-2636-E-194-001.

References

1. Arias, A. C., J. D. MacKenzie, I. McCulloch, J. Rivnay, and A. Salleo. (2010). Materials and applications for large area electronics: Solution-based approaches, *Chem. Rev.*, **110**, pp. 3–24.

2. Liao, C., M. Zhang, M. Y. Yao, T. Hua, L. Li, and F. Yan. (2015). Flexible organic electronics in biology: Materials and devices, *Adv. Mater.*, **27**, pp. 7493–7527.

3. Janneck, R., N. Pilet, S. P. Bommanaboyena, B. Watts, P. Heremans, J. Genoe, and C. Rolin. (2017). Highly crystalline C8-BTBT thin-film transistors by lateral homo-epitaxial growth on printed templates, *Adv. Mater.*, **29**, pp. 1703864.

4. Allard, S., M. Forster, B. Souharce, H. Thiem, and U. Scherf. (2008). Organic semiconductors for solution-processable field-effect transistors (OFETs), *Angew. Chem. Int. Ed.*, **47**, pp. 4070–4098.

5. Russew, M.-M. and S. Hecht. (2010). Photoswitches: From molecules to materials, *Adv. Mater.*, **22**, pp. 3348–3360.

6. Klauk, H. (2010). Organic thin-film transistors, *Chem. Soc. Rev.*, **39**, pp. 2643–2666.

7. Mei, J., Y. Diao, A. L. Appleton, L. Fang, and Z. Bao. (2013). Integrated materials design of organic semiconductors for field-effect transistors, *J. Am. Chem. Soc.*, **135**, pp. 6724–6746.

8. Wang, X., Y. Gu, Z. Xiong, Z. Cui, and T. Zhang. (2014). Silk-molded flexible, ultrasensitive, and highly stable electronic skin for monitoring human physiological signals, *Adv. Mater.*, **26**, pp. 1336–1342.

9. Gsänger, M., D. Bialas, L. Huang, M. Stolte, and F. Würthner. (2016). Organic semiconductors based on dyes and color pigments, *Adv. Mater.*, **28**, pp. 3615–3645.

10. Richter, L. J., D. M. DeLongchamp, and A. Amassian. (2017). Morphology development in solution-processed functional organic blend films: An in situ viewpoint, *Chem. Rev.*, **117**, pp. 6332–6366.

11. Huang, H., L. Yang, A. Facchetti, and T. J. Marks. (2017). Organic and polymeric semiconductors enhanced by noncovalent conformational locks, *Chem. Rev.*, **117**, pp. 10291–10318.

12. Wang, B., W. Huang, L. Chi, M. Al-Hashimi, T. J. Marks, and A. Facchetti. (2018). High-k gate dielectrics for emerging flexible and stretchable electronics, *Chem. Rev.*, **118**, pp. 5690–5754.

13. Liu, A., H. Zhu, H. Sun, Y. Xu, and Y.-Y. Noh. (2018). Solution processed metal oxide high-κ dielectrics for emerging transistors and circuits, *Adv. Mater.*, **30**, pp. 1706364.

14. Głowacki, E. D., M. Irimia-Vladu, M. Kaltenbrunner, J. Gsiorowski, M. S. White, U. Monkowius, G. Romanazzi, G. P. Suranna, P. Mastrorilli, T. Sekitani, S. Bauer, T. Someya, L. Torsi, and N. S. Sariciftci. (2013). Hydrogen-bonded semiconducting pigments for air-stable field-effect transistors, *Adv. Mater.*, **25**, pp. 1563–1569.

15. Kline, R. J., M. D. McGehee, E. N. Kadnikova, J. Liu, J. M. J. Fréchet, and M. F. Toney. (2005). Dependence of regioregular poly(3-hexylthiophene) film morphology and field-effect mobility on molecular weight, *Macromolecules*, **38**, pp. 3312–3319.

16. Ong, B. S., Y. Wu, P. Liu, and S. Gardner. (2004). High-performance semiconducting polythiophenes for organic thin-film transistors, *J. Am. Chem. Soc.*, **126**, pp. 3378–3379.

17. Sirringhaus, H., R. J. Wilson, R. H. Friend, M. Inbasekaran, W. Wu, E. P. Woo, M. Grell, and D. D. C. Bradley. (2000). Mobility enhancement in conjugated polymer field-effect transistors through chain alignment in a liquid-crystalline phase, *Appl. Phys. Lett.*, **77**, pp. 406–408.

18. Lin, Y., D. J. Gundlach, S. F. Nelson, and T. N. Jackson. (1997). Stacked pentacene layer organic thin-film transistors with improved characteristics, *IEEE Electron. Device Lett.*, **18**, pp. 606–608.

19. Anthony, J. E., J. S. Brooks, D. L. Eaton, and S. R. Parkin. (2001). Functionalized pentacene: Improved electronic properties from control of solid-state order, *J. Am. Chem. Soc.*, **123**, pp. 9482–9483.

20. Takimiya, K., H. Ebata, K. Sakamoto, T. Izawa, T. Otsubo, and Y. Kunugi. (2006). 2,7-Diphenyl[1]benzothieno[3,2-b]benzothiophene, a new organic semiconductor for air-stable organic field-effect transistors with mobilities up to 2.0 cm^2 V-1 s-1, *J. Am. Chem. Soc.*, **128**, pp. 12604–12605.

21. Yamamoto, T. and K. Takimiya. (2007). Facile synthesis of highly π-extended heteroarenes, dinaphtho[2,3-b:2′,3′-f] chalcogenopheno[3,2-b]chalcogenophenes, and their application to field-effect transistors, *J. Am. Chem. Soc.*, **129**, pp. 2224–2225.

22. Sirringhaus, H., P. J. Brown, R. H. Friend, M. M. Nielsen, K. Bechgaard, B. M. W. Langeveld-Voss, A. J. H. Spiering, R. A. J. Janssen, E. W. Meijer, P. Herwig, and D. M. de Leeuw. (1999). Two-dimensional charge transport in self-organized, high-mobility conjugated polymers, *Nature*, **401**, pp. 685.

23. Newman, C. R., C. D. Frisbie, D. A. da Silva Filho, J.-L. Brédas, P. C. Ewbank, and K. R. Mann. (2004). Introduction to organic thin film transistors and design of n-channel organic semiconductors, *Chem. Mater.*, **16**, pp. 4436–4451.

24. Chua, L.-L., J. Zaumseil, J.-F. Chang, E. C. W. Ou, P. K. H. Ho, H. Sirringhaus, and R. H. Friend. (2005). General observation of n-type field-effect behaviour in organic semiconductors, *Nature*, **434**, pp. 194.

25. de Leeuw, D. M., M. M. J. Simenon, A. R. Brown, and R. E. F. Einerhand (1997). Stability of n-type doped conducting polymers and consequences for polymeric microelectronic devices, *Synth. Met.*, **87**, pp. 53–59.

26. Babel, A. and S. A. Jenekhe. (2003). High electron mobility in ladder polymer field-effect transistors, *J. Am. Chem. Soc.*, **125**, pp. 13656–13657.

27. Babel, A. and S. A. Jenekhe. (2002). Electron transport in thin-film transistors from an n-type conjugated polymer, *Adv. Mater.*, **14**, pp. 371–374.

28. Chen, Z., Y. Zheng, H. Yan, and A. Facchetti. (2009). Naphthalenedicarboximide- vs perylenedicarboximide-based copolymers. Synthesis and semiconducting properties in bottom-gate N-channel organic transistors, *J. Am. Chem. Soc.*, **131**, pp. 8–9.

29. Diao, Y., L. Shaw, Z. Bao, and S. C. B. Mannsfeld. (2014). Morphology control strategies for solution-processed organic semiconductor thin films, *Energ. Environ. Sci.*, **7**, pp. 2145–2159.

30. Mei, J., D. H. Kim, A. L. Ayzner, M. F. Toney, and Z. Bao. (2011). Siloxane-terminated solubilizing side chains: Bringing conjugated polymer backbones closer and boosting hole mobilities in thin-film transistors, *J. Am. Chem. Soc.*, **133**, pp. 20130–20133.

31. Knopfmacher, O., M. L. Hammock, A. L. Appleton, G. Schwartz, J. Mei, T. Lei, J. Pei, and Z. Bao. (2014). Highly stable organic polymer field-effect transistor sensor for selective detection in the marine environment, *Nat. Commun.*, **5**, pp. 2954.

32. Wang, C.-H., C.-Y. Hsieh, and J.-C. Hwang. (2011). Flexible organic thin-film transistors with silk fibroin as the gate dielectric, *Adv. Mater.*, **23**, pp. 1630–1634.

33. Lee, B. H., B. B. Y. Hsu, S. N. Patel, J. Labram, C. Luo, G. C. Bazan, and A. J. Heeger. (2016). Flexible organic transistors with controlled nanomorphology, *Nano Lett.*, **16**, pp. 314–319.

34. Maculan, G., A. D. Sheikh, A. L. Abdelhady, M. I. Saidaminov, M. A. Haque, B. Murali, E. Alarousu, O. F. Mohammed, T. Wu, and O. M. Bakr. (2015). $CH_3NH_3PbCl_3$ single crystals: Inverse temperature crystallization and visible-blind UV-photodetector, *J. Phys. Chem. Lett.*, **6**, pp. 3781–3786.

35. He, M., Y. Chen, H. Liu, J. Wang, X. Fang, and Z. Liang. (2015). Chemical decoration of $CH_3NH_3PbI_3$ perovskites with graphene oxides for photodetector applications, *Chem. Commun.*, **51**, pp. 9659–9661.

36. Wang, H., R. Haroldson, B. Balachandran, A. Zakhidov, S. Sohal, J. Y. Chan, A. Zakhidov, and W. Hu. (2016). Nanoimprinted perovskite

nanograting photodetector with improved efficiency, *ACS Nano*, **10**, pp. 10921–10928.

37. Lee, Y., J. Kwon, E. Hwang, C.-H. Ra, W. J. Yoo, J.-H. Ahn, J. H. Park, and J. H. Cho. (2015). High-performance perovskite–graphene hybrid photodetector, *Adv. Mater.*, **27**, pp. 41–46.

38. Chen, S., C. Teng, M. Zhang, Y. Li, D. Xie, and G. Shi. (2016). A flexible UV–vis–NIR photodetector based on a perovskite/conjugated-polymer composite, *Adv. Mater.*, **28**, pp. 5969–5974.

39. Ma, C., Y. Shi, W. Hu, M.-H. Chiu, Z. Liu, A. Bera, F. Li, H. Wang, L.-J. Li, and T. Wu. (2016). Heterostructured $WS_2/CH_3NH_3PbI_3$ photoconductors with suppressed dark current and enhanced photodetectivity, *Adv. Mater.*, **28**, pp. 3683–3689.

40. Liu, X., E. K. Lee, D. Y. Kim, H. Yu, and J. H. Oh. (2016). Flexible organic phototransistor array with enhanced responsivity via metal–ligand charge transfer, *ACS Appl. Mater. Interfaces*, **8**, pp. 7291–7299.

41. Chou, W.-Y., Y.-S. Lin, L.-L. Kuo, S.-J. Liu, H.-L. Cheng, and F.-C. Tang. (2014). Light sensing in photosensitive, flexible n-type organic thin-film transistors, *J. Mater. Chem. C*, **2**, pp. 626–632.

42. Gu, P., Y. Yao, L. Feng, S. Niu, and H. Dong. (2015). Recent advances in polymer phototransistors, *Polym. Chem.*, **6**, pp. 7933–7944.

43. Shockley, W., M. Sparks, and G. K. Teal. (1951). p-n Junction transistors, *Phys. Rev.*, **83**, pp. 151–162.

44. Xia, H., K. Xie, and G. Zou. (2017). Advances in spiropyrans/spirooxazines and applications based on fluorescence resonance energy transfer (FRET) with fluorescent materials, *Molecules*, **22**, pp. 2236.

45. Xu, H., J. Li, B. H. K. Leung, C. C. Y. Poon, B. S. Ong, Y. Zhang, and N. Zhao. (2013). A high-sensitivity near-infrared phototransistor based on an organic bulk heterojunction, *Nanoscale*, **5**, pp. 11850–11855.

46. Sze, S. M. (1985). *Semiconductor Devices, Physics and Technology*. New York: Wiley.

47. Kim, M., S. B. Jo, J. H. Park, and K. Cho. (2015). Flexible lateral organic solar cells with core–shell structured organic nanofibers, *Nano Energy*, **18**, pp. 97–108.

48. Leydecker, T., M. Herder, E. Pavlica, G. Bratina, S. Hecht, E. Orgiu, and P. Samorì. (2016). Flexible non-volatile optical memory thin-film transistor device with over 256 distinct levels based on an organic bicomponent blend, *Nat. Nanotech.*, **11**, pp. 769.

49. Ercole, F., T. P. Davis, and R. A. Evans. (2010). Photo-responsive systems and biomaterials: Photochromic polymers, light-triggered self-assembly, surface modification, fluorescence modulation and beyond, *Polym. Chem.*, **1**, pp. 37–54.

50. Peng, H., Y. Yan, Y. Yang, L. Zhou, W. Wu, Q. Sun, J. Zhuang, S.-T. Han, C.-C. Ko, Z. Xu, X. Xie, R. K. Y. Li, and V. A. L. Roy. (2018). Interface engineering via photopolymerization-induced phase separation for flexible UV-responsive phototransistors, *ACS Appl. Mat. Interfaces*, **10**, pp. 7487–7496.

51. Miyata, A., A. Mitioglu, P. Plochocka, O. Portugall, J.T.-W. Wang, S. D. Stranks, H. J. Snaith, and R. J. Nicholas. (2015). Direct measurement of the exciton binding energy and effective masses for charge carriers in organic-inorganic tri-halide perovskites, *Nat. Phys.*, **11**, pp. 582–587.

52. Johnston, M. B. and L. M. Herz. (2016). Hybrid perovskites for photovoltaics: Charge-carrier recombination, diffusion, and radiative efficiencies, *Acc. Chem. Res.*, **49**, pp. 146–154.

53. Xu, W., H. Cho, Y.-H. Kim, Y.-T. Kim, C. Wolf, C.-G. Park, and T.-W. Lee. (2016). Organometal halide perovskite artificial synapses, *Adv. Mater.*, **28**, pp. 5916–5922.

54. Chen, J.-Y., Y.-C. Chiu, Y.-T. Li, C.-C. Chueh, and W.-C. Chen. (2017). Nonvolatile perovskite-based photomemory with a multilevel memory behavior, *Adv. Mater.*, **29**, pp. 1702217.

55. Qin, A., J. W. Y. Lam, and B. Z. Tang. (2010). Click polymerization: Progresses, challenges, and opportunities, *Macromolecules*, **43**, pp. 8693–8702.

56. Kolb, H. C., M. G. Finn, and K. B. Sharpless. (2001). Click chemistry: Diverse chemical function from a few good reactions, *Angew. Chem. Int. Ed.*, **40**, pp. 2004–2021.

Chapter 2

Flexible and Organic Solar Cells

Bing Cao
Department of Chemistry, 11227 Saskatchewan Drive,
University of Alberta, Edmonton, Alberta T6G 2N4, Canada
bcao@ualberta.ca

2.1 Introduction

Organic solar cells (OSCs) have attracted considerable attention as clean-energy technology. With the growth in world population and economic development, more and more energy is needed. While the current energy supply relies heavily on fossil fuel sources, such as oil, coal, and natural gas, there are associated problems. First, these energy resources are finite, which cannot be recovered; second, combusting fossil fuels releases carbon dioxide, which leads to greenhouse effect and global warming; third, the combustion of fossil fuels also leads to air pollution, which is a severe problem in developing countries now. To solve these problems and meet the world's future energy demand, renewable and clean energy is needed. Currently, solar, wind, biomass, and hydropower are the main renewable energy resources. Among them, solar energy is

Flexible and Stretchable Electronics: Materials, Designs, and Devices
Edited by Run-Wei Li and Gang Liu

ISBN 978-981-4800-46-4 (Hardcover), 978-0-429-05890-5 (eBook)
www.jennystanford.com

the most abundant renewable energy that could solve the energy demands in the future. The solar energy that strikes the earth in 1 h can provide enough energy to power the entire world for a whole year [1]. Among the solar cell technologies, OSCs have emerged as a promising candidate that can greatly improve the economic and environmental performance of the solar industry. OSCs are made from abundant, nontoxic organic materials in a layered structure, which can be processed onto plastic rolls through solution-phase large-scale printing techniques such as roll-to-roll printing, ink injecting, and spray coating, etc. [2–4]. The cost of OSCs is, hence, much lower than that of the inorganic solar cells [5]. Organic materials are light weight and color-tunable, allowing their application in wearable electronics. OSC technologies have also shown the potential for lowering the environmental impact compared to conventional silicon-based solar cell technologies [5]. These unique features make them a major player in the energy marketplace [4, 6–8]. Up to date, the world record of OSC efficiency is reported to be 14.2% in literature [9], and 13.2% in commercialized industry by Heliatek [10].

This chapter will provide an overview of OSCs from the basics to the state-of-art research frontiers. The structures and operation principle of OSCs will be introduced. The development of new materials for high-performance OSCs is the focus of this chapter and will be discussed in details. The current progress, advantages, and challenges of flexible OSCs will be presented and discussed as well.

2.2 Basic Solar Cell Concepts

2.2.1 Structure of Organic Solar Cells

OSCs or organic photovoltaics (OPVs) are fabricated in a layered structure. An OSC device consists of three important components: photoactive layer, electrodes, and interfacial layers [11, 12]. A typical solar cell device has a transparent indium-tin-oxide-coated glass as the anode, a noble metal top contact as the cathode, and a photon-absorbing active layer sandwiched in between the two electrodes, as shown in Fig. 2.1. The early stage of OSCs adopted the bilayer structure as the photoactive layer [13], as shown in

Fig. 2.1a; the photon-absorbing molecule/polymer is deposited by thermal evaporation, followed by acceptor molecule. While less than 1% efficiency was obtained using PcCu and C_{60} bilayer structure [13], the limitation of this structure is that the diffusion length is greatly limited [14, 15]. Later in 1990s, the concept of bulk heterojunction (BHJ) structure was introduced, in which the donor and acceptor materials are mixed homogeneously to form an interpercolated mixture matrix as the photoactive layer [14, 15]. The BHJ is dominating in the field nowadays. Besides the electrodes and photoactive material, the interfacial layers that sit between photoactive layer and electrodes also play critical roles, such as blocking undesired charge and aligning energy level, to guarantee OSC devices work properly [16–18].

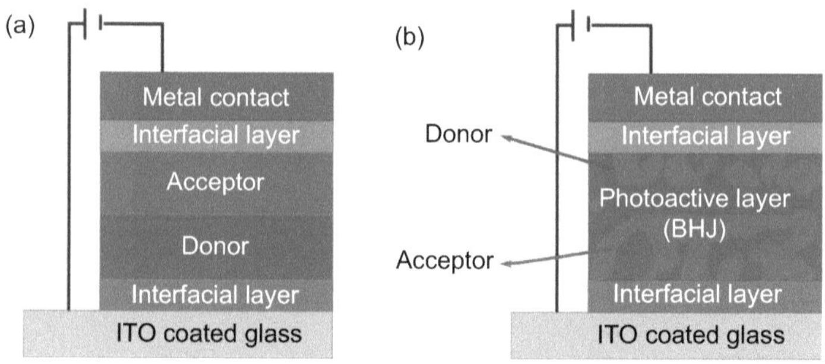

Figure 2.1 Typical solar cell structures of (a) bilayer and (b) BHJ devices.

2.2.2 Operation Principle of Organic Solar Cells

The core of a solar cell device is the photoactive BHJ layer. The donor material is a p-type organic semiconductor that absorbs photons, and the acceptor material is an n-type semiconductor that accepts photogenerated electrons [19]. The photovoltaic process occurs through several steps [20, 21]: (1) When light shines through an OSC device, photons with energy equal to or greater than the bandgap of the donor polymers/molecules can be absorbed. This high-energy photon will excite one electron in the highest occupied molecular orbitals (HOMO) of the polymer to the lowest unoccupied molecular orbitals (LUMO) level, leaving a hole behind, generating an electron–hole pair (which is called exciton). (2) This closely bonded electron–hole pair (exciton) is then diffused to the interface of donor and acceptor. (3) At the interface, the exciton is dissociated. The electron is transferred to the acceptor phase, while the hole remains in the donor phase. (4) These separated electrons and holes are called charge carriers and will migrate through the acceptor/donor phase to the cathode/anode to be collected. The process is described in Fig. 2.2.

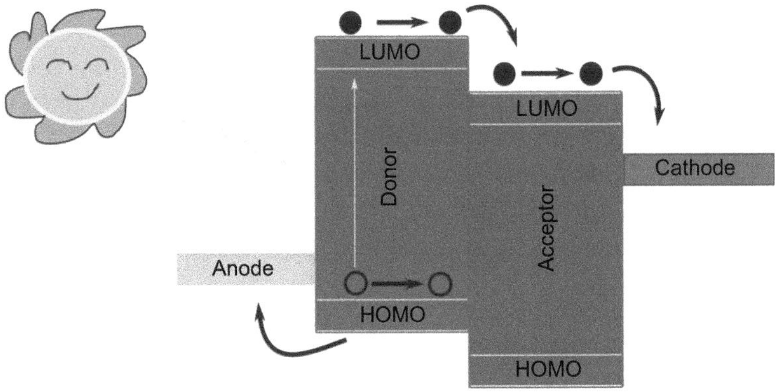

Figure 2.2 Photovoltaic process in OSCs.

2.2.3 Photovoltaic Parameters

When no light is present, the OSCs work as a diode. When light shines, photocurrent will be generated and the solar cell works as a current source in parallel with a diode [22]. In ideal cells, the total current equals to the photocurrent minus the diode current. In practice, shunt resistance (R_{SH}) and series resistance (R_S) exist as internal parasitic resistances in solar cell devices. The equivalent circuit model is depicted in Fig. 2.3. The total generated current can be expressed in the following equation:

$$I = I_L - I_0 \left(e^{\frac{q(V+I^*R_S)}{n^*k^*T}} - 1 \right) - \frac{V + I^*R_S}{R_{SH}} \tag{2.1}$$

where I is the total current, I_L is the photocurrent, I_0 is the saturation current of the diode, n is the diode ideality factor, q is the elementary charge of 1.6×10^{19} C, k is a constant and equals to 1.38×10^{-23} J/K, T is the cell temperature, V is the measured cell voltage, R_S is the series resistance, and R_{SH} is the shunt resistance.

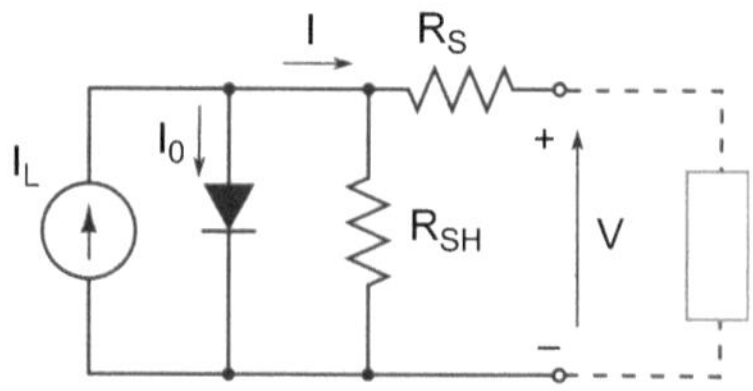

Figure 2.3 Simplified equivalent circuit model for a solar cell device.

To evaluate the performance of OSCs, a current density–voltage (J–V) curve is obtained under simulated solar light conditions. The standard simulation condition for solar cell testing is air mass 1.5 global, which is equivalent to an integrated power of 100 mW/cm^2 at 25°C. A voltage is applied to the solar cell devices, and the current response is measured. This voltage–current density relationship is plotted as the J–V curve. Photovoltaic parameters can be extracted from the J–V curve, as shown in Fig. 2.4. J_{SC} is the short-circuit current that occurs when the voltage is 0; V_{OC} is the open-circuit voltage that occurs when no current is passing the solar cell device; *FF* is the fill factor, which measures the maximum power that can be extracted from the J–V curve. *FF* measures the quality of a solar cell. Graphically, *FF* is the largest possible rectangle area which could be fitted in the J–V curve and expressed as:

$$FF = \frac{P_{MAX}}{P_T} = \frac{J_{MP}V_{MP}}{J_{SC}V_{OC}} \tag{2.2}$$

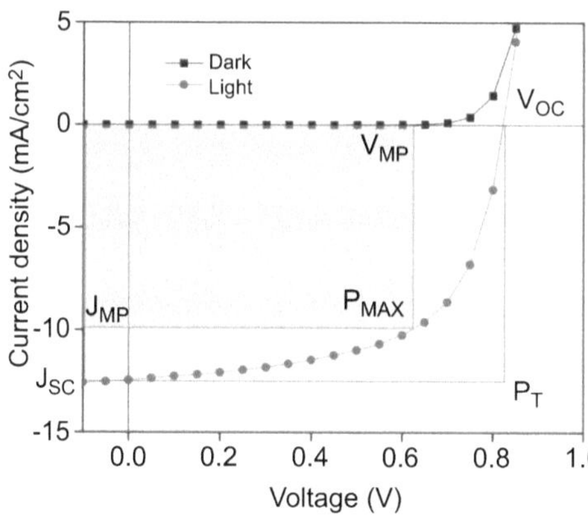

Figure 2.4 An example of J–V curve.

The efficiency of a solar cell is evaluated by power conversion efficiency (PCE), which is the ratio of the maximum power generated by the solar cell to the input power. PCE can be expressed with the J_{SC}, V_{OC}, and FF, as described in Eq. (2.3):

$$PCE = \frac{P_{MAX}}{P_{in}} = \frac{J_{SC} {}^{*} V_{OC} {}^{*} FF}{P_{in}} \tag{2.3}$$

where P_{in} is the incident light power. R_S and R_{SH} can also be extracted from the J–V curve by taking the inverse slope of the J–V curve at V_{OC} and J_{SC} point. J_{SC}, V_{OC}, FF, PCF, R_S, and R_{SH} are important photovoltaic parameters that depict the performance of a solar cell.

2.3 Donor Materials Development

The semiconducting organic materials that can absorb photons are the key to OSCs. The conductivity of organic molecules is fulfilled by the conjugated system, which has delocalized π-electrons on alternating single and double bonds [23]. Conjugated polymers and conjugated small molecules are both well studied. A high-performance semiconducting polymer or small-molecule donor should have several desired properties, i.e., narrow bandgap, proper energy levels, high hole mobility, optimal morphology, and good stability [24, 25]. The bandgap of donor material determines the theoretical upper limit for J_{SC}. Roughly 70% of sunlight energy comes from the wavelength range of 300–900 nm, so an ideal donor material should have a bandgap of 1.4–1.5 eV. V_{OC} is determined by the HOMO of the donor and the LUMO of the acceptor. Generally, the relationship is expressed as

$$V_{OC} = \frac{1}{e}\left(\left|E^{\text{Donor}}\text{HOMO}\right| - \left|E^{\text{Acceptor}}\text{LUMO}\right|\right) - 0.3\ \text{V} \tag{2.4}$$

where e is the elementary charge. A minimum energy difference of ~0.3 eV between the LUMO energy levels of the donor polymer and the acceptor is required to facilitate efficient exciton splitting and charge dissociation [26]. In theory, a deep-lying HOMO donor can exhibit higher V_{OC}. The energy levels (HOMO, LUMO) and electronic properties can be manipulated by structural modification [27–29]. Development of donor materials with regard to conjugated system design, substitutes, and side chain will be discussed in the following context.

2.3.1 Conjugated Polymers

Even the early-stage OSC research started from investigating vacuum-evaporated small-molecule materials [13], conjugated polymers as donor dominated the field since 1990s. Impressive progress has been achieved using solution-processed conjugated polymers as donor material. The first well-studied polymer donors are poly(phenylenevinylene) (PPV) and its derivatives [19]. For example, poly[2-methoxy-5-(2-ethylhexyloxy)]-1,4-phenylenevinylene (MEH-PPV) was first discovered to have ultrafast electron transfer to C_{60} fullerene, suggesting the use of conjugated polymer as donor and fullerene as acceptor in OSC devices [30, 31]. After the concept of BHJ was introduced, the conjugated polymer donors and fullerene acceptors were mixed to fabricate BHJ OSCs [14, 32]. The introduced BHJ provides remarkable PCE improvement of PPV-based OSCs, marking a milestone in the area [14]. The best performed PPV derivative as donor material is poly-[2-methoxy-5-(30,70-dimethyloctyloxy)-1,4-phenylenevinylene] (MDMO-PPV), which was mixed with phenyl-C61-butyric acid methyl ester (PCBM) to form a junction and exhibits PCE of 3.3% [33]. The chemical structures of PPV-based polymers are presented in Fig. 2.5. The limiting factor is the low J_{SC}, resulting from the large bandgap of the PPV-based donor.

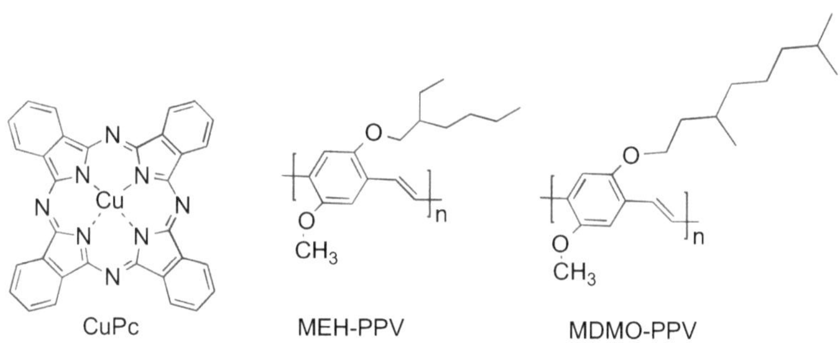

Figure 2.5 Chemical structures of CuPc, MEH-PPV, and MDMO-PPV.

The second generation of the well-studied conjugated polymer is poly[3-hexylthiophene-2,5-diyl] (P3HT). P3HT emerged as a predominant donor due to its relatively narrow bandgap of 1.9 eV compared to PPV polymers [15, 28, 34, 35]. P3HT also has a higher hole mobility due to the increase in π stacking and crystallinity [36, 37]. Intense research has been done on P3HT:PCBM-based OSCs to study the structure–property relationships. Research showed that the morphological control is the key to achieve high PCE. When processing conditions of thermal annealing [36], solvent annealing [38], and additives [39, 40] are well optimized, an impressive 5% PCE was achieved when blended with $PC_{61}BM$ (Fig. 2.6) [36].

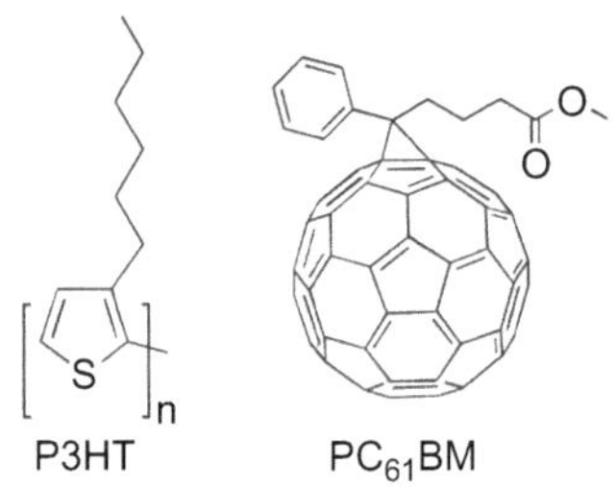

Figure 2.6 Chemical structures of P3HT and $PC_{61}BM$.

In order to further improve the efficiency, Eq. (2.3) suggests that both J_{SC} and V_{OC} need to be optimized. One strategy to achieve this goal is to design a polymer that has a deep-lying HOMO while keeping a narrow bandgap. A new generation of copolymers is then introduced, in which alternating electron-rich (i.e., donor) and electron-deficient (i.e., acceptor) moieties are incorporated in the polymer backbone [26, 41, 42]. This donor–acceptor (D–A) approach (in some cases named as push–pull) enables the construction of a low-bandgap polymer with tunable energy levels. Typically, the electron-rich fragment has high-lying HOMO levels and the acceptor has low-lying LUMO level. The repeating D–A units determine the bandgap and energy levels of the polymer. A simplified molecular orbital theory is depicted in Fig. 2.7 [27]. Hence, the HOMO and LUMO of the polymer can be tuned individually by selecting the desired donor and acceptor units [42, 43]. This D–A approach has made great success in designing low-bandgap polymers and dominates the development of new OSC materials in the community. A variety of D–A polymers are developed based on this concept. The most well-known donor units include fluorene, cyclopentadithiophene (CPDT), oligothiophene, benzodithiophene (BDT), indacenodithiophene (IDT), etc. The most well-known acceptor units include benzothiadiazole (BT), thieno[3,4-b] thiophene (TTP), 4,7-di[2-thienyl]-2,1,3-benzothiadiazole (DTBT), dithienyl-diketopyrrolopyrrole (DTDPP), thienothiophene (TT), thienopyrroledione (TPD), isoindigo (IID) moieties, etc. The chemical structures of these moieties are summarized in Fig. 2.8.

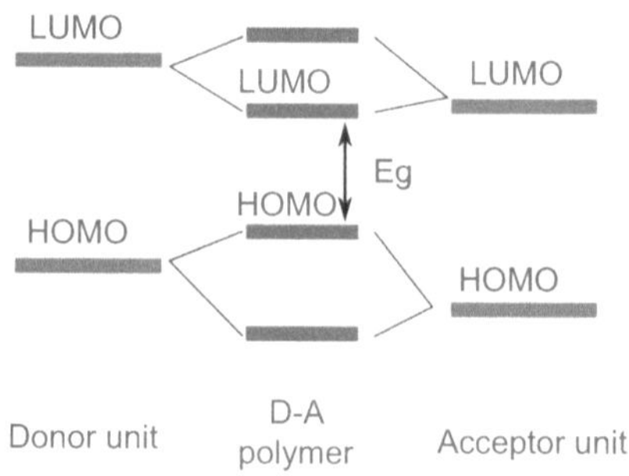

Figure 2.7 Schematic illustration of LUMO, HOMO, and bandgap of D–A polymer.

One of the earliest D–A polymers is PFDTBT, which consists of fluorene moiety as the donor unit and BT moiety as the acceptor unit in the backbone. It was blended with $PC_{61}BM$ to form BHJ OSCs and exhibits 2.2% PCE [44]. Better performing D–A polymers were developed over years, including PCDTBT [45], PBDTTPD [46], PTB7 [25, 47], PffB4T4-2OD [48, 49], PDTB-EF-T [9], etc. These D–A polymers exhibit 6–14% efficiency with proper processing and device engineering [25]. Figure 2.9 lists representative D–A polymers that will be discussed in this chapter.

A common feature of the D–A repeating units is the existence of quinoidal structure (Fig. 2.10). The quinoidal structure increases the planarity of the conjugated backbone, enabling efficient π delocalization, hence lowering the bandgap [27]. If a polymer-conjugated backbone naturally favors quinoidal form, the bandgap is lower. Based on this concept, a successful approach was using alternative TTP and BDT units [50–52]. A series of D–A polymers (PTB1, PTB2, PTB3, PTB4, PTB5, PTB6) based on the two units was synthesized, and >6% PCE was obtained [53]. This work eventually led to the synthesis of PTB7, which initially exhibits PCE of 7.7% [54], and further optimized to 9.2% with inverted structure engineering [55]. Based on the success of PTB7, the same group further synthesized PTB7-based 2D polymers by adding aromatic groups to the BDT unit to enhance the planarity of backbone. PTB7-Th was one of those 2D polymers that exhibited 9.9% PCE in inverted devices [56].

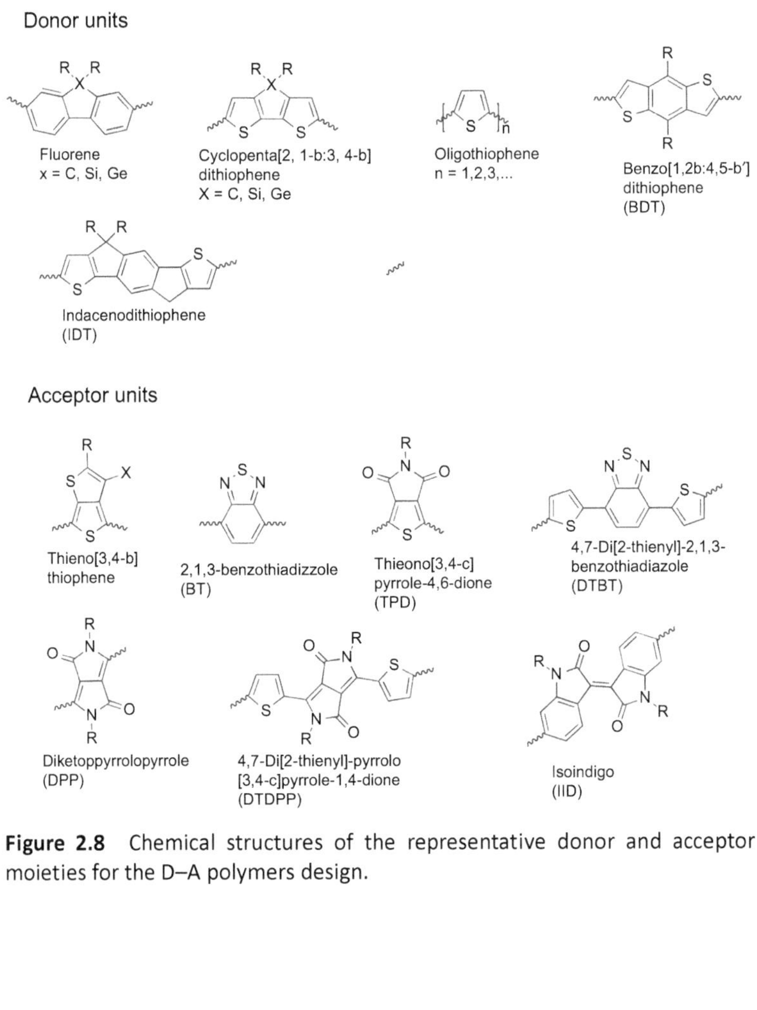

Figure 2.8 Chemical structures of the representative donor and acceptor moieties for the D–A polymers design.

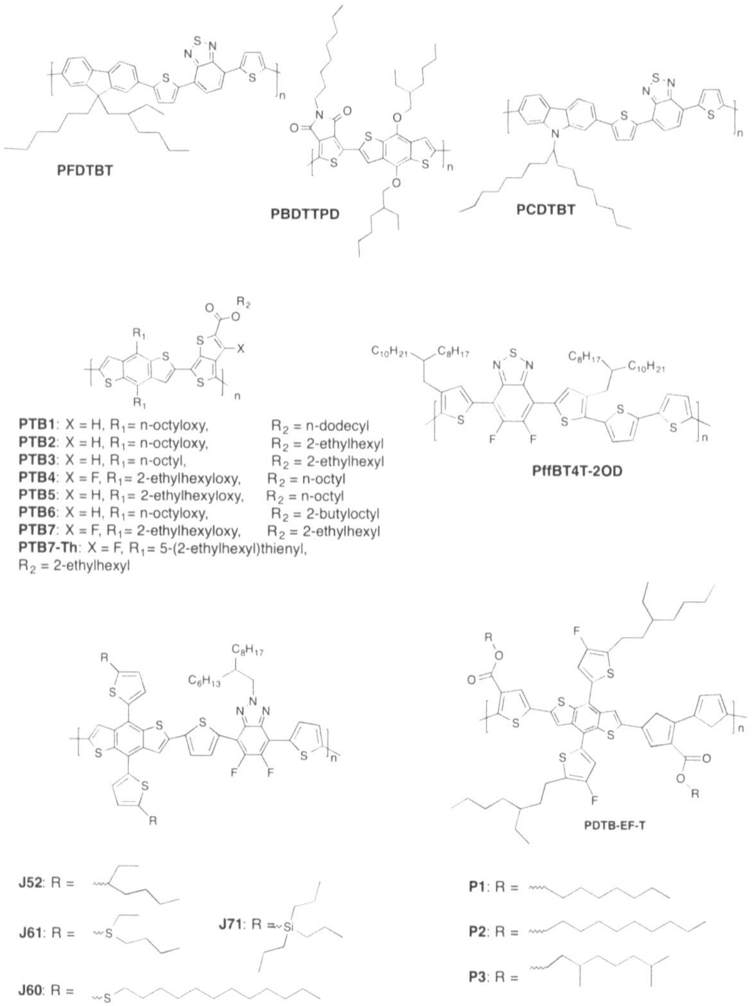

Figure 2.9 Chemical structures of selected high-performance D–A conjugated polymers.

Further optimization of D–A polymer is aimed at enhancing charge carrier mobility. OSCs suffer from the low charge carrier mobility issue as compared to their inorganic counterparts [5, 57–59]. Adding high-mobility-inducing fragments is an effective way to improve the charge carrier mobility. For example, DPP showed high hole mobility when incorporated into a D–A polymer because of the planarity structure and strong electron-withdrawing ability of the DPP unit. That is, PDPP-TT-T-TT has a hole mobility of 1.42 $cm^2/V{\cdot}s$ [60]; PDPPT4T-2OD, with a hole mobility of $1.5–3.0 \times 10^{-2}$ $cm^2/V{\cdot}s$, exhibits 10.5% PCE when fabricated into devices [61].

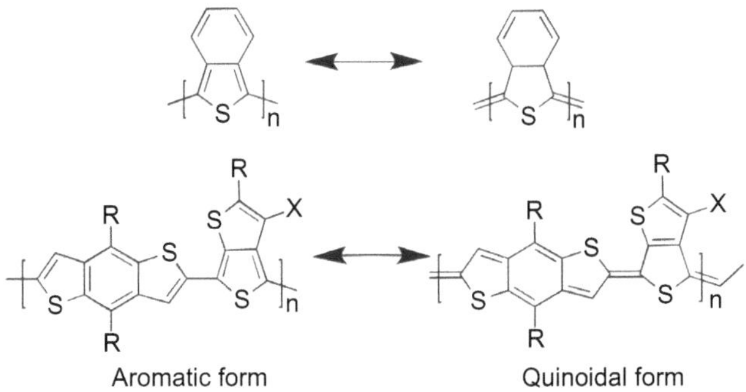

Figure 2.10 Aromatic and quinoidal forms of poly(benzo[c]thiophene) and PTB polymers.

In the conjugated polymer, the side chain not only determines the solubility of the polymer but also plays an important role in determining the intermolecular interactions. For example, Li et al. synthesized and studied a group of polymers with different alkylthio side chains, namely, J52, J60, J61, and J71 (Fig. 2.10) [62, 63]. These polymers are paired with a small-molecule acceptor (which will be discussed later in Section 2.4.2) to form BHJ solar cells. The side chain conformation on the photovoltaic performance is studied. Among them, J61, which has a linear alkylthio side chain, exhibits a PCE of 9.5% with enhanced J_{SC} of 17.43 mA/cm^2 [63]. Closely following this work, the same group synthesized polymer J71 with a star-shaped alkylthio side chain and obtained 11.4% PCE [62, p. 4].

In addition to side chain modification, functional groups, such as fluorine, cyano group, and heteroatoms (e.g., Si and Se), can significantly influence the energy levels and physical properties. Fluorination is one of the most studied and widely applied methods to alter the energy levels [64, 65]. Sitting in the top-right corner of the periodic table, fluorine atom is the most electronegative element, while the atom is small in size. This allows fluorination to lower the HOMO and LUMO levels without introducing much steric hindrance [64]. Besides, fluorination changes the charge distribution along the polymer and enhances the intermolecular interaction to improve polymer crystallinity [64, 66]. A group of PTB-based donor polymers were synthesized with fluorine atoms at a different position on the backbone to study the effect of fluorination [67]. The performance of the solar cells strongly depends on the position and number of the fluorine atoms. By substituting 0, 1, 2, and 3 fluorine atoms, 5.1%, 7.4%, 3.2%, and 2.3% PCE were obtained. The results also indicate that fluorination lowered the HOMO and LUMO levels, enhance the crystallinity of the polymer, and increase phase separation between polymer and PCBM. In a recent study, Hou et al. designed a deep HOMO polymer PDTB-EF-T (Fig. 2.9) to match up with a fluorinated, non-fullerene acceptor named IT-4F (which will be discussed in Section 2.4.2) [9]. In this system, the fluorination of PDTB-EF-T successfully achieved a deep HOMO level (5.5 eV) and obtained a high V_{OC} of 0.90 eV and a record-breaking PCE of 14.2%.

2.3.2 Conjugated Small Molecules

Even though polymer donors have achieved great success in achieving >10% efficiency, polymers have their drawbacks. First, the synthesis of polymers is usually a multistep, low-yield process [68]. Second, polymers suffer from batch-to-batch variation [69]. Subtle difference in average molecular weight and polydispersity could affect the solar cell fabrication process by changing the polymer properties such as solubility, mixing, and nanoscale phase segregation. Hence, this batch-to-batch variation could influence the final BHJ morphology and hence the PCE. It limits the reproducibility of polymer solar cells and is not ideal for large-scale production.

Semiconducting small molecules are designed to solve this problem. Compared to polymers, small molecules are synthesized in fewer steps, which enables high yield. Small molecules have well-defined chemical structures with fixed and uniform molecular weight so that they do not suffer the batch-to-batch variation [70–74].

CuPc was the first small molecule used as donor in solar cells [13]. In 1986, Tang et al. fabricated bilayer CuPc-based solar cells with <1% PCE. Inspired by this work, Pc-based OSCs have been explored [75–79]. However, the PCE of Pc-based molecules is still low as compared to polymer-based OSCs. For years, research on small-molecule donors lagged behind the development of conjugated polymer donors. In the 2010s, use of solution-process small molecules as donor became an active area of research. In a few years, the PCE of small-molecule OSCs quickly improved and reached over 11% in 2018 [65, 80, 81].

Inspired by the successful D–A design of conjugated polymer donors, small molecules with alternating D–A structures have been developed. One common feature of the high-performing molecules is their axisymmetric nature. The D–A units are often designed to be symmetric about a central core unit, such as A–D–A, A1–D1–D2–D1–A, D1–A–D2–A–D, etc. Similar to conjugated polymers, there are common donor and acceptor units. For example, fluorinated benzodiathiazole (FBT), rhodanine (RH) and derivatives, and DPP are the most adopted acceptor units, while dithien- osilole (DTS), IDT, and BDT are the most adopted donor units.

Similar to conjugated polymers, the selection of the donor and acceptor units, the conjugation length, the shape and position of the side chain, and the functional group are critical in determining the electronic and physical properties of the small-molecule donors [73, 82, 83]. The similar strategies are shared between designing conjugated small molecules and polymers, and we will discuss and classify the D–A small molecules based on the most common moieties.

Thiophene is the most common moiety of OSC materials. Oligothiophene-based small molecules emerged as efficient donor material in OSCs, some of which is presented in Fig. 2.11. The first example shown in Fig. 2.11 is a series of six-thiophenes oligomers [84]. These molecules are designed with furan, thiophene, and selenophene as linkers to study the structure–property relationship. It was found that electron linkers with different electron-donating and polarization properties have an effective influence toward the energy levels. Among the three molecules, the highest PCE of 6.2% was achieved for the T3:$PC_{71}BM$-based OSC device. Continuous improvement was done by modification of the end group. In 2015, Zhang et al. modified the end group of a seven-unit oligothiophene [82]. RH and its derivatives with additional electron-withdrawing groups substituted at the C=S bond were used as the end-cap group to make DRCN7T and DERHD7T. Efficiencies of 6.0% and 9.3% were obtained by DRCN7T- and DERHD7T-based OSC devices, respectively. Since then RH became an essential end group in small-molecule donors. Besides the side chain and end group, the length of the oligothiophene core affects the molecule properties as well. A series of DRCNnT oligothiophene with 4, 5, 6, 7, 8, and 9 thiophene rings was synthesized and studied. It was found that odd-numbered oligothiophene chains outperformed those of even number, and the best efficiency was obtained by DRCN5T with the PCE of 10.0% [85].

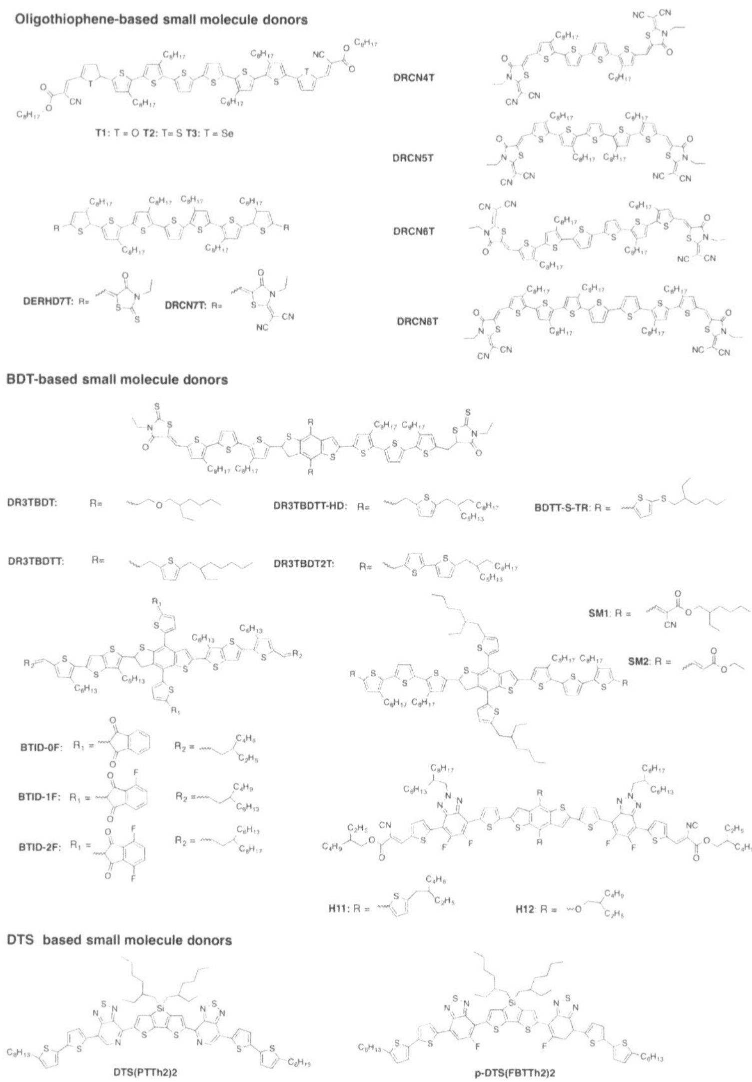

Figure 2.11 Chemical structures of conjugated small-molecule donors.

As discussed earlier, BDT unit has made great success in D–A polymer, e.g., PTB7. It has been intensively adopted in small-molecule design as well. BDT-based small molecules such as DR3TBDT, BDT-S-TR, SM1, SM2, H11, and H12 are presented in Fig. 2.11. The functions of side chains, end groups have been intensively studied. For instance, DR3TBDT, DR3TBDTT-HD, DR3TBDTT, and DR3TBDT2T (Fig. 2.11) are synthesized based on the same conjugated backbone but with different side chains [70]. PCEs ranging from 6.9% to 8.1% were obtained. Later, a follow-up work adopted the same structure with alkylthio-thienyl side chains, namely, BDTT-S-TR, and achieving 9.2% PCE [86]. Different film-forming properties were observed for BDTT-S-TR. Unlike other small-molecule counterparts, the processing does not require thermal or solvent annealing treatment. Further improvement was done in altering the end group. A recent work has synthesized SM1 and SM2 (Fig. 2.11) with weak electron-withdrawing ester end groups [87]. SM1 with cyano group (CN) on the ester end group wins over SM2, and 10.1% PCE was obtained. Along with side chain and end group modification, addition of functional atoms and groups is also important in determining the physical and electronic properties of small molecules. For instance, BTID-0F, BTID-1F, and BTID-2F were synthesized with different number of –F atoms [65]. An 11.3% PCE was obtained by BTID-2F-based OSC devices. In another work, Bin et al. designed BDT-based small molecules with fluorobenzotriazole as the acceptor unit, namely, H11 and H12 [88]. A PCE of 9.7% was obtained in the H11-based all-small-molecule OSC device.

Donors based on dithienosilole (DTS) have received much attention in recent molecular OSCs as well. Remarkable efficiencies above 8% have been achieved. For example, $DTS(PTTh_2)_2$ was designed with a DTS unit and a PT electron-withdrawing unit. When paired with $PC_{71}BM$, 6.7% PCE was obtained [69]. Further improvement with fluorination leads to the synthesis of *p*-$DTS(FBTTh_2)_2$, which exhibits 8.9% PCE with proper device engineering [89].

Besides molecule design, intensive research has been done with regard to morphological control and device engineering to get high-performance small-molecule OSCs [90–93].

2.4 Acceptor Materials Development

2.4.1 Fullerene Derivatives

Fullerene derivatives such as $PC_{61}BM$ and $PC_{71}BM$ were predominant acceptors for OSCs in the past decades. In 1992, Sariciftci et al. observed ultrafast electron transfer from MEH-PPV to C_{60} [30]. Based on this discovery, fullerene is widely studied as an electron acceptor in OSCs. However, the solubility of C_{60} in most organic solvents is low, limiting its application in BHJ OSCs. A C_{60} derivative $PC_{61}BM$ was synthesized to improve the solubility in common organic solvents [94]. The structure of $PC_{61}BM$ is shown in Fig. 2.12. The addition of alkyne chains improves the solubility of $PC_{61}BM$ to 35 mg/ml in o-DCB [95]. It has been intensively used in OSCs as fullerene acceptor.

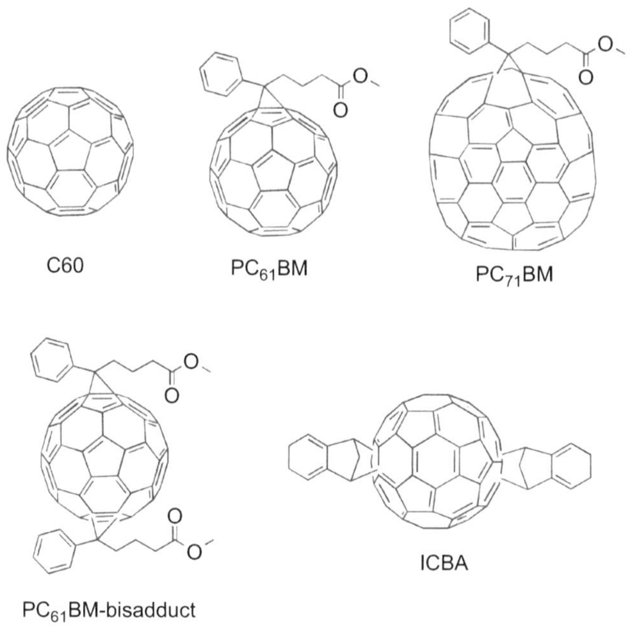

Figure 2.12 Fullerene-based acceptors for OSCs.

However, one drawback of $PC_{61}BM$ is the limited absorption in the visible light region. $PC_{71}BM$, a C_{70} derivative, was synthesized to enhance the absorption in the visible region [96]. The solubility of $PC_{71}BM$ is also greatly improved to be as high as 80 mg/ml in chloroform. $PC_{71}BM$ became the most common acceptor material and has been approved to match well with several low-bandgap polymers, such as PBDTTPD, PCDTBT, PffBT4T-2OD, exhibiting PCE up to 12% [97]. Further improvement in fullerene derivatives as acceptor material is mainly focused on structural modification on the substituent for the purpose of higher solubility, or for facilitating synthesis, purification, and processing [95].

One limiting factor of PCBM-based acceptor is the limited V_{OC}. As expressed in Eq. (2.4), V_{OC} is determined by the difference between the LUMO of acceptor and the HOMO of donor. In the case of P3HT:PCBM OSCs, since the HOMO of P3HT is −5.1 eV and the LUMO of $PC_{61}BM$ is −4.2 eV, V_{OC} of the system is restricted around 0.6 eV [24]. In order to obtain the up-shifted LUMO level of the acceptor material, fullerene bis-adduct and multi-adduct derivatives were synthesized to up-shift the LUMO level. These include the PCBM-based bis-adducts/multi-adducts [98–100] and indene-C_{60} bis-adduct and indene-C_{70} bis-adduct [101–103]. These approaches have successfully improved the V_{OC} from 0.6 to 0.84 eV when paired with P3HT [95].

Despite the beneficial properties of fullerene-based acceptors, there are considerable limitations and problems with them. (1) PCBM has poor absorption in the visible and near-infrared (NIR) regions. (2) Limited chemical modifications can be made to PCBM to tune the energy levels. This not only increases the synthetic complexity but also makes fullerenes less likely to gain complementary light harvesting to the polymers. (3) PCBM has poor stability in blend films when made into devices [104, 105]. Under irradiation, the fullerenes easily diffuse and excessively aggregate to big clusters (crystallization) and can lead to very rapid losses in PCE over a short time period [106]. These limitations and challenges triggered the development of non-fullerene acceptors (NFAs). There are polymers- and small-molecule-based NFAs; this chapter refers NFAs to small-molecule acceptors unless specified.

2.4.2 Non-fullerene Small Molecules

In the past 5 years, small-molecule acceptors have developed fast. NFAs win over PCBM with distinct features: (1) tunable bandgap and energy levels; (2) broadened absorption in the visible–NIR region; (3) flexibility of structural modification to control the planarity and crystallinity [41, 107–111].

The development of NFAs can be traced back to the 1980s. Indeed, the first reported bilayer OSCs used perylene-based acceptor [13]. Many NFAs have been synthesized and showed great potential for tuning energy levels, planarity, and crystallinity. These include perylene diimide (PDI) [112, 113], naphthalene diimide (NDI), BT-based NFAs, and so on. Although at the time these NFA-based OSCs generally exhibit lower efficiencies than the fullerene-based OSCs, these researches provide helpful insight into designing high-performance NFAs. Since 2015, NFAs have become a major research focus and rapid progress has been achieved. A vast variety of NFAs have been examined in BHJ OSCs. PCEs for NFA-based OSCs increased from 6% to 14% in the literature.

The first well-studied NFAs are PDI-based molecules (Fig. 2.13). PDI derivatives were first designed to replace $PC_{61}BM$. For example, a PDI dimer-based NFA, SF-PDI_2, was synthesized and paired with P3HT to make fullerene-free OSC devices [114]. However, the PCE of this P3HT:PDI_2-based solar cell was only 2.4%. The reason for the low efficiency is that SF-PDI_2 has a bandgap of 2.0 eV and P3HT has a bandgap of 1.9 eV. The wide bandgap for both donor and polymer limits the photon harvesting of the OSC device. The immediate strategy to increase the absorption of BHJ in the visible–NIR region is to use a readily low-bandgap polymer instead of P3HT. For example, a low-bandgap polymer PTB7-Th (bandgap 1.6 eV) was paired with a wide-bandgap PDI-based NFA, hPDI_4 (bandgap 2.35 eV) to make BHJ OSCs, and achieved 8.3% PCE [115]. Further research studied chemical structure and property relationships and their effect on the PCE. Two PDI trimmers, Ta-PDI and Ph-PDI, were synthesized with different cores. Even though Ta-PDI and Ph-PDI have a similar bandgap of 2.2 eV, Ta-PDI OSC demonstrates a PCE of 9.2% when paired with PTB7-Th. It wins over the Ph-PDI-based devices because its core is more electron deficient [116]. A strategy to continue the increase in PCE is selecting donor and acceptor with complementary absorption in the visible–NIR region. Impressive achievement has been obtained using this strategy [114, 116]. For example, in 2016, an NFA OSC with 9.5% efficiency was demonstrated by using a low-bandgap polymer P3TEA (1.7 eV) and SF-PDI_2. This BHJ has broader absorption due to the less overlap between donor and acceptor absorption, exhibits boosted J_{SC} and nearly 90% internal quantum efficiency [117].

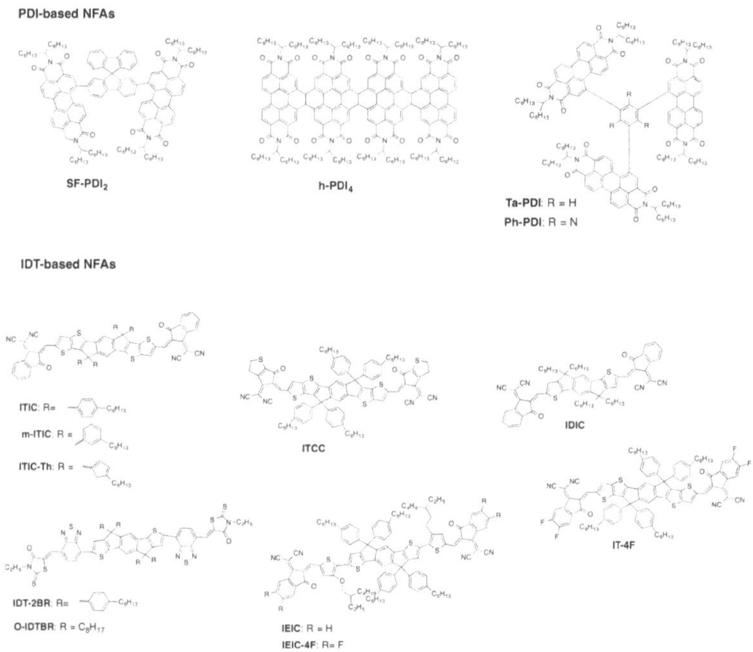

Figure 2.13 Chemical structures of representative non-fullerene acceptors for OSCs.

Despite the success of using low-bandgap polymer with PDI-based NFAs in OSCs, the PDI-based molecules generally have wide bandgap greater than 2.0 eV, reduction of which is required for further improvements. Learning from the design of low-bandgap polymers, the concept of D–A design has been adopted when designing low-bandgap NFAs. In 2015, ITIC was synthesized as an A–D–A small-molecule acceptor, which has a bandgap of only 1.6 eV [118]. The chemical structure is depicted in Fig. 2.13; it has an IDT moiety as the core and 2-(3-oxo-2,3-dihydroinden-1-ylidene) malononitrile (INCN) moieties as the end groups. The coplanar structure of the IDT [119, 120] core provides good hole mobility and broadens absorption, while the electron-withdrawing INCN [121] unit provides the availability of deep LUMO levels. When paired with PTB7-Th, PTB7-Th:ITIC OSC device exhibits 6.8% PCE [118]. Later, ITIC was paired with a wide bandgap 2D polymer, named J71, exhibiting a PCE of 11.8% [62]. The high PCE is benefitted from the broadened absorption region of BHJ and the complementary absorption of the J71 and ITIC. Since then, a number of works have been done with regard to ITIC structure modification. For example, ITIC-Th [122] and m-ITIC [123] are synthesized by modifying the side chain. In the report, ITIC-Th was paired with PDBT-T1, exhibiting a PCE of 9.6% [122]. m-ITIC was paired with J61, exhibiting a PCE of 11.7% [123].

Further modification tuned the energy level of ITIC to achieve the desired electronic properties. For example, ITCC was synthesized by replacing the phenyl end group with thienyl group [124]. The corresponding LUMO of ITCC was up-shifted by 0.14 eV as compared to ITIC, leading to a high V_{OC} of 1.01 eV when paired with PBDB-T. A PCE of 11.4% was obtained in PBDB-T:ITCC OSCs [124]. Besides, the number of fused rings in the core plays a role in determining the properties as well [41, 122, 125]. IDIC, which has five fused rings (IDIC) as the core unit, outperforms the seven-ring ITIC [126]. PCE as high as 11% was obtained from the IDIC-based OSCs. Based on this achievement, continued modification was done to IDIC with regard to the side chain engineering. For example, IDT-2BR [127] and O-IDTBR [128] were synthesized by adding a phenyl side chain and a longer alkyl chain, respectively. PCE from 6% to 8% was obtained when blended with donor polymers. In another work, IEIC was synthesized by adding a bridging thiophene between the IDT core unit and the end groups; a bandgap of 1.57 eV was obtained and a PCE of 6.3% was obtained when paired with PTB7-Th [129].

Exciting and rapid progress has been made recently with regard to further lowering the bandgap of ITIC-based fused-ring NFAs. NFAs with extremely narrow bandgap below 1.5 eV were reported. For example, IHIC adopts a six-fused-ring IDT core and exhibits broad and strong absorption with a bandgap of 1.38 eV [130]. A 9.7% PCE was obtained when paired with PTB7-Th. Similar to conjugated D–A polymers/small molecules, adding a functional group greatly affects the energy levels and bandgap. Fluorination of IEIC and ITIC leads to extremely narrow bandgaps of 1.24 and 1.22 eV for IEICO-4F [131] and IT-4F [132, 133], respectively. IEICO-4F was paired with mixed PTB7-Th:J52 donor, exhibiting approximately 11% PCE [131]. IT-4F was paired with PDTB-EF-T to make OSCs and exhibited a new world record efficiency of 14.2% [9]. To date, the best NFA solar cells are based on ITIC-type acceptors. With proper optimization, screening donor small molecules/polymers, BHJ morphology control, and device engineering, it is possible to obtain higher PCE.

Besides polymer:NFA OSCs, exciting progress has been made with all-small-molecule OSCs. Profound efficiencies in the range of 6–10% were obtained using small molecules as donor materials and NFA as acceptors [87, 134–137]. This area has attracted intense research interest and has great potential in the future.

2.5 Interfacial Materials and Device Engineering

Interfacial materials play several important roles [18, 138–140]. First, an interfacial layer works as an electron/hole-blocking layer between BHJ and electrode. Second, an interfacial layer can modify the workfunction of the substrates and tune the energy level alignment at the BHJ/electrode interfaces. Third, an interfacial layer works as a buffer layer to prevent the migration of metal electrode diffusion into the organic layer. Fourth, it can work as an optical spacer to modulate the light absorption in BHJ. Fifth, it can alter the surface energy of the substrate, which as a result alters the morphology and phase distribution in the BHJ.

According to the charge flow direction, OSCs can be divided into conventional and inverted devices. The feasibility of choosing interfacial layers with different workfuntions allows the device structure engineering possible. The forward and inverted structure can be built upon the selection of a proper electron/hole-selective interfacial layer. For example, in the conventional forward structure solar cell, the general structure is ITO/PEDOT:PSS/BHJ/LiF/Al [15], while ITO is the anode collecting holes and Al collecting electrons. PEDOT:PSS works as a hole-selective layer while blocking electrons. If we change the interfacial layers and make it to ITO/ZnO/BHJ/MoOx/Ag [141], ZnO works as hole-blocking layers to make ITO a cathode to collect electrons. In this case, an inverted structure is built. Using interfacial layers to determine the polarity of a solar cell device allows feasibility of device structure engineering, especially in unconventional structures and tandem solar cells.

Tremendous work has been done in the design of interfacial layers. Among the family of electron-selecting interfacial layer, n-type metal oxides with deep-lying energy level are currently dominant in the field. ZnO [142–144], TiO_x [145–147], Nb_2O_5 [148–150], and ZrO_2 [151] are well-developed electron-selective materials. Organic materials, including polymers and small molecules, have often been used in the recent years to avoid thermal treatment and improve compatibility with organic layers. Figure 2.14 presents the chemical structures of several well-known organic interfacial molecules. These include inexpensive nonconjugated polyelectrolytes such as poly(ethylenimine)-ethoxylated (PEIE) [152], poly(vinylpyrrolidone) (PVP) [153], polyethyleneimine (PEI) [154], and polyallylamine (PAA) [154] as well as conjugated polymers such as poly[(9,9-bis(3′-(*N*,*N*-dimethylamino)-propyl)-2,7-fluorene)-alt-2,7-(9,9- dioctylfluorene)] (PFN) [55] and its derivatives (Fig. 2.14). PFN was used as an electron-selective layer in an inverted PTB7:$PC_{71}BM$, and 9.2% PCE was obtained [55]. Small molecules, such as bathocuproine (BCP) [155, 156], bathophenanthroline (Bphen) [157], pyridinium salt-based molecules, triazine- and pyridinium-based small molecules [158], have also been adopted as effective electron-selective layers for OSCs. Recent research also indicates other types of electron-selective materials, such as low-workfunction metal salts, carbon-based materials, inorganic–organic hybrid composites, and so on [16].

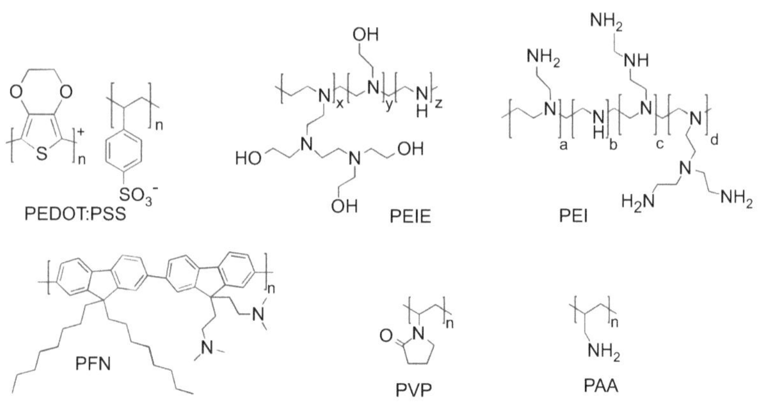

Figure 2.14 Representatives of polymeric interfacial materials.

Hole-selective interfacial layers should have high workfunctions to match the HOMO level of donor polymer. PEDOT:PSS is one of the most studied and adopted hole-selective interfacial layer [159]. It has advantages such as good solubility in water, transparency in the visible–NIR region, good conductivity, and proper workfunction (5.1 eV) that matches well with many polymer donors to form good Ohmic contact. However, it also has drawbacks such as its acidic nature and water-absorbing problems, which impairs the device stability [160–162]. To overcome these problems, new conducting polymers are developed to replace PEDOT:PSS, such as PANI [163] and HAPAN [164]. Efforts have also been done by adding a modifier layer on top of PEDOT:PSS, in order to migrate the disadvantages while keeping the favorable properties [165–168]. Other classes of materials are also developed to replace PEDOT:PSS. These includes metal oxides and sulfides (MoO_3 [169–171], MoS_2 [172, p. 2; 173], NiO_x [174], CrO_x [175], etc.), graphene oxides [176–179], and conjugated polyelectrolytes/small molecules [180, 181].

A lot of studies have focused on exploring interfacial layers. Sophisticated reviews can be found for more detailed progress of the emerging interfacial materials [16, 18, 138]. There are common rules to follow when designing new interfacial layers. A proper interfacial layer should satisfy several requirements: (1) transparent in visible and NIR regions; (2) proper energy level; (3) preferentially select charges (hole or electron); (4) good conductivity and stability; (5) low-cost and easy to process. The vast library of interfacial materials allows state-of-the-art device engineering, especially in tandem cells and flexible devices [18, 138, 182].

2.6 Flexible and Organic Solar Cells

The distinct advantages of OSCs over other solar cell technologies are solution processability and potential for large-scale production. As introduced in previous sections, OSCs adopt a layered structure with organic polymers and small molecules. These polymers/molecules can be processed in solution onto flexible foils to enable fast, cost-effective, and large-scale production, such as roll-to-roll printing

[183–186]. This low-energy-cost, high-throughput manufacturing can eventually make the price of OSCs competitive in the market [6, 7, 187].

Figure 2.15 Example of large-scale OSC installation. Reprinted from Ref. [7], Copyright 2017, with permission from Elsevier.

Compared to Si-based solar cells, the organic components in OSCs are nontoxic, earth abundant, light weighted, and low cost. Researchers have developed a massive library of organic OSC materials for desired functions. For example, the color of the OSC can be tuned by customizing the chemical structure of photoactive materials. For building-integration OSCs, it can be made semi-transparent by tailoring the absorption range of photoactive materials. The light-weighted and flexible OSCs can be rolled or folded during transportation, which saves the cost for bulky protective packing. It is also suitable for large-scale installation and building integrated applications [188, 189]. The BHJ structure enables the active layer to have sufficient light harvesting with a very small thickness (~100 nm); OSC modules can be fabricated with a total thickness of 2 μm, which is appealing to wearable and portable applications [190, 191].

Although ITO-coated glass substrate is still dominating in lab research, many efforts have been made to replace it for flexible

OSCs. Emerging candidates such as metal grid/conducting polymer composites [192], carbon nanotube/graphene networks [193, 194], ultra-thin metals [184], and metal nanowire networks [195] have been developed. OSC scientists have invested tremendous efforts into this area. Krebs et al. denoted a great deal of work in developing and studying the fully solution-processed flexible OSCs by large-scale manufacturing [196–202]. The group is leading large-scale flexible OSC research and founded a startup named infinityPV in 2014 to commercialize flexible OSCs. In one of their reports, they fabricated flexible OSC devices by roll-coating fabrication on two kinds of flexible substrates: polyethylene terephthalate (PET) foil with printed Ag grid and stripe-patterned thin ITO-coated PET substrate (Fig. 2.16) [203]. Flexible devices based on PTB7:IEIC and PTB7:$PC_{61}BM$ were fabricated, and module efficiencies of 2.2% and 2.4% were obtained [203]. In another work, transparent flexible OSCs with 6.8% efficiency were obtained by an all-solution-process manufacturing using a TiO_2-coated PET substrate [184]. The efficiency of the flexible OSCs still has a lot of room to improve. Currently, the reported efficiencies of flexible OSC module still lag behind the lab efficiency [7, 186]. More efforts need to be devoted to improve the PCEs of OSCs produced at a large scale. Understanding the effect of processing parameters on the morphology, crystallinity, stability, and PCE is a critical task in the future [202, 204–209]. All these challenges need to be addressed for long-term application and successful commercialization. More work is needed for large-scale preparation, installation, and operation of large-polymer solar cell panels in a grid-tied system.

The future market potential of OSCs has been analyzed and evaluated. Table 2.1 compares the cost of different solar cell technologies. As shown in the table, even though OSCs (OPV) are behind in stability and efficiency, the speed of manufacturing and installation of OSCs is way faster, and the energy payback time (EPBT) is remarkably shorter. These features lead to an overall competitive commercial price for OSCs, especially in the ideal condition that PCE is greater than 7%.

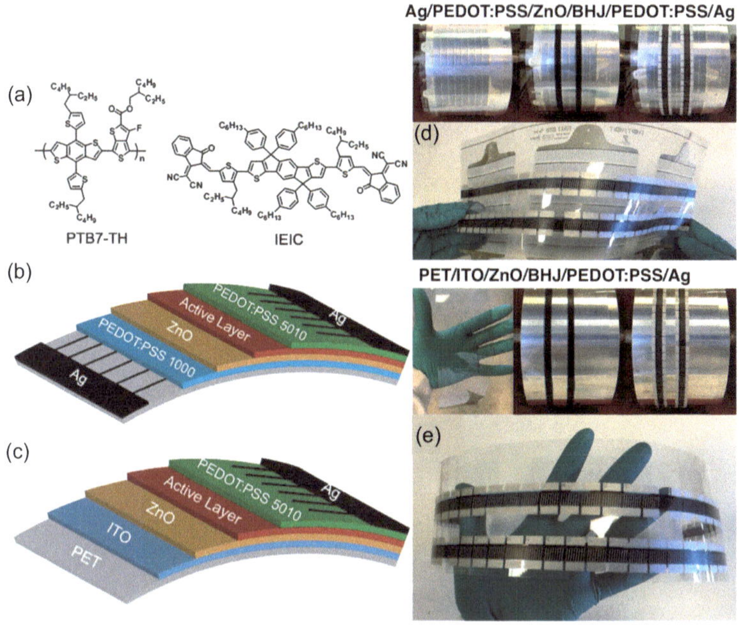

Figure 2.16 PTB7-Th:IEIC-based (a) OSC devices roll coated on two different flexible electrodes: (b, d) PEDOT:PSS and ZnO-coated PET flexible substrate with an Ag grid; and (c, f) stripe-patterned ITO on PET substrate. Reconstructed with permission from Ref. [203], Copyright 2015, Royal Society of Chemistry.

Table 2.1 Comparison of scaled cost and installation between photovoltaic/solar cell technologies

	€/W_p	EPBT (days)	Manufacturing speed (W_p/h)	Installation speed (W_p/s)	Lifetime T_{80} (years)	PCE (%)
OPV (this work)	1	60	~1 kW_p	8	2	4
OPV (limit)	0.05	1	>1,000,000 kW_p	20	10	>7
c-Si	0.35	<400	4500 kW_p	<1	>25	21
CdTe	0.57	<300	300 kW_p	<1	>25	19.5

Note: Reconstructed from Ref. [7] with permission, copyright © 2017 Elsevier Inc.

In summary, OSCs have unique properties over other solar cell technologies. The flexibility, transparency, light weight, tunable color, potential of low-cost large-scale manufacturing, and low environmental impact make OSCs a competitive player in the future renewable energy marketplace. With the new record of 14.2% in academia [9] and 13.2% in commercialized industry by Heliatek [10], OSCs have enormous potential in contributing to our sustainable future.

References

1. Lewis, N. S. and Nocera, D. G. (2006). Powering the planet: Chemical challenges in solar energy utilization. *Proc. Natl. Acad. Sci.*, **103**, pp. 15729–15735.
2. Reale, A., La Notte, L., Salamandra, L., Polino, G., Susanna, G., Brown, T. M., Brunetti, F., and Di Carlo, A. (2015). Spray coating for polymer solar cells: An up-to-date overview. *Energy Technol.*, **3**, pp. 385–406.
3. Krebs, F. C., Espinosa, N., Hösel, M., Søndergaard, R. R., and Jørgensen, M. (2014). 25th Anniversary Article: Rise to power: OPV-based solar parks. *Adv. Mater.*, **26**, pp. 29–39.
4. Nielsen, T. D., Cruickshank, C., Foged, S., Thorsen, J., and Krebs, F. C. (2010). Business, market and intellectual property analysis of polymer solar cells. *Sol. Energy Mater. Sol. Cells*, **94**, pp. 1553–1571.
5. Darling, S. B. and You, F. (2013). The case for organic photovoltaics. *RSC Adv.*, **3**, pp. 17633–17648.
6. Espinosa, N., Hösel, M., Angmo, D., and Krebs, F. C. (2012). Solar cells with one-day energy payback for the factories of the future. *Energy Environ. Sci.*, **5**, pp. 5117–5132.
7. Carlé, J. E., Helgesen, M., Hagemann, O., Hösel, M., Heckler, I. M., Bundgaard, E., Gevorgyan, S. A., Søndergaard, R. R., Jørgensen, M., García-Valverde, R., Chaouki-Almagro, S., Villarejo, J. A., and Krebs, F. C. (2017). Overcoming the scaling lag for polymer solar cells. *Joule*, **1**, pp. 274–289.
8. Machui, F., Hösel, M., Li, N., Spyropoulos, G. D., Ameri, T., Søndergaard, R. R., Jørgensen, M., Scheel, A., Gaiser, D., Kreul, K., Lenssen, D., Legros, M., Lemaitre, N., Vilkman, M., Välimäki, M., Nordman, S., Brabec, C. J., and Krebs, F. C. (2014). Cost analysis of roll-to-roll fabricated ITO free single and tandem organic solar modules based on data from manufacture. *Energy Environ. Sci.*, **7**, pp. 2792–2802.

9. Li, S., Ye, L., Zhao, W., Yan, H., Yang, B., Liu, D., Li, W., Ade, H., and Hou, J. (2018). A wide band gap polymer with a deep highest occupied molecular orbital level enables 14.2% efficiency in polymer solar cells. *J. Am. Chem. Soc.*, **140**, pp. 7159–7167.

10. Heliatek sets new organic photovoltaic world record efficiency of 13.2%. https://www.heliatek.com/en/press/press-releases/details/heliatek-sets-new-organic-photovoltaic-world-record-efficiency-of-13-2.

11. Mazzio, K. A. and Luscombe, C. K. (2014). The future of organic photovoltaics. *Chem. Soc. Rev.*, **44**, pp. 78–90.

12. Brabec, C. J., Sariciftci, N. S., and Hummelen, J. C. (2001). Plastic solar cells. *Adv. Funct. Mater.*, **11**, pp. 15–26.

13. Tang, C. W. (1986). Two-layer organic photovoltaic cell. *Appl. Phys. Lett.*, **48**, pp. 183–185.

14. Yu, G., Gao, J., Hummelen, F. W., and Heeger, A. J. (1995). Polymer photovoltaic cells: Enhanced efficiencies via a network of internal donor-acceptor heterojunctions. *Science*, **270**, pp. 1789–1791.

15. Padinger, F., Rittberger, R. S., and Sariciftci, N. S. (2003). Effects of postproduction treatment on plastic solar cells. *Adv. Funct. Mater.*, **13**, pp. 85–88.

16. Yin, Z., Wei, J., and Zheng, Q. (2016). Interfacial materials for organic solar cells: Recent advances and perspectives. *Adv. Sci.*, **3**, pp. 1500362.

17. Kim, H., Lim, K.-G., and Lee, T.-W. (2016). Planar heterojunction organometal halide perovskite solar cells: Roles of interfacial layers. *Energy Environ. Sci.*, **9**, pp. 12–30.

18. Chueh, C.-C., Li, C.-Z., and Jen, A.K.-Y. (2015). Recent progress and perspective in solution-processed interfacial materials for efficient and stable polymer and organometal perovskite solar cells. *Energy Environ. Sci.*, **8**, pp. 1160–1189.

19. Pivrikas, A., Sariciftci, N. S., Juška, G., and Österbacka, R. (2007). A review of charge transport and recombination in polymer/fullerene organic solar cells. *Prog. Photovolt. Res. Appl.*, **15**, pp. 677–696.

20. Li, G., Zhu, R., and Yang, Y. (2012). Polymer solar cells. *Nat. Photonics*, **6**, pp. 153–161.

21. Cao, W. and Xue, J. (2014). Recent progress in organic photovoltaics: Device architecture and optical design. *Energy Environ. Sci.*, **7**, pp. 2123–2144.

22. Lindholm, F. A., Fossum, J. G., and Burgess, E. L. (1979). Application of the superposition principle to solar-cell analysis. *IEEE Trans. Electron Devices*, **26**, pp. 165–171.

23. Guo, X., Zhou, N., Lou, S. J., Smith, J., Tice, D. B., Hennek, J. W., Ortiz, R. P., Navarrete, J. T. L., Li, S., Strzalka, J., Chen, L. X., Chang, R. P. H., Facchetti, A., and Marks, T. J. (2013). Polymer solar cells with enhanced fill factors. *Nat. Photonics*, **7**, pp. 825–833.

24. Zhou, H., Yang, L., and You, W. (2012). Rational design of high performance conjugated polymers for organic solar cells. *Macromolecules*, **45**, pp. 607–632.

25. Jung, J. W., Jo, J. W., Jung, E. H., and Jo, W. H. (2016). Recent progress in high efficiency polymer solar cells by rational design and energy level tuning of low bandgap copolymers with various electron-withdrawing units. *Org. Electron.*, **31**, pp. 149–170.

26. Scharber, M. C., Mühlbacher, D., Koppe, M., Denk, P., Waldauf, C., Heeger, A. J., and Brabec, C. J. (2006). Design rules for donors in bulk-heterojunction solar cells: Towards 10% energy-conversion efficiency. *Adv. Mater.*, **18**, pp. 789–794.

27. Lu, L., Zheng, T., Wu, Q., Schneider, A. M., Zhao, D., and Yu, L. (2015). Recent advances in bulk heterojunction polymer solar cells. *Chem. Rev.*, **115**, pp. 12666–12731.

28. Dennler, G., Scharber, M. C., and Brabec, C. J. (2009). Polymer-fullerene bulk-heterojunction solar cells. *Adv. Mater.*, **21**, pp. 1323–1338.

29. Zhang, S., Ye, L., Zhao, W., Liu, D., Yao, H., and Hou, J. (2014). Side chain selection for designing highly efficient photovoltaic polymers with 2D-conjugated structure. *Macromolecules*, **47**, pp. 4653–4659.

30. Sariciftci, N. S., Smilowitz, L., Heeger, A. J., and Wudl, F. (1992). Photoinduced electron transfer from a conducting polymer to buckminsterfullerene. *Science*, **258**, pp. 1474–1476.

31. Kraabel, B., Lee, C. H., McBranch, D., Moses, D., Sariciftci, N. S., and Heeger, A. J. (1993). Ultrafast photoinduced electron transfer in conducting polymer—buckminsterfullerene composites. *Chem. Phys. Lett.*, **213**, pp. 389–394.

32. Halls, J. J. M., Walsh, C. A., Greenham, N. C., Marseglia, E. A., Friend, R. H., Moratti, S. C., and Holmes, A. B. (1995). Efficient photodiodes from interpenetrating polymer networks. *Nature*, **376**, pp. 498–500.

33. Brabec, C. J., Shaheen, S. E., Winder, C., Sariciftci, N. S., and Denk, P. (2002). Effect of LiF/metal electrodes on the performance of plastic solar cells. *Appl. Phys. Lett.*, **80**, pp. 1288–1290.

34. Schilinsky, P., Waldauf, C., and Brabec, C. J. (2002). Recombination and loss analysis in polythiophene based bulk heterojunction photodetectors. *Appl. Phys. Lett.*, **81**, pp. 3885–3887.

35. Kim, J. Y., Kim, S. H., Lee, H.-H., Lee, K., Ma, W., Gong, X., and Heeger, A. J. (2006). New architecture for high-efficiency polymer photovoltaic cells using solution-based titanium oxide as an optical spacer. *Adv. Mater.*, **18**, pp. 572–576.

36. Ma, W., Yang, C., Gong, X., Lee, K., and Heeger, A. J. (2005). Thermally stable, efficient polymer solar cells with nanoscale control of the interpenetrating network morphology. *Adv. Funct. Mater.*, **15**, pp. 1617–1622.

37. Reyes-Reyes, M., Kim, K., and Carroll, D. L. (2005). High-efficiency photovoltaic devices based on annealed poly(3-hexylthiophene) and 1-(3-methoxycarbonyl)-propyl-1-phenyl-(6,6)C61 blends. *Appl. Phys. Lett.*, **87**, pp. 083506.

38. Verploegen, E., Mondal, R., Bettinger, C. J., Sok, S., Toney, M. F., and Bao, Z. (2010). Effects of thermal annealing upon the morphology of polymer–fullerene blends. *Adv. Funct. Mater.*, **20**, pp. 3519–3529.

39. Lou, S. J., Szarko, J. M., Xu, T., Yu, L., Marks, T. J., and Chen, L. X. (2011). Effects of additives on the morphology of solution phase aggregates formed by active layer components of high-efficiency organic solar cells. *J. Am. Chem. Soc.*, **133**, pp. 20661–20663.

40. Lee, J. K., Ma, W. L., Brabec, C. J., Yuen, J., Moon, J. S., Kim, J. Y., Lee, K., Bazan, G. C., and Heeger, A. J. (2008). Processing additives for improved efficiency from bulk heterojunction solar cells. *J. Am. Chem. Soc.*, **130**, pp. 3619–3623.

41. Lin, H., Chen, S., Li, Z., Lai, J. Y. L., Yang, G., McAfee, T., Jiang, K., Li, Y., Liu, Y., Hu, H., Zhao, J., Ma, W., Ade, H., and Yan, H. (2015). High-performance non-fullerene polymer solar cells based on a pair of donor–acceptor materials with complementary absorption properties. *Adv. Mater.*, **27**, pp. 7299–7304.

42. Zhou, H., Yang, L., Stoneking, S., and You, W. (2010). A weak donor–strong acceptor strategy to design ideal polymers for organic solar cells. *ACS Appl. Mater. Interfaces*, **2**, pp. 1377–1383.

43. Brédas, J.-L., Norton, J. E., Cornil, J., and Coropceanu, V. (2009). Molecular understanding of organic solar cells: The challenges. *Acc. Chem. Res.*, **42**, pp. 1691–1699.

44. Svensson, M., Zhang, F., Veenstra, S. C., Verhees, W. J. H., Hummelen, J. C., Kroon, J. M., Inganäs, O., and Andersson, M. R. (2003). High-

performance polymer solar cells of an alternating polyfluorene copolymer and a fullerene derivative. *Adv. Mater.*, **15**, pp. 988–991.

45. Park, S. H., Roy, A., Beaupré, S., Cho, S., Coates, N., Moon, J. S., Moses, D., Leclerc, M., Lee, K., and Heeger, A. J. (2009). Bulk heterojunction solar cells with internal quantum efficiency approaching 100%. *Nat. Photonics*, **3**, pp. 297–302.

46. Zou, Y., Najari, A., Berrouard, P., Beaupré, S., Réda Aïch, B., Tao, Y., and Leclerc, M. (2010). A thieno[3,4-c]pyrrole-4,6-dione-based copolymer for efficient solar cells. *J. Am. Chem. Soc.*, **132**, pp. 5330–5331.

47. Liang, Y., Xu, Z., Xia, J., Tsai, S.-T., Wu, Y., Li, G., Ray, C., and Yu, L. (2010). For the bright future: Bulk heterojunction polymer solar cells with power conversion efficiency of 7.4%. *Adv. Mater.*, **22**, pp. E135–E138.

48. Zhang, Y., Parnell, A. J., Pontecchiani, F., Cooper, J. F. K., Thompson, R. L., Jones, R. A. L., King, S. M., Lidzey, D. G., and Bernardo, G. (2017). Understanding and controlling morphology evolution via DIO plasticization in PffBT4T-2OD/PC71BM devices. *Sci. Rep.*, **7**, pp. 44269.

49. Ma, W., Yang, G., Jiang, K., Carpenter, J. H., Wu, Y., Meng, X., McAfee, T., Zhao, J., Zhu, C., Wang, C., Ade, H., and Yan, H. (2015). Influence of processing parameters and molecular weight on the morphology and properties of high-performance PffBT4T-2OD:PC71BM organic solar cells. *Adv. Energy Mater.*, **5**, pp. 1501400.

50. Chu, T.-Y., Alem, S., Verly, P. G., Wakim, S., Lu, J., Tao, Y., Beaupré, S., Leclerc, M., Bélanger, F., Désilets, D., Rodman, S., Waller, D., and Gaudiana, R. (2009). Highly efficient polycarbazole-based organic photovoltaic devices. *Appl. Phys. Lett.*, **95**, pp. 063304.

51. Liang, Y., Wu, Y., Feng, D., Tsai, S.-T., Son, H.-J., Li, G., and Yu, L. (2009). Development of new semiconducting polymers for high performance solar cells. *J. Am. Chem. Soc.*, **131**, pp. 56–57.

52. Chen, H.-Y., Hou, J., Zhang, S., Liang, Y., Yang, G., Yang, Y., Yu, L., Wu, Y., and Li, G. (2009). Polymer solar cells with enhanced open-circuit voltage and efficiency. *Nat. Photonics*, **3**, pp. 649–653.

53. Liang, Y., Feng, D., Wu, Y., Tsai, S.-T., Li, G., Ray, C., and Yu, L. (2009). Highly efficient solar cell polymers developed via fine-tuning of structural and electronic properties. *J. Am. Chem. Soc.*, **131**, pp. 7792–7799.

54. Hu, Z., Ying, L., Huang, F., and Cao, Y. (2017). Towards a bright future: Polymer solar cells with power conversion efficiencies over 10%. *Sci. China Chem.*, **60**, pp. 571–582.

55. He, Z., Zhong, C., Su, S., Xu, M., Wu, H., and Cao, Y. (2012). Enhanced power-conversion efficiency in polymer solar cells using an inverted device structure. *Nat. Photonics*, **6**, pp. 591–595.

56. He, Z., Xiao, B., Liu, F., Wu, H., Yang, Y., Xiao, S., Wang, C., Russell, T. P., and Cao, Y. (2015). Single-junction polymer solar cells with high efficiency and photovoltage. *Nat. Photonics*, **9**, pp. 174–179.

57. Mihailetchi, V. D., Duren, J. K. J. van, Blom, P. W. M., Hummelen, J. C., Janssen, R. A. J., Kroon, J. M., Rispens, M. T., Verhees, W. J. H., and Wienk, M. M. (2003). Electron transport in a methanofullerene. *Adv. Funct. Mater.*, **13**, pp. 43–46.

58. Mihailetchi, V. D., Xie, H. X., de Boer, B., Koster, L. J. A., and Blom, P. W. M. (2006). Charge transport and photocurrent generation in poly(3-hexylthiophene): Methanofullerene bulk-heterojunction solar cells. *Adv. Funct. Mater.*, **16**, pp. 699–708.

59. Mihailetchi, V. D., Koster, L. J. A., Hummelen, J. C., and Blom, P. W. M. (2004). Photocurrent generation in polymer-fullerene bulk heterojunctions. *Phys. Rev. Lett.*, **93**, pp. 216601.

60. Bronstein, H., Chen, Z., Ashraf, R. S., Zhang, W., Du, J., Durrant, J. R., Shakya Tuladhar, P., Song, K., Watkins, S. E., Geerts, Y., Wienk, M. M., Janssen, R. A. J., Anthopoulos, T., Sirringhaus, H., Heeney, M., and McCulloch, I. (2011). Thieno[3,2-b]thiophene–diketopyrrolopyrrole-containing polymers for high-performance organic field-effect transistors and organic photovoltaic devices. *J. Am. Chem. Soc.*, **133**, pp. 3272–3275.

61. Liu, Y., Zhao, J., Li, Z., Mu, C., Ma, W., Hu, H., Jiang, K., Lin, H., Ade, H., and Yan, H. (2014). Aggregation and morphology control enables multiple cases of high-efficiency polymer solar cells. *Nat. Commun.*, **5**, pp. 5293.

62. Bin, H., Gao, L., Zhang, Z.-G., Yang, Y., Zhang, Y., Zhang, C., Chen, S., Xue, L., Yang, C., Xiao, M., and Li, Y. (2016). 11.4% Efficiency non-fullerene polymer solar cells with trialkylsilyl substituted 2D-conjugated polymer as donor. *Nat. Commun.*, **7**, pp. 13651.

63. Bin, H., Zhang, Z.-G., Gao, L., Chen, S., Zhong, L., Xue, L., Yang, C., and Li, Y. (2016). Non-fullerene polymer solar cells based on alkylthio and fluorine substituted 2D-conjugated polymers reach 9.5% efficiency. *J. Am. Chem. Soc.*, **138**, pp. 4657–4664.

64. Reichenbächer, K., Süss, H. I., and Hulliger, J. (2005). Fluorine in crystal engineering: "The little atom that could." *Chem. Soc. Rev.*, **34**, pp. 22–30.

65. Deng, D., Zhang, Y., Zhang, J., Wang, Z., Zhu, L., Fang, J., Xia, B., Wang, Z., Lu, K., Ma, W., and Wei, Z. (2016). Fluorination-enabled optimal morphology leads to over 11% efficiency for inverted small-molecule organic solar cells. *Nat. Commun.*, **7**, pp. 13740.

66. Wang, Y., Parkin, S. R., Gierschner, J., and Watson, M. D. (2008). Highly fluorinated benzobisbenzothiophenes. *Org. Lett.*, **10**, pp. 3307–3310.

67. Son, H. J., Wang, W., Xu, T., Liang, Y., Wu, Y., Li, G., and Yu, L. (2011). Synthesis of fluorinated polythienothiophene-co-benzodithiophenes and effect of fluorination on the photovoltaic properties. *J. Am. Chem. Soc.*, **133**, pp. 1885–1894.

68. Boudreault, P.-L.T., Najari, A., and Leclerc, M. (2011). Processable low-bandgap polymers for photovoltaic applications. *Chem. Mater.*, **23**, pp. 456–469.

69. Sun, Y., Welch, G. C., Leong, W. L., Takacs, C. J., Bazan, G. C., and Heeger, A. J. (2012). Solution-processed small-molecule solar cells with 6.7% efficiency. *Nat. Mater.*, **11**, pp. 44–48.

70. Zhou, J., Zuo, Y., Wan, X., Long, G., Zhang, Q., Ni, W., Liu, Y., Li, Z., He, G., Li, C., Kan, B., Li, M., and Chen, Y. (2013). Solution-processed and high-performance organic solar cells using small molecules with a benzodithiophene unit. *J. Am. Chem. Soc.*, **135**, pp. 8484–8487.

71. Walker, B., Kim, C., and Nguyen, T.-Q. (2011). Small molecule solution-processed bulk heterojunction solar cells. *Chem. Mater.*, **23**, pp. 470–482.

72. Mishra, A. and Bäuerle, P. (2012). Small molecule organic semiconductors on the move: Promises for future solar energy technology. *Angew. Chem. Int. Ed.*, **51**, pp. 2020–2067.

73. Li, Y., Guo, Q., Li, Z., Pei, J., and Tian, W. (2010). Solution processable D-A small molecules for bulk-heterojunction solar cells. *Energy Environ. Sci.*, **3**, pp. 1427–1436.

74. Chen, Y., Wan, X., and Long, G. (2013). High performance photovoltaic applications using solution-processed small molecules. *Acc. Chem. Res.*, **46**, pp. 2645–2655.

75. Roncali, J. (2009). Molecular bulk heterojunctions: An emerging approach to organic solar cells. *Acc. Chem. Res.*, **42**, pp. 1719–1730.

76. Peumans, P., Yakimov, A., and Forrest, S. R. (2003). Small molecular weight organic thin-film photodetectors and solar cells. *J. Appl. Phys.*, **93**, pp. 3693–3723.

77. Walzer, K., Maennig, B., Pfeiffer, M., and Leo, K. (2007). Highly efficient organic devices based on electrically doped transport layers. *Chem. Rev.*, **107**, pp. 1233–1271.

78. de Bettignies, R., Nicolas, Y., Blanchard, P., Levillain, E., Nunzi, J.-M., and Roncali, J. (2003). Planarized star-shaped oligothiophenes as a new class of organic semiconductors for heterojunction solar cells. *Adv. Mater.*, **15**, pp. 1939–1943.

79. Uhrich, C., Schueppel, R., Petrich, A., Pfeiffer, M., Leo, K., Brier, E., Kilickiran, P., and Baeuerle, P. (2007). Organic thin-film photovoltaic cells based on oligothiophenes with reduced bandgap. *Adv. Funct. Mater.*, **17**, pp. 2991–2999.

80. Nian, L., Gao, K., Jiang, Y., Rong, Q., Hu, X., Yuan, D., Liu, F., Peng, X., Russell, T. P., and Zhou, G. (2017). Small-molecule solar cells with simultaneously enhanced short-circuit current and fill factor to achieve 11% efficiency. *Adv. Mater.*, **29**, pp. 1700616.

81. Wan, J., Xu, X., Zhang, G., Li, Y., Feng, K., and Peng, Q. (2017). Highly efficient halogen-free solvent processed small-molecule organic solar cells enabled by material design and device engineering. *Energy Environ. Sci.*, **10**, pp. 1739–1745.

82. Zhang, Q., Kan, B., Liu, F., Long, G., Wan, X., Chen, X., Zuo, Y., Ni, W., Zhang, H., Li, M., Hu, Z., Huang, F., Cao, Y., Liang, Z., Zhang, M., Russell, T. P., and Chen, Y. (2015). Small-molecule solar cells with efficiency over 9%. *Nat. Photonics*, **9**, pp. 35–41.

83. Ilmi, R., Haque, A., and Khan, M. S. (2018). High efficiency small molecule-based donor materials for organic solar cells. *Org. Electron.*, **58**, pp. 53–62.

84. Liu, Y., Yang, Y. (Michael), Chen, C.-C., Chen, Q., Dou, L., Hong, Z., Li, G., and Yang, Y. (2013). Solution-processed small molecules using different electron linkers for high-performance solar cells. *Adv. Mater.*, **25**, pp. 4657–4662.

85. Kan, B., Li, M., Zhang, Q., Liu, F., Wan, X., Wang, Y., Ni, W., Long, G., Yang, X., Feng, H., Zuo, Y., Zhang, M., Huang, F., Cao, Y., Russell, T. P., and Chen, Y. (2015). A series of simple oligomer-like small molecules based on oligothiophenes for solution-processed solar cells with high efficiency. *J. Am. Chem. Soc.*, **137**, pp. 3886–3893.

86. Cui, C., Guo, X., Min, J., Guo, B., Cheng, X., Zhang, M., Brabec, C. J., and Li, Y. (2015). High-performance organic solar cells based on a small molecule with alkylthio-thienyl-conjugated side chains without extra treatments. *Adv. Mater.*, **27**, pp. 7469–7475.

87. Qiu, B., Xue, L., Yang, Y., Bin, H., Zhang, Y., Zhang, C., Xiao, M., Park, K., Morrison, W., Zhang, Z.-G., and Li, Y. (2017). All-small-molecule nonfullerene organic solar cells with high fill factor and high efficiency over 10%. *Chem. Mater.*, **29**, pp. 7543–7553.

88. Bin, H., Yang, Y., Zhang, Z.-G., Ye, L., Ghasemi, M., Chen, S., Zhang, Y., Zhang, C., Sun, C., Xue, L., Yang, C., Ade, H., and Li, Y. (2017). 9.73% Efficiency nonfullerene all organic small molecule solar cells with absorption-complementary donor and acceptor. *J. Am. Chem. Soc.*, **139**, pp. 5085–5094.

89. Kyaw, A. K. K., Wang, D. H., Wynands, D., Zhang, J., Nguyen, T.-Q., Bazan, G. C., and Heeger, A. J. (2013). Improved light harvesting and improved efficiency by insertion of an optical spacer (ZnO) in solution-processed small-molecule solar cells. *Nano Lett.*, **13**, pp. 3796–3801.

90. Miao, J., Chen, H., Liu, F., Zhao, B., Hu, L., He, Z., and Wu, H. (2015). Efficiency enhancement in solution-processed organic small molecule: Fullerene solar cells via solvent vapor annealing. *Appl. Phys. Lett.*, **106**, pp. 183302.

91. Ran, N. A., Kuik, M., Love, J. A., Proctor, C. M., Nagao, I., Bazan, G. C., and Nguyen, T.-Q. (2014). Understanding the charge-transfer state and singlet exciton emission from solution-processed small-molecule organic solar cells. *Adv. Mater.*, **26**, pp. 7405–7412.

92. Li, L., Xiao, L., Qin, H., Gao, K., Peng, J., Cao, Y., Liu, F., Russell, T. P., and Peng, X. (2015). High-efficiency small molecule-based bulk-heterojunction solar cells enhanced by additive annealing. *ACS Appl. Mater. Interfaces*, **7**, pp. 21495–21502.

93. Love, J. A., Proctor, C. M., Liu, J., Takacs, C. J., Sharenko, A., Poll, T. S. van der, Heeger, A. J., Bazan, G. C., and Nguyen, T.-Q. (2013). Film morphology of high efficiency solution-processed small-molecule solar cells. *Adv. Funct. Mater.*, **23**, pp. 5019–5026.

94. Hummelen, J. C., Knight, B. W., LePeq, F., Wudl, F., Yao, J., and Wilkins, C. L. (1995). Preparation and characterization of fulleroid and methanofullerene derivatives. *J. Org. Chem.*, **60**, pp. 532–538.

95. He, Y. and Li, Y. (2011). Fullerene derivative acceptors for high performance polymer solar cells. *Phys. Chem. Chem. Phys.*, **13**, pp. 1970–1983.

96. Wienk, M. M., Kroon, J. M., Verhees, W. J. H., Knol, J., Hummelen, J. C., van Hal, P. A., and Janssen, R. A. J. (2003). Efficient methano[70]fullerene/MDMO-PPV bulk heterojunction photovoltaic cells. *Angew. Chem. Int. Ed Engl.*, **42**, pp. 3371–3375.

97. Troshin, P. A., Hoppe, H., Renz, J., Egginger, M., Mayorova, J. Y., Goryachev, A. E., Peregudov, A. S., Lyubovskaya, R. N., Gobsch, G., Sariciftci, N. S., and Razumov, V. F. (2009). Material solubility-photovoltaic performance relationship in the design of novel fullerene derivatives for bulk heterojunction solar cells. *Adv. Funct. Mater.*, **19**, pp. 779–788.

98. Lenes, M., Wetzelaer, G.-J.A.H., Kooistra, F. B., Veenstra, S. C., Hummelen, J. C., and Blom, P. W. M. (2008). Fullerene bisadducts for enhanced open-circuit voltages and efficiencies in polymer solar cells. *Adv. Mater.*, **20**, pp. 2116–2119.

99. Lenes, M., Shelton, S. W., Sieval, A. B., Kronholm, D. F., Hummelen, J. C., and Blom, P. W. M. (2009). Electron trapping in higher adduct fullerene-based solar cells. *Adv. Funct. Mater.*, **19**, pp. 3002–3007.

100. Cates, N. C., Gysel, R., Beiley, Z., Miller, C. E., Toney, M. F., Heeney, M., McCulloch, I., and McGehee, M. D. (2009). Tuning the properties of polymer bulk heterojunction solar cells by adjusting fullerene size to control intercalation. *Nano Lett.*, **9**, pp. 4153–4157.

101. He, Y., Zhao, G., Peng, B., and Li, Y. (2010). High-yield synthesis and electrochemical and photovoltaic properties of indene-C70 bisadduct. *Adv. Funct. Mater.*, **20**, pp. 3383–3389.

102. Zhao, G., He, Y., and Li, Y. (2010). 6.5% Efficiency of polymer solar cells based on poly(3-hexylthiophene) and indene-C60 bisadduct by device optimization. *Adv. Mater.*, **22**, pp. 4355–4358.

103. He, Y., Chen, H.-Y., Hou, J., and Li, Y. (2010). Indene–C60 bisadduct: A new acceptor for high-performance polymer solar cells. *J. Am. Chem. Soc.*, **132**, pp. 5532–5532.

104. Jorgensen, M., Norrman, K., Gevorgyan, S. A., Tromholt, T., Andreasen, B., and Krebs, F. C. (2012). Stability of polymer solar cells. *Adv. Mater.*, **24**, pp. 580–612.

105. Cheng, P. and Zhan, X. (2016). Stability of organic solar cells: Challenges and strategies. *Chem. Soc. Rev.*, **45**, pp. 2544–2582.

106. Morse, G. E., Tournebize, A., Rivaton, A., Chassé, T., Taviot-Gueho, C., Blouin, N., Lozman, O. R., and Tierney, S. (2015). The effect of polymer solubilizing side-chains on solar cell stability. *Phys. Chem. Chem. Phys.*, **17**, pp. 11884–11897.

107. Nielsen, C. B., Holliday, S., Chen, H.-Y., Cryer, S. J., and McCulloch, I. (2015). Non-fullerene electron acceptors for use in organic solar cells. *Acc. Chem. Res.*, **48**, pp. 2803–2812.

108. Lopez, S. A., Sanchez-Lengeling, B., de Goes Soares, J., and Aspuru-Guzik, A. (2017). Design principles and top non-fullerene acceptor candidates for organic photovoltaics. *Joule*, **1**, pp. 857–870.

109. Cheng, P., Li, G., Zhan, X., and Yang, Y. (2018). Next-generation organic photovoltaics based on non-fullerene acceptors. *Nat. Photonics*, **12**, pp. 131–142.

110. Yan, C., Barlow, S., Wang, Z., Yan, H., Jen, A.K.-Y., Marder, S. R., and Zhan, X. (2018). Non-fullerene acceptors for organic solar cells. *Nat. Rev. Mater.*, **3**, pp. 18003.

111. Zhang, G., Zhao, J., Chow, P. C. Y., Jiang, K., Zhang, J., Zhu, Z., Zhang, J., Huang, F., and Yan, H. (2018). Nonfullerene acceptor molecules for bulk heterojunction organic solar cells. *Chem. Rev.*, **118**, pp. 3447–3507.

112. Zhan, X., Tan, Z., Domercq, B., An, Z., Zhang, X., Barlow, S., Li, Y., Zhu, D., Kippelen, B., and Marder, S. R. (2007). A high-mobility electron-transport polymer with broad absorption and its use in field-effect transistors and all-polymer solar cells. *J. Am. Chem. Soc.*, **129**, pp. 7246–7247.

113. Zhong, Y., Trinh, M. T., Chen, R., Wang, W., Khlyabich, P. P., Kumar, B., Xu, Q., Nam, C.-Y., Sfeir, M. Y., Black, C., Steigerwald, M. L., Loo, Y.-L., Xiao, S., Ng, F., Zhu, X.-Y., and Nuckolls, C. (2014). Efficient organic solar cells with helical perylene diimide electron acceptors. *J. Am. Chem. Soc.*, **136**, pp. 15215–15221.

114. Yan, Q., Zhou, Y., Zheng, Y.-Q., Pei, J., and Zhao, D. (2013). Towards rational design of organic electron acceptors for photovoltaics: A study based on perylenediimide derivatives. *Chem. Sci.*, **4**, pp. 4389–4394.

115. Zhong, Y., Trinh, M. T., Chen, R., Purdum, G. E., Khlyabich, P. P., Sezen, M., Oh, S., Zhu, H., Fowler, B., Zhang, B., Wang, W., Nam, C.-Y., Sfeir, M. Y., Black, C. T., Steigerwald, M. L., Loo, Y.-L., Ng, F., Zhu, X.-Y., and Nuckolls, C. (2015). Molecular helices as electron acceptors in high-performance bulk heterojunction solar cells. *Nat. Commun.*, **6**, pp. 8242.

116. Duan, Y., Xu, X., Yan, H., Wu, W., Li, Z., and Peng, Q. (2017). Pronounced effects of a triazine core on photovoltaic performance–efficient organic solar cells enabled by a PDI trimer-based small molecular acceptor. *Adv. Mater.*, **29**, pp. 1605115.

117. Liu, J., Chen, S., Qian, D., Gautam, B., Yang, G., Zhao, J., Bergqvist, J., Zhang, F., Ma, W., Ade, H., Inganäs, O., Gundogdu, K., Gao, F., and Yan, H. (2016). Fast charge separation in a non-fullerene organic solar cell with a small driving force. *Nat. Energy*, **1**, pp. 16089.

118. Lin, Y., Wang, J., Zhang, Z.-G., Bai, H., Li, Y., Zhu, D., and Zhan, X. (2015). An electron acceptor challenging fullerenes for efficient polymer solar cells. *Adv. Mater.*, **27**, pp. 1170–1174.

119. Zhang, Y., Zou, J., Yip, H.-L., Chen, K.-S., Zeigler, D. F., Sun, Y., and Jen, A.K.-Y. (2011). Indacenodithiophene and quinoxaline-based conjugated polymers for highly efficient polymer solar cells. *Chem. Mater.*, **23**, pp. 2289–2291.

120. Chen, C.-P., Chan, S.-H., Chao, T.-C., Ting, C., and Ko, B.-T. (2008). Low-bandgap poly(thiophene-phenylene-thiophene) derivatives with broaden absorption spectra for use in high-performance bulk-heterojunction polymer solar cells. *J. Am. Chem. Soc.*, **130**, pp. 12828–12833.

121. He, G., Li, Z., Wan, X., Zhou, J., Long, G., Zhang, S., Zhang, M., and Chen, Y. (2013). Efficient small molecule bulk heterojunction solar cells with high fill factors via introduction of π-stacking moieties as end group. *J. Mater. Chem. A*, **1**, pp. 1801–1809.

122. Lin, Y., Zhao, F., He, Q., Huo, L., Wu, Y., Parker, T. C., Ma, W., Sun, Y., Wang, C., Zhu, D., Heeger, A. J., Marder, S. R., and Zhan, X. (2016). High-performance electron acceptor with thienyl side chains for organic photovoltaics. *J. Am. Chem. Soc.*, **138**, pp. 4955–4961.

123. Yang, Y., Zhang, Z.-G., Bin, H., Chen, S., Gao, L., Xue, L., Yang, C., and Li, Y. (2016). Side-chain isomerization on an n-type organic semiconductor ITIC acceptor makes 11.77% high efficiency polymer solar cells. *J. Am. Chem. Soc.*, **138**, pp. 15011–15018.

124. Yao, H., Ye, L., Hou, J., Jang, B., Han, G., Cui, Y., Su, G. M., Wang, C., Gao, B., Yu, R., Zhang, H., Yi, Y., Woo, H. Y., Ade, H., and Hou, J. (2017). Achieving highly efficient nonfullerene organic solar cells with improved intermolecular interaction and open-circuit voltage. *Adv. Mater.*, **29**, pp. 1700254.

125. Lin, Y., He, Q., Zhao, F., Huo, L., Mai, J., Lu, X., Su, C.-J., Li, T., Wang, J., Zhu, J., Sun, Y., Wang, C., and Zhan, X. (2016). A facile planar fused-ring electron acceptor for as-cast polymer solar cells with 8.71% efficiency. *J. Am. Chem. Soc.*, **138**, pp. 2973–2976.

126. Lin, Y., Zhao, F., Wu, Y., Chen, K., Xia, Y., Li, G., Prasad, S. K. K., Zhu, J., Huo, L., Bin, H., Zhang, Z.-G., Guo, X., Zhang, M., Sun, Y., Gao, F., Wei, Z., Ma, W., Wang, C., Hodgkiss, J., Bo, Z., Inganäs, O., Li, Y., and Zhan, X. (2016). Mapping polymer donors toward high-efficiency fullerene free organic solar cells. *Adv. Mater.*, **29**, pp. 1604155.

127. Wu, Y., Bai, H., Wang, Z., Cheng, P., Zhu, S., Wang, Y., Ma, W., and Zhan, X. (2015). A planar electron acceptor for efficient polymer solar cells. *Energy Environ. Sci.*, **8**, pp. 3215–3221.

128. Holliday, S., Ashraf, R. S., Wadsworth, A., Baran, D., Yousaf, S. A., Nielsen, C. B., Tan, C.-H., Dimitrov, S. D., Shang, Z., Gasparini, N., Alamoudi, M., Laquai, F., Brabec, C. J., Salleo, A., Durrant, J. R., and McCulloch, I. (2016). High-efficiency and air-stable P3HT-based polymer solar cells with a new non-fullerene acceptor. *Nat. Commun.*, **7**, pp. 11585.

129. Lin, Y., Zhang, Z.-G., Bai, H., Wang, J., Yao, Y., Li, Y., Zhu, D., and Zhan, X. (2015). High-performance fullerene-free polymer solar cells with 6.31% efficiency. *Energy Environ. Sci.*, **8**, pp. 610–616.

130. Wang, W., Yan, C., Lau, T.-K., Wang, J., Liu, K., Fan, Y., Lu, X., and Zhan, X. (2017). Fused hexacyclic nonfullerene acceptor with strong near-infrared absorption for semitransparent organic solar cells with 9.77% efficiency. *Adv. Mater.*, **29**, pp. 1701308.

131. Yao, H., Cui, Y., Yu, R., Gao, B., Zhang, H., and Hou, J. (2017). Design, synthesis, and photovoltaic characterization of a small molecular acceptor with an ultra-narrow band gap. *Angew. Chem. Int. Ed.*, **56**, pp. 3045–3049.

132. Zhao, W., Li, S., Yao, H., Zhang, S., Zhang, Y., Yang, B., and Hou, J. (2017). Molecular optimization enables over 13% efficiency in organic solar cells. *J. Am. Chem. Soc.*, **139**, pp. 7148–7151.

133. Li, W., Ye, L., Li, S., Yao, H., Ade, H., and Hou, J. (2018). A high-efficiency organic solar cell enabled by the strong intramolecular electron push–pull effect of the nonfullerene acceptor. *Adv. Mater.*, **30**, pp. 1707170.

134. Xin, R., Feng, J., Zeng, C., Jiang, W., Zhang, L., Meng, D., Ren, Z., Wang, Z., and Yan, S. (2017). Nonfullerene-acceptor all-small-molecule organic solar cells based on highly twisted perylene bisimide with an efficiency of over 6%. *ACS Appl. Mater. Interfaces*, **9**, pp. 2739–2746.

135. Hadmojo, W. T., Yim, D., Sinaga, S., Lee, W., Ryu, D. Y., Jang, W.-D., Jung, I. H., and Jang, S.-Y. (2018). Near-infrared harvesting fullerene-free all-small-molecule organic solar cells based on porphyrin donors. *ACS Sustain. Chem. Eng.*, **6**, pp. 5306–5313.

136. Wang, Y., Chang, M., Kan, B., Wan, X., Li, C., and Chen, Y. (2018). All-small-molecule organic solar cells based on pentathiophene donor and alkylated indacenodithiophene-based acceptors with efficiency over 8%. *ACS Appl. Energy Mater.*, **1**, pp. 2150–2156.

137. Zhang, J., Zhu, L., and Wei, Z. (2017). Toward over 15% power conversion efficiency for organic solar cells: Current status and perspectives. *Small Methods*, **1**, pp. 1700258.

138. Yip, H.-L. and Jen, A.K.-Y. (2012). Recent advances in solution-processed interfacial materials for efficient and stable polymer solar cells. *Energy Environ. Sci.*, **5**, pp. 5994–6011.

139. Zeng, H., Zhu, X., Liang, Y., and Guo, X. (2015). Interfacial layer engineering for performance enhancement in polymer solar cells. *Polymers*, **7**, pp. 333–372.

140. Hau, S. K., Yip, H.-L., Acton, O., Baek, N. S., Ma, H., and Jen, A.K.-Y. (2008). Interfacial modification to improve inverted polymer solar cells. *J. Mater. Chem.*, **18**, pp. 5113–5119.

141. Sun, Y., Seo, J. H., Takacs, C. J., Seifter, J., and Heeger, A. J. (2011). Inverted polymer solar cells integrated with a low-temperature-annealed sol-gel-derived ZnO film as an electron transport layer. *Adv. Mater.*, **23**, pp. 1679–1683.

142. Yuan, Z., Fu, M., and Huang, W. (2013). Air-stable inverted ZnO nanorod arrays/polymer hybrid solar cell. *Synth. Met.*, **185–186**, pp. 133–136.

143. Noh, Y.-J., Na, S.-I., and Kim, S.-S. (2013). Inverted polymer solar cells including ZnO electron transport layer fabricated by facile spray pyrolysis. *Sol. Energy Mater. Sol. Cells*, **117**, pp. 139–144.

144. White, M. S., Olson, D. C., Shaheen, S. E., Kopidakis, N., and Ginley, D. S. (2006). Inverted bulk-heterojunction organic photovoltaic device using a solution-derived ZnO underlayer. *Appl. Phys. Lett.*, **89**, pp. 143517.

145. Xiong, J., Yang, B., Zhou, C., Yang, J., Duan, H., Huang, W., Zhang, X., Xia, X., Zhang, L., Huang, H., and Gao, Y. (2014). Enhanced efficiency and stability of polymer solar cells with TiO_2 nanoparticles buffer layer. *Org. Electron.*, **15**, pp. 835–843.

146. Lee, K., Kim, J. Y., Park, S. H., Kim, S. H., Cho, S., and Heeger, A. J. (2007). Air-stable polymer electronic devices. *Adv. Mater.*, **19**, pp. 2445–2449.

147. Waldauf, C., Morana, M., Denk, P., Schilinsky, P., Coakley, K., Choulis, S. A., and Brabec, C. J. (2006). Highly efficient inverted organic photovoltaics using solution based titanium oxide as electron selective contact. *Appl. Phys. Lett.*, **89**, pp. 233517.

148. Hamada, K., Murakami, N., Tsubota, T., and Ohno, T. (2013). Solution-processed amorphous niobium oxide as a novel electron collection layer for inverted polymer solar cells. *Chem. Phys. Lett.*, **586**, pp. 81–84.

149. Siddiki, M. K., Venkatesan, S., Galipeau, D., and Qiao, Q. (2013). Kelvin probe force microscopic imaging of the energy barrier and energetically favorable offset of interfaces in double-junction organic solar cells. *ACS Appl. Mater. Interfaces*, **5**, pp. 1279–1286.

150. Siddiki, M. K., Venkatesan, S., and Qiao, Q. (2012). Nb_2O_5 as a new electron transport layer for double junction polymer solar cells. *Phys. Chem. Chem. Phys.*, **14**, pp. 4682–4686.

151. Vasilopoulou, M., Georgiadou, D. G., Soultati, A., Boukos, N., Gardelis, S., Palilis, L. C., Fakis, M., Skoulatakis, G., Kennou, S., Botzakaki, M., Georga, S., Krontiras, C. A., Auras, F., Fattakhova-Rohlfing, D., Bein, T., Papadopoulos, T. A., Davazoglou, D., and Argitis, P. (2014). Atomic-layer-deposited aluminum and zirconium oxides for surface passivation of TiO_2 in high-efficiency organic photovoltaics. *Adv. Energy Mater.*, **4**, pp. 1400214.

152. Zhou, Y., Fuentes-Hernandez, C., Shim, J., Meyer, J., Giordano, A. J., Li, H., Winget, P., Papadopoulos, T., Cheun, H., Kim, J., Fenoll, M., Dindar, A., Haske, W., Najafabadi, E., Khan, T. M., Sojoudi, H., Barlow, S., Graham, S., Brédas, J.-L., Marder, S. R., Kahn, A., and Kippelen, B. (2012). A universal method to produce low-work function electrodes for organic electronics. *Science*, **336**, pp. 327–332.

153. Wang, H., Zhang, W., Xu, C., Bi, X., Chen, B., and Yang, S. (2013). Efficiency enhancement of polymer solar cells by applying poly(vinylpyrrolidone) as a cathode buffer layer via spin coating or self-assembly. *ACS Appl. Mater. Interfaces*, **5**, pp. 26–34.

154. Kang, H., Hong, S., Lee, J., and Lee, K. (2012). Electrostatically self-assembled nonconjugated polyelectrolytes as an ideal interfacial layer for inverted polymer solar cells. *Adv. Mater.*, **24**, pp. 3005–3009.

155. Gommans, H., Verreet, B., Rand, B. P., Muller, R., Poortmans, J., Heremans, P., and Genoe, J. (2008). On the role of bathocuproine in organic photovoltaic cells. *Adv. Funct. Mater.*, **18**, pp. 3686–3691.

156. Vogel, M., Doka, S., Breyer, C., Lux-Steiner, M. C., and Fostiropoulos, K. (2006). On the function of a bathocuproine buffer layer in organic photovoltaic cells. *Appl. Phys. Lett.*, **89**, pp. 163501.

157. Jin, F., Chu, B., Li, W., Su, Z., Zhao, B., Yan, X., Zhang, F., Fan, D., Zhang, T., Gao, Y., Lee, C. S., and Wang, J. (2013). Improvement in power conversion efficiency and long-term lifetime of organic photovoltaic cells by using bathophenanthroline/molybdenum oxide as compound cathode buffer layer. *Sol. Energy Mater. Sol. Cells*, **117**, pp. 189–193.

158. Ye, H., Hu, X., Jiang, Z., Chen, D., Liu, X., Nie, H., Su, S.-J., Gong, X., and Cao, Y. (2013). Pyridinium salt-based molecules as cathode interlayers for enhanced performance in polymer solar cells. *J. Mater. Chem. A*, **1**, pp. 3387–3394.

159. Sun, K., Zhang, S., Li, P., Xia, Y., Zhang, X., Du, D., Isikgor, F. H., and Ouyang, J. (2015). Review on application of PEDOTs and PEDOT:PSS in energy

conversion and storage devices. *J. Mater. Sci. Mater. Electron.*, **26**, pp. 4438–4462.

160. Lee, E. J., Han, J. P., Jung, S. E., Choi, M. H., and Moon, D. K. (2016). Improvement in half-life of organic solar cells by using a blended hole extraction layer consisting of PEDOT:PSS and conjugated polymer electrolyte. *ACS Appl. Mater. Interfaces*, **8**, pp. 31791–31798.

161. Vitoratos, E., Sakkopoulos, S., Dalas, E., Paliatsas, N., Karageorgopoulos, D., Petraki, F., Kennou, S., and Choulis, S. A. (2009). Thermal degradation mechanisms of PEDOT:PSS. *Org. Electron.*, **10**, pp. 61–66.

162. Grossiord, N., Kroon, J. M., Andriessen, R., and Blom, P. W. M. (2012). Degradation mechanisms in organic photovoltaic devices. *Org. Electron.*, **13**, pp. 432–456.

163. Silva, W. J. da, Kim, H. P., Yusoff, A. R. bin M., and Jang, J. (2013). Transparent flexible organic solar cells with 6.87% efficiency manufactured by an all-solution process. *Nanoscale*, **5**, pp. 9324–9329.

164. Zhao, W., Ye, L., Zhang, S., Fan, B., Sun, M., and Hou, J. (2014). Ultrathin polyaniline-based buffer layer for highly efficient polymer solar cells with wide applicability. *Sci. Rep.*, **4**, pp. 6570.

165. Cao, B., He, X., Fetterly, C. R., Olsen, B. C., Luber, E. J., and Buriak, J. M. (2016). Role of interfacial layers in organic solar cells: Energy level pinning versus phase segregation. *ACS Appl. Mater. Interfaces.*, **8**, pp. 18238–18248.

166. Meng, B., Fu, Y., Xie, Z., Liu, J., and Wang, L. (2014). Phosphonate-functionalized donor polymer as an underlying interlayer to improve active layer morphology in polymer solar cells. *Macromolecules*, **47**, pp. 6246–6251.

167. Xu, Q., Wang, F., Qian, D., Tan, Z., Li, L., Li, S., Tu, X., Sun, G., Hou, X., Hou, J., and Li, Y. (2013). Construction of planar and bulk integrated heterojunction polymer solar cells using cross-linkable D-A copolymer. *ACS Appl. Mater. Interfaces*, **5**, pp. 6591–6597.

168. Zhang, W., Zhao, B., He, Z., Zhao, X., Wang, H., Yang, S., Wu, H., and Cao, Y. (2013). High-efficiency ITO-free polymer solar cells using highly conductive PEDOT:PSS/surfactant bilayer transparent anodes. *Energy Environ. Sci.*, **6**, pp. 1956–1964.

169. Jasieniak, J. J., Seifter, J., Jo, J., Mates, T., and Heeger, A. J. (2012). A solution-processed MoOx anode interlayer for use within organic photovoltaic devices. *Adv. Funct. Mater.*, **22**, pp. 2594–2605.

170. Xie, F., Choy, W. C. H., Wang, C., Li, X., Zhang, S., and Hou, J. (2013). Low-temperature solution-processed hydrogen molybdenum and

vanadium bronzes for an efficient hole-transport layer in organic electronics. *Adv. Mater.*, **25**, pp. 2051–2055.

171. Villers, B. J. T. de, MacKenzie, R. C. I., Jasieniak, J. J., Treat, N. D., and Chabinyc, M. L. (2013). Linking vertical bulk-heterojunction composition and transient photocurrent dynamics in organic solar cells with solution-processed MoOx contact layers. *Adv. Energy Mater.*, **4**, pp. 1301290.

172. Gu, X., Cui, W., Li, H., Wu, Z., Zeng, Z., Lee, S.-T., Zhang, H., and Sun, B. (2013). A solution-processed hole extraction layer made from ultrathin MoS_2 nanosheets for efficient organic solar cells. *Adv. Energy Mater.*, **3**, pp. 1262–1268.

173. Qin, P., Fang, G., Ke, W., Cheng, F., Zheng, Q., Wan, J., Lei, H., and Zhao, X. (2014). In situ growth of double-layer MoO_3/MoS_2 film from MoS_2 for hole-transport layers in organic solar cell. *J. Mater. Chem. A*, **2**, pp. 2742–2756.

174. Bai, S., Cao, M., Jin, Y., Dai, X., Liang, X., Ye, Z., Li, M., Cheng, J., Xiao, X., Wu, Z., Xia, Z., Sun, B., Wang, E., Mo, Y., Gao, F., and Zhang, F. (2013). Low-temperature combustion-synthesized nickel oxide thin films as hole-transport interlayers for solution-processed optoelectronic devices. *Adv. Energy Mater.*, **4**, pp. 1301460.

175. Tu, X., Wang, F., Li, C., Tan, Z., and Li, Y. (2014). Solution-processed and low-temperature annealed CrOx as anode buffer layer for efficient polymer solar cells. *J. Phys. Chem. C*, **118**, pp. 9309–9317.

176. Chuang, M.-K., Lin, S.-W., Chen, F.-C., Chu, C.-W., and Hsu, C.-S. (2014). Gold nanoparticle-decorated graphene oxides for plasmonic-enhanced polymer photovoltaic devices. *Nanoscale*, **6**, pp. 1573–1579.

177. Stratakis, E., Stylianakis, M. M., Koudoumas, E., and Kymakis, E. (2013). Plasmonic organic photovoltaic devices with graphene based buffer layers for stability and efficiency enhancement. *Nanoscale*, **5**, pp. 4144–4150.

178. Murray, I. P., Lou, S. J., Cote, L. J., Loser, S., Kadleck, C. J., Xu, T., Szarko, J. M., Rolczynski, B. S., Johns, J. E., Huang, J., Yu, L., Chen, L. X., Marks, T. J., and Hersam, M. C. (2011). Graphene oxide interlayers for robust, high-efficiency organic photovoltaics. *J. Phys. Chem. Lett.*, **2**, pp. 3006–3012.

179. Li, S.-S., Tu, K.-H., Lin, C.-C., Chen, C.-W., and Chhowalla, M. (2010). Solution-processable graphene oxide as an efficient hole transport layer in polymer solar cells. *ACS Nano*, **4**, pp. 3169–3174.

180. Lee, B. H., Lee, J.-H., Jeong, S. Y., Park, S. B., Lee, S. H., and Lee, K. (2014). Broad work-function tunability of p-type conjugated polyelectrolytes for efficient organic solar cells. *Adv. Energy Mater.*, **5**, pp. 1401653.

181. Choi, H., Kim, H.-B., Ko, S.-J., Kim, J. Y., and Heeger, A. J. (2015). An organic surface modifier to produce a high work function transparent electrode for high performance polymer solar cells. *Adv. Mater.*, **27**, pp. 892–896.

182. Lu, L., Xu, T., Jung, I. H., and Yu, L. (2014). Match the interfacial energy levels between hole transport layer and donor polymer to achieve high solar cell performance. *J. Phys. Chem. C*, **118**, pp. 22834–22839.

183. Angmo, D. and Krebs, F. C. (2013). Flexible ITO-free polymer solar cells. *J. Appl. Polym. Sci.*, **129**, pp. 1–14.

184. Silva, W. J. da, Kim, H. P., Yusoff, A. R. bin M., and Jang, J. (2013). Transparent flexible organic solar cells with 6.87% efficiency manufactured by an all-solution process. *Nanoscale*, **5**, pp. 9324–9329.

185. Tong, G., Jia, Z., and Chang, J. (2018). Flexible hybrid electronics: Review and challenges. *2018 IEEE Int. Symp. Circuits Syst. ISCAS*, pp. 1–5.

186. Zou, M., Ma, Y., Yuan, X., Hu, Y., Liu, J., and Jin, Z. (2018). Flexible devices: From materials, architectures to applications. *J. Semicond.*, **39**, pp. 011010.

187. Espinosa, N., García-Valverde, R., and Krebs, F. C. (2011). Life-cycle analysis of product integrated polymer solar cells. *Energy Environ. Sci.*, **4**, pp. 1547–1557.

188. Shaheen, S. E., Ginley, D. S., and Jabbour, G. E. (2005). Organic-based photovoltaics: Toward low-cost power generation. *MRS Bull.*, **30**, pp. 10–19.

189. Forrest, S. R. (2004). The path to ubiquitous and low-cost organic electronic appliances on plastic. *Nature*, **428**, pp. 911–918.

190. Kaltenbrunner, M., White, M. S., Głowacki, E. D., Sekitani, T., Someya, T., Sariciftci, N. S., and Bauer, S. (2012). Ultrathin and lightweight organic solar cells with high flexibility. *Nat. Commun.*, **3**, pp. 770.

191. Jean, J., Wang, A., and Bulović, V. (2016). In situ vapor-deposited parylene substrates for ultra-thin, lightweight organic solar cells. *Org. Electron.*, **31**, pp. 120–126.

192. Aernouts, T., Vanlaeke, P., Geens, W., Poortmans, J., Heremans, P., Borghs, S., Mertens, R., Andriessen, R., and Leenders, L. (2004). Printable anodes for flexible organic solar cell modules. *Thin Solid Films*, **451–452**, pp. 22–25.

193. Ostfeld, A. E., Catheline, A., Ligsay, K., Kim, K.-C., Chen, Z., Facchetti, A., Fogden, S., and Arias, A. C. (2014). Single-walled carbon nanotube transparent conductive films fabricated by reductive dissolution and

spray coating for organic photovoltaics. *Appl. Phys. Lett.*, **105**, pp. 253301.

194. Pasquier, A. D., Unalan, H. E., Kanwal, A., Miller, S., and Chhowalla, M. (2005). Conducting and transparent single-wall carbon nanotube electrodes for polymer-fullerene solar cells. *Appl. Phys. Lett.*, **87**, pp. 203511.

195. De, S., Higgins, T. M., Lyons, P. E., Doherty, E. M., Nirmalraj, P. N., Blau, W. J., Boland, J. J., and Coleman, J. N. (2009). Silver nanowire networks as flexible, transparent, conducting films: Extremely high DC to optical conductivity ratios. *ACS Nano*, **3**, pp. 1767–1774.

196. Pastorelli, F., Schmidt, T., Hösel, M., Søndergaard, R., Jørgensen, M., and Krebs, F. (2016). The organic power transistor: Roll-to-roll manufacture, thermal behavior, and power handling when driving printed electronics. *Adv. Eng. Mater.*, **18**, pp. 51–55.

197. Dam, H., Andersen, T., Madsen, M. K., Mortensen, T. F., Pedersen, M., Nielsen, U., and Krebs, F. (2015). Roll and roll-to-roll process scaling through development of a compact flexo unit for printing of back electrodes. *Sol. Energy Mater. Sol. Cells*, **140**, pp. 187–192.

198. Gevorgyan, S., Madsen, M., Roth, B., Corazza, M., Hösel, M., Søndergaard, R., Jørgensen, M., and Krebs, F. (2015). Lifetime of organic photovoltaics: Status and predictions. *Adv. Energy Mater.*, **6**, pp. 1501208.

199. Roth, B., Søndergaard, R., and Krebs, F. (2015). *Handbook of Flexible Organic Electronics: Materials, Manufacturing and Applications*, 1st Ed. Chapter 7 "Roll-to-roll printing and coating techniques for manufacturing large-area flexible organic electronics," (Woodhead Publishing), pp. 171–197.

200. Kim, Y., Bradley, D., Hou, Y., Krebs, F., and McNeill, R. C. (2016). Introduction to the issue on organic nanophotonics. *IEEE J. Sel. Top. Quantum Electron.*, **22**, pp. 1–3.

201. Garcia-Valverde, R., Villarejo, J., Hösel, M., Madsen, M., Søndergaard, R., Jørgensen, M., and Krebs, F. (2016). Scalable single point power extraction for compact mobile and stand-alone solar harvesting power sources based on fully printed organic photovoltaic modules and efficient high voltage DC/DC conversion. *Sol. Energy Mater. Sol. Cells*, **144**, pp. 48–54.

202. Rossander, L., Dam, H., Carlé, J., Helgesen, M., Rajkovic, I., Corazza, M., Krebs, F., and Wenzel Andreasen, J. (2017). In-line, roll-to-roll morphology analysis of organic solar cell active layers. *Energy Env. Sci*, **10**, pp. 2411–2419.

203. Liu, K., Larsen-Olsen, T. T., Lin, Y., Beliatis, M., Bundgaard, E., Jørgensen, M., Krebs, F. C., and Zhan, X. (2016). Roll-coating fabrication of flexible organic solar cells: Comparison of fullerene and fullerene-free systems. *J. Mater. Chem. A*, **4**, pp. 1044–1051.

204. Corazza, M., Simonsen, S., Gnaegi, H., Thydén, K., Krebs, F., and Gevorgyan, S. (2016). Comparison of ultramicrotomy and focused-ion-beam for the preparation of TEM and STEM cross section of organic solar cells. *Appl. Surf. Sci.*, **389**, pp. 462–488.

205. Gevorgyan, S., Espinosa, N., Ciammaruchi, L., Roth, B., Livi, F., Tsopanidis, S., Zuefle, S., Queirós, S., Gregori, A., Benatto, G., Corazza, M., Madsen, M., Hösel, M. J., Beliatis, M., Trofod, T., Pastorelli, F., Castro, A., Mingorance, A., Lenzi, V., and Krebs, F. (2016). Baselines for lifetime of organic solar cells. *Adv. Energy Mater.*, **6**, pp. 1600910.

206. Benatto, G., Corazza, M., Roth, B., Schütte, F., Rengenstein, M., Gevorgyan, S., and Krebs, F. (2016). Inside or outside? Linking outdoor and indoor lifetime tests of ITO-free organic photovoltaic devices for greenhouse applications. *Energy Technol.* **5**, pp. 338–344.

207. Torto, L., Cester, A., Rizzo, A., Wrachien, N., Gevorgyan, S., Corazza, M., and Krebs, F. (2016). Model of organic solar cell photocurrent including the effect of charge accumulation at interfaces and non-uniform carrier generation. *IEEE J. Electron Devices Soc.*, **4**, pp. 387–395.

208. Gevorgyan, S., Maria Heckler, I., Bundgaard, E., Corazza, M., Hösel, M., Søndergaard, R., Benatto, G., Jørgensen, M., and Krebs, F. (2017). Improving, characterizing and predicting the lifetime of organic photovoltaics. *J. Phys. Appl. Phys.*, **50**, pp. 103001.

209. Pastorelli, F., Accanto, N., Jørgensen, M., van Hulst, N., and Krebs, F. (2017). In situ electrical and thermal monitoring of printed electronics by two-photon mapping OPEN. *Sci. Rep.*, **7**, pp. 3837.

Chapter 3

Flexible Parylene-C Material and Its Applications in MOSFETs, RRAMs, and Sensors

Yimao Cai, Min Lin, and Qingyu Chen
Institude of Microelectronics, Peking University, Beijing 100871, China
caiyimao@pku.edu.cn

Poly(para-xylylene) (parylene) is a transparent and flexible polymer. With good mechanical and electrical properties, for the more common types of –N, –C, –D, –HT, parylene has been investigated for a variety of applications, such as interlayer dielectric for the advanced VLSI interconnect technology [52], moisture barriers [5], electrets [16], and gate dielectric for metal-oxide semiconductor field-effect transistors (MOSFETs) [26]. This chapter mainly focuses on the properties of parylene-C and its applications in MOSFETs, resistive random access memories (RRAMs), and sensors.

Flexible and Stretchable Electronics: Materials, Designs, and Devices
Edited by Run-Wei Li and Gang Liu

ISBN 978-981-4800-46-4 (Hardcover), 978-0-429-05890-5 (eBook)
www.jennystanford.com

3.1 An Introduction to Parylene

3.1.1 Types and Growth of Parylene Thin Films

Parylene is the generic name for members of a unique family of thermoplastic polymers that are deposited by using the dimer of para-xylylene (di-para-xylylene, or DPXN). Commonly available parylene variants are parylene-N, -C, -D, -HT, as shown in Fig. 3.1. Each type of parylene with its unique structure has different properties so that these four types of parylene can be used in different applications. Among them, parylene-C provides excellent combination of electrical and physical properties, especially low permeability in moist and corrosive environment. Therefore, we will focus on the properties and applications of parylene-C in the following sections.

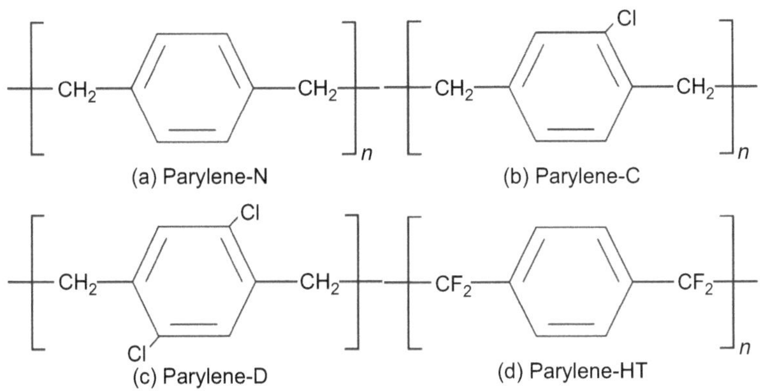

Figure 3.1 Molecules for the different types of parylene.

Parylene-C can be deposited by the polymer chemical vapor deposition (CVD) process at room temperature. The process can be divided into three distinct steps, as shown in Fig. 3.2. First, parylene-C is vaporized from its solid dimer form by the application of heat under vacuum. Then, with a high-temperature tube furnace, the gaseous form of the dimer cleaves into a monomer. Finally, the polymerization of the gaseous monomer occurs at room temperature as parylene-C in the vacuum chamber. Parylene-C deposition is not line-of-sight, and all sides of an object to be encapsulated are uniformly impinged by the gaseous monomer so that the truly conformal nature of the coating is obtained, which is a great advantage over other deposition techniques. In addition, the ability of forming pinhole-free parylene-C films above 100 nm [60] by the polymer CVD process is also very useful.

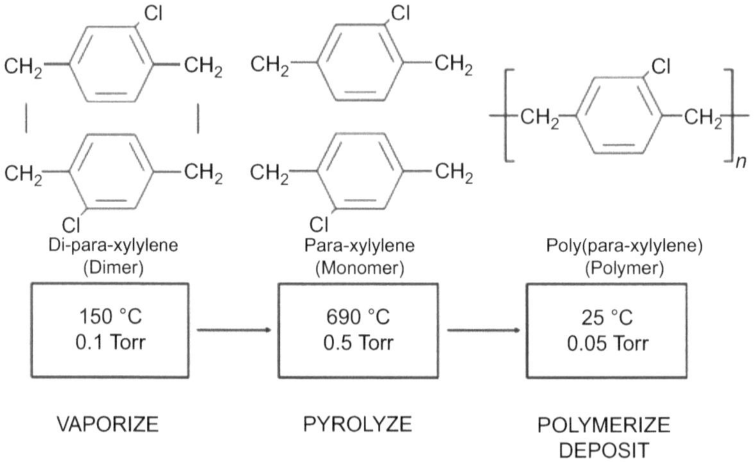

Figure 3.2 Parylene-C deposition process.

3.1.2 Properties of Parylene-C Thin Films

Parylene-C has good physical and mechanical properties with stable thermal expansion coefficient and tensile strength, high dielectric constant, and satisfactory hydrophobicity.

Parylene-C is a transparent, flexible, and chemically stable polymer and can be deposited by polymer CVD, which could avoid the contamination of solvent and by-products compared to the spin-coating process. In particular, parylene-C is immune to common organic solvents and can be patterned by standard photolithography process. In addition, the biocompatibility of materials to body tissue, body fluids, moisture, electrolytes, etc. is very important for medical implants and surgical components. Parylene-C is a Food and Drug Administration (FDA)-approved and biofriendly material, which has been widely investigated for use as implantable microdevices, such as microeletrodes [21], pacemakers [27], and microelectromechanical devices [7] for intraocular pressure monitoring.

In summary, with good chemical inertness, excellent biological compatibility, low permeability to moisture and gases, and conformal coverage with various substrates, parylene-C has been extensively used in various applications. In the following sections, we will give a detailed description of the applications of parylene-C in MOSFETs, RRAMs, and sensors.

3.2 Application of Parylene-C in MOSFETs

Parylene-C is a flexible and transparent polymer with conformal pinhole-free coverage and water-free and oxygen-free surface, which has been chosen as the gate dielectric, substrate, and passivation material to improve the performance of MOSFETs. Compared with the commonly used materials, MOSFETs using parylene-C material have achieved some promising results.

3.2.1 Gate Dielectric

Although the semiconductor and its contact have traditionally attracted considerable attention in organic field-effect transistors (OFETs), the gate dielectric also plays an increasingly important role in improving the performance of MOSFETs. Proper choice of the gate dielectric, which would form a trap-free interface with active materials, is crucial. Parylene-C, with excellent dielectric and mechanical properties, has been widely used as gate dielectric to increase mobility, yield n-type conduction, and reduce hysteresis in MOSFETs.

Podzorov et al. demonstrated an OFET on the surface of single crystals of rubrene by using parylene-C as gate dielectric [61]. The OFET showed typical p-type conductivity with the field-effect mobility up to 1 cm^2/Vs and the on/off ratio up to 10^4 at room temperature. In order to further study the effect of gate dielectrics on device performance, Stassen et al. performed a comparative study of rubrene single-crystal FETs fabricated using Ta_2O_5, Al_2O_3, SiO_2, and parylene-C as gate dielectrics [64]. The mobility of rubrene single-crystal FETs was different for devices fabricated with the different gate dielectrics. The device based on parylene-C as gate dielectric indicated the maximum mobility ranging from 6 to 10

cm^2/Vs when compared with other devices. This result illustrates that the mobility measured in OFETs is not only a property of the organic materials used, but also an intrinsic property of the organic–dielectric interface.

With the strong trapping of electrons, OFETs typically show p-type, but not n-type, conduction even with the appropriate low-workfunction electrodes, which greatly limits their application in organic complementary metal-oxide semiconductor (CMOS) circuits. The reason why n-type conduction is so elusive has been attributed to the electrochemical trapping of electrons by SiOH silanol groups at the semiconductor–dielectric interface in the case of the commonly used SiO_2 dielectric [9]. Chua et al. successfully yielded n-channel FET conduction in most conjugated polymers by using an appropriate hydroxyl-free gate dielectric, such as divinyl-tetramethylsiloxane-bis(benzocyclobutene) derivative (BCB), polyethylene, poly(methyl methacrylate), and parylene-N [9]. Similarly, Aguirre et al. also investigated the suppression of electron conduction in carbon nanotube field-effect transistors (CFETs), but they got different conclusions from Chua et al. [1]. They assessed the electrical transport characteristics in air for CFETs fabricated with both SiO_2 and parylene-C gate dielectrics. The I–V curve showed typical unipolar p-type conduction behavior of individual carton nanotube device with SiO_2 gate dielectric, independent of the metal used as the source and drain contacts. Unlike SiO_2-CFETs, ambipolar operation was achieved in parylene-C CFETs with both negative and positive gate voltages (Fig. 3.3). By investigating the impact of ambient adsorbates on the electrical behavior of CFET devices fabricated with both SiO_2 and parylene-C gate dielectrics, they found that an electrochemical charge transfer reaction between the semiconducting channel and the aqueous oxygen redox couple was the underlying phenomenon governing the suppression of electron conduction in CFET with SiO_2 gate dielectric and an adsorbed oxygen/water layer was a prerequisite to the electron transfer process. The hydrophilic gate dielectrics such as SiO_2 and other oxygen-containing dielectrics would exacerbate the electron transfer process and thus resulted in the suppression of electron conduction. However, parylene-C, with good chemical inertness and conformal pinhole-free surface, was an oxygen-free and hydrophobic polymer, which led to ambipolar behavior in CFETs.

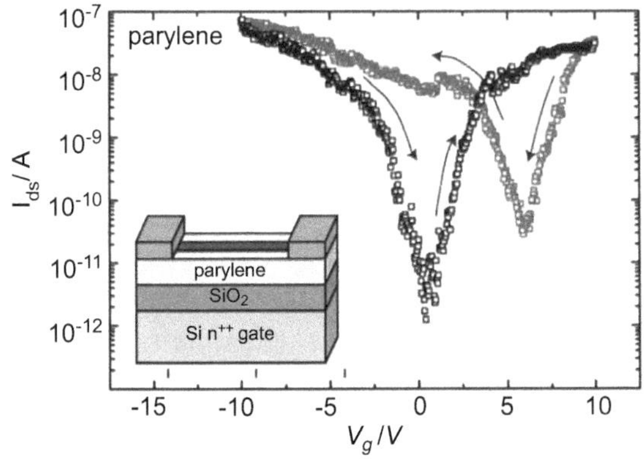

Figure 3.3 Transfer characteristics for parylene-C-based CFET in air with ambipolar behavior. Reprinted with permission from Ref. [1], Copyright 2009, John Wiley & Sons.

It was remarkable that current hysteresis was often observed between the forward and reverse gate voltage sweeps with both SiO_2 and other gate dielectrics. Due to the excellent properties of parylene-C, the magnitude of gate voltage hysteresis observed for parylene-C CFETs was approximately half of that observed for SiO_2 CFETs [1]. Sabri et al. also reported hysteresis behavior in graphene FETs with SiO_2 and parylene-C gate dielectrics [65]. Hysteresis with gate voltage was significant in oxide-gated FETs (about 9 V) and minimal in parylene-C-gated FETs. The larger hysteresis in oxide-gated FETs was attributed to the fast trapping and detrapping of electrons in the oxide layer. The hydrophobic parylene-C surface reduced ambient doping (water and oxygen) and thus reduced hysteresis due to electron trapping when compared with the most commonly used SiO_2 gate dielectric. In addition, Kawasaki et al. also reported a flexible picene thin-film FET with parylene-C gate dielectric fabricated in a top-contact structure [29]. A hysteresis-free transfer curve was achieved, as shown in Fig. 3.4. In fact, they had already observed a reduced hysteresis in picene thin-film FETs with SiO_2 gate dielectrics coating by hydrophobic polymers, Cytop^{TM}, and polystyrene in comparison with those merely employing an SiO_2 gate dielectric [30, 39]. Nevertheless, a small hysteresis still remained in the transfer curves in spite of the better hydrophobicity of Cytop^{TM} than parylene-C. They explained that the negligible hysteresis in a picene FET with parylene-C gate dielectric might be due to both the hydrophobic parylene-C surface and an improved growth of piece on the parylene-C surface.

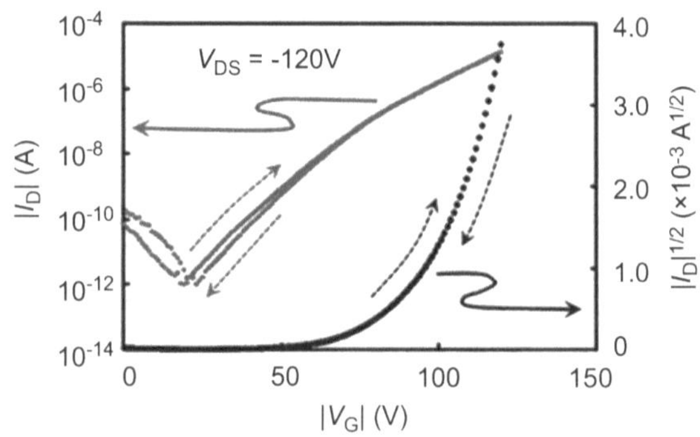

Figure 3.4 Transfer characteristics of a picene FET based on parylene-C gate dielectric with a negligible current hysteresis. Reprinted with permission from Ref. [29], Copyright 2010, AIP Publishing.

3.2.2 Substrate

With the development of flexible displays [31, 53, 77], wearable electronics [40, 66, 78], and biomedical devices [62, 67, 68], flexible devices and systems are getting more attractive to academia and industry. Among various flexible materials, parylene-C has been widely used as a substrate in implantable biomedical devices because of its excellent safety and biocompatibility.

Park et al. fabricated a flexible graphene FET with a bottom-gate coplanar structure on a parylene-C substrate for flexible implantable biomedical applications [63]. The fabricated graphene FET on parylene-C substrate could be released from the silicon wafer in cold water due to the hydrophobic property of parylene-C. The graphene FET on parylene-C substrate demonstrated good flexibility and transparency (Fig. 3.5), indicating great advantages in the field of flexible devices. In addition, Marszalek et al. further investigated the influence of bending on device performance for OFETs based on tetrathiafulvalene (TTF) derivatives as active material and parylene-C as dielectric and substrate [54]. The flexible OFET was bent using rollers, and the electrical characteristics of the OFET with different radii of curvature were measured in detail, as shown in Fig. 3.6. It can be seen that the device could work normally even if the radius reached 5 mm. These results indicate that the OFET fabricated on a parylene-C substrate not only has flexibility but also maintains good electrical performance under bending conditions.

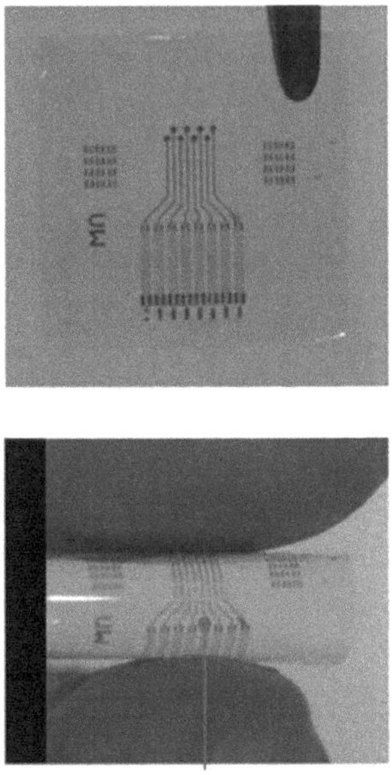

Figure 3.5 Flexible graphene transistors on parylene-C substrate. Reprinted with permission from Ref. [63], Copyright 2016, AIP Publishing.

Chamlagain et al. compared the electrical properties of $MoSe_2$ FETs fabricated on SiO_2 and parylene-C substrates [10]. In spite of the qualitatively similar transfer and output characteristics between devices on SiO_2 and parylene-C substrates, the $MoSe_2$ device on parylene-C substrate showed a mobility close to the mobility of bulk $MoSe_2$ (100–160 cm^2/Vs), which was significantly higher than that on SiO_2 substrate (about 50 cm^2/Vs). A detailed temperature-dependent electrical study of ultrahigh crystalline-quality multilayer $MoSe_2$ FETs of varied thickness (5–15 nm) on SiO_2 and parylene-C substrates indicated that the surface polar optical phonon scattering originating from the SiO_2 substrate was responsible for the reduced mobility of $MoSe_2$ FETs on SiO_2 substrate. The SiO_2 surface polar optical phonon mode with an energy of about 60 meV could be easily excited by thermal energy at room temperature, while parylene-C with its lowest energy optical phonon mode of 130 meV (corresponding to the vibrational stretch of C–Cl bond) could not be easily excited at room temperature, which was more suitable to be used as a substrate material.

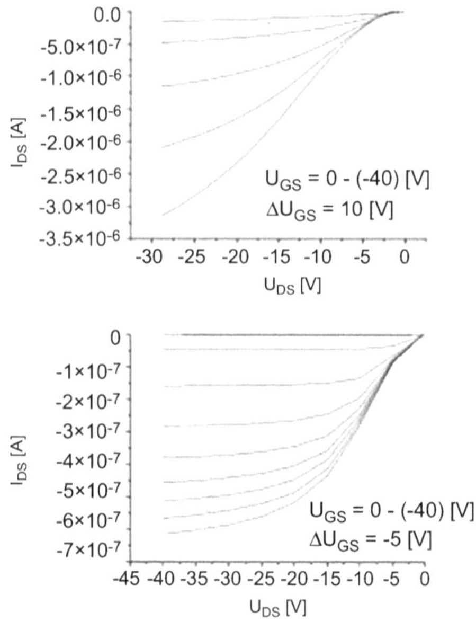

Figure 3.6 Output characteristic of TTF-4SC22 FET on parylene-C during bending test with curvature radius *r* = 25 mm (top) and *r* = 5 mm (bottom). Reprinted with permission from Ref. [54], Copyright 2012, Elsevier.

3.2.3 Encapsulation Gate Dielectric

Encapsulation is often used to protect MOSFETs from the moisture and gases in the environment and ensure their electrical functionality. Therefore, high-quality encapsulation materials are necessary to possess good barrier properties and protect the characteristics of MOSFETs. A variety of encapsulation materials have been widely studied with limited success. For example, Si_3N_4 (catalytic CVD at 270 °C) and Al_2O_3 (atomic layer deposition at 300 °C) have been used as passivation layers for CFETs. However, the high-temperature process used in Si_3N_4 and Al_2O_3 deposition process limited their applications [32], especially in flexible devices. A 1.7 μm PMMA coating for passivating transistors has also been introduced, but the electrical characteristics of the fabricated transistors changed due to the permeation of water molecules through PMMA [33]. In addition, it has also been reported that gas molecules could diffuse through the SU-8/PMMA layer when exposing these devices to an NO_2 environment for 1 h and cause changes in device performance [79].

Parylene-C is a pinhole-free film and can provide an excellent barrier to both liquids and gases. It can be deposited by polymer CVD with advantages of high step coverage, good purity without solvent contamination, and excellent compatibility with various substrates. Therefore, parylene-C is a very promising material as a passivation layer. Feili et al. created a flexible OFET with pentacene used as the active layer and parylene-C used as the encapsulation material [17]. However, ultraviolet radiation could influence the charge carrier transfer mechanism in pentacene when exposing these devices in the surrounding environment, which had an adverse effect on the drain current and mobility of the OFET. After encapsulation with parylene-C, the ultraviolet radiation with a wavelength below 400 nm could be absorbed by the 15 μm parylene-C layer so as to avoid the effect on pentacene. In addition, a slight increase in drain current and mobility was observed after covering parylene-C film due to the doping of pentacene during parylene-C deposition process. Besides, the deposited parylene-C might extract O_2 and H_2O molecules and brought better contact between pentacene molecules in crystal form, which was also responsible for the increase in drain current.

Similarly, Selvarasah et al. also demonstrated a parylene-C passivation approach for CFETs fabricated on a flexible substrate [69]. The device was fabricated with parylene-C back gate on a parylene-C substrate (Fig. 3.7a), exhibiting p-type conductivity. After encapsulation, the electrical properties of the CFETs were re-measured and compared with the values obtained before encapsulation (Fig. 3.7b). A slight decrease in the ON current of these devices after encapsulation was noticed, which is caused by the desorption of oxygen from the sidewalls of the CFETs during polymerization of parylene-C. Due to the suppression of water molecules and other trapped charges after the parylene-C passivation, the gate hysteresis widths of the CFETs were significantly reduced.

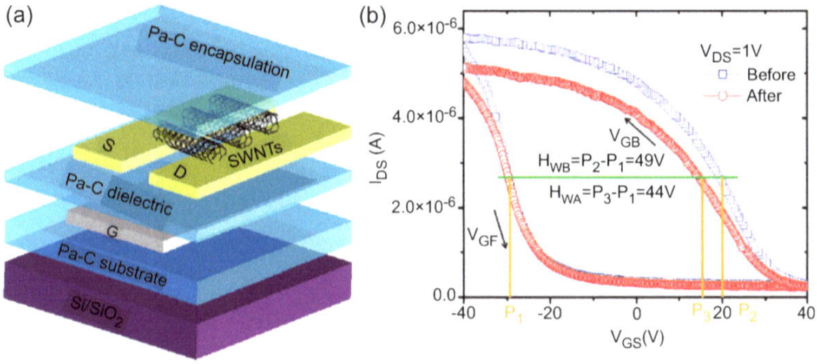

Figure 3.7 (a) The schematic drawing of the bottom-gated CFET. (b) Transfer characteristics of the transistor measured before and after 3 μm parylene-C deposition with a slight decrease in the ON current. Reprinted with permission from Ref. [69], Copyright 2010, AIP Publishing.

With excellent flexibility, improved mobility, restored ambipolar behavior, and negligible hysteresis, MOSFETs based on parylene-C as gate dielectric, substrate, or passivating material have shown great potential for application in CMOS circuits. This will greatly extend the range of suitable materials for CMOS circuits.

3.3 Application of Parylene-C in RRAM

Conventional Si-based memory technology is facing theoretical and physical limits of downscaling, so novel memory technologies have attracted significant interest from the academia and industry as a promising alternative or supplementary memory devices, such as ferroelectric random access memory [70], phase-change random access memory [22], magnetoresistive random access memory [2], and RRAM [74]. In particular, with low cost, ease of processing, flexibility, and good scalability, organic RRAM is considered to be a viable candidate for next-generation nonvolatile memory devices. Organic RRAMs with a two-terminal device structure can store data by defining the two states of electrical bistability (low-resistance state (LRS) and high-resistance state (HRS)) as "1" and "0" in response to an applied electric filed. Various organic materials have demonstrated resistive switching behaviors, including small molecules [14, 15, 41], polymers [11, 23, 42], and composites

[43–45]. Compared with other organic materials, polymers show relatively uniform and robust film qualities and are more suitable for organic RRAM applications.

Due to its high chemical stability, good mechanical flexibility, and excellent compatibility with various substrates, parylene-C-based organic RRAM devices have been investigated in detail. Cai et al. reported a flexible organic RRAM device based on single-layer parylene-C with brilliant mechanical and electrical stabilities for wearable biomedical applications [12]. Parylene-C was used not only as an active layer but also as a flexible substrate in a sandwich structure of Al/parylene-C/W/parylene-C substrate, as shown in Fig. 3.8a. Thanks to the good hydrophobicity of parylene-C, the fabricated devices with about 5 μm flexible parylene-C substrate could be torn easily from the top of a silicon wafer by soaking in cold water (Fig. 3.8b). The fabricated device was initially in an HRS and exhibited forming-free bipolar resistance characteristics. The reliability of the organic RRAM was verified by a robust endurance of about 500 cycles and a retention time of more than 10^5 s (Figs. 3.8c and 3.8d). Due to the good mechanical flexibility of parylene-C, the fabricated RRAM device showed stable memory operations up to a bending radius of 10 mm, indicating superior bending stability and corresponding electrical reliability (Figs. 3.8e and 3.8f). These results reveal the significant potential of parylene-C-based flexible RRAM devices for wearable biomedical applications.

The switching mechanism of parylene-C-based RRAM devices was attributed to the formation and dissolution of a metal bridge related to the redox reaction of the active metal electrode. The random growth and dissolution of conductive filaments resulted in the poor uniformity of operation voltages and HRS/LRS state resistances. Liu et al. proposed an HfO_x/parylene-C hybrid RRAM device to improve the uniformity [46]. The device indicated a tight operation voltage distribution, which was an important consideration in order to satisfy the requirements of practical applications. The comparison of resistance distribution of HRS/LRS between Pt/HfO_x/parylene-C/Al device and Al/parylene-C/W device is shown in Fig. 3.9, indicating dramatically improved resistance uniformity with a resistance ratio >1000. The main reason for the improved performance lay in its switching mechanism that localized conducting filaments were formed and ruptured during switching by inserting an HfO_x layer between the active electrode and parylene-C. With tight distribution of resistance and large window, the hybrid device is very suitable for multilevel storage.

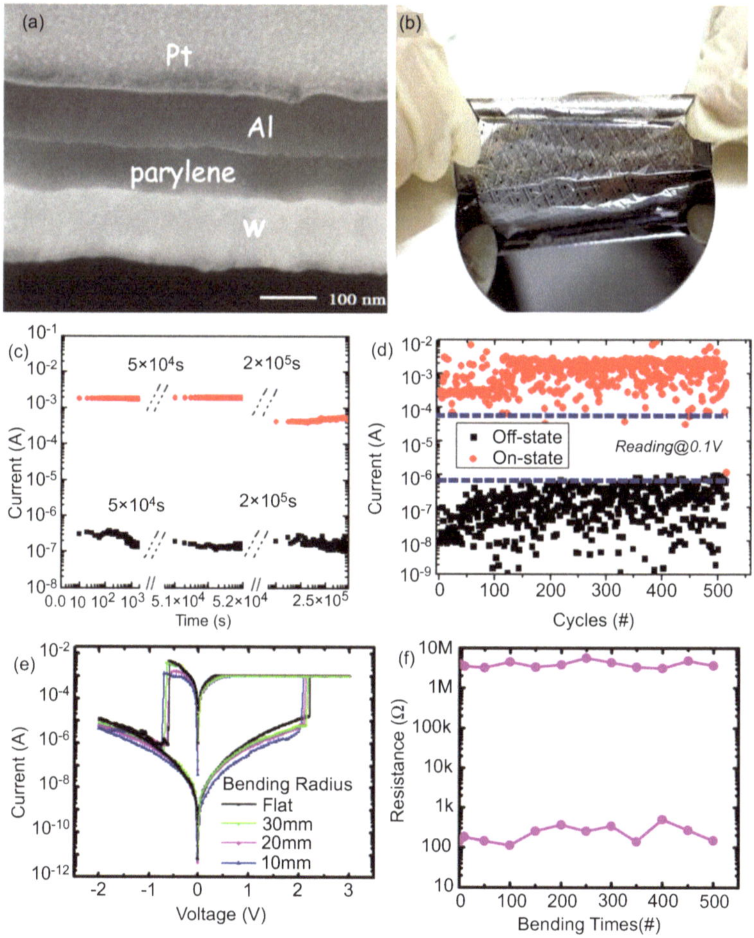

Figure 3.8 (a) Cross-sectional SEM image of parylene-C-based RRAM device on a flexible substrate; (b) procedure of tearing fabricated flexible devices from the holding silicon water; (c) retention behavior of the flexible device; (d) cycling characteristics of the flexible device; (e) switching I–V curves of the flexible device as a function of bending radius; (f) resistance ratio between the HRS and LRS as a function of bending times. Reprinted with permission from Ref. [12], Copyright 1990, IOP Publishing.

It has been reported that a porous parylene-C film can be formed when its thickness is less than 100 nm. So when a 40 nm thick single-layer parylene-C film is used as the active layer in RRAM cell, the conductive filaments tend to grow randomly through pinholes between top and bottom electrodes. In addition, metal atoms may diffuse into the parylene-C layer during the deposition of the top electrode and lead to the formation of thicker conductive filaments. Therefore, large set and reset currents and programming power consumption are obtained in Al/parylene-C/W device, which become the major impediments to future high-density and low-power applications. In order to address these problems, Bai et al. introduced a novel organic RRAM device based on double-layer parylene-C with bipolar switching behavior and good nonvolatile memory characteristics (Fig. 3.10a) [3]. The extremely low reset current of sub-20 nA and set current of 0.15 μA were obtained for the first time in the field of organic RRAM, almost 10^5 times lower than that of the single-layer parylene-C RRAM device, as shown in Figs. 3.10c and 3.10d. With 0.39 μW set power and extremely low reset power of 66.6 nW, the double-layer parylene-C RRAM device exhibited great potentials for high-density and low-power applications.

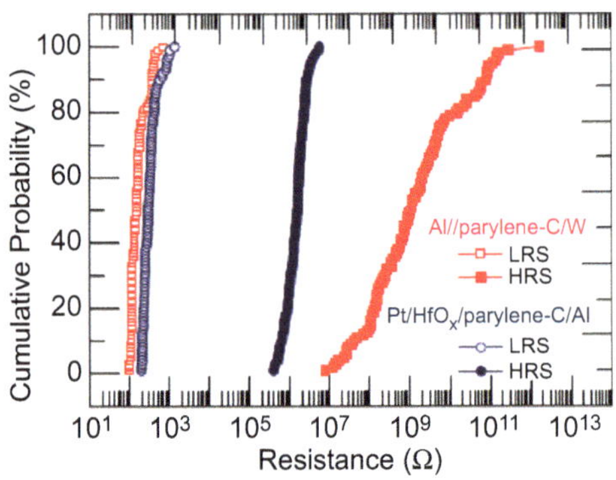

Figure 3.9 Cumulative probability of both LRS and HRS of Pt/HfO_x/parylene-C/Al and all-parylene-C RRAM devices. Reprinted with permission from Ref. [46], Copyright 2014, IEEE.

In recent years, wearable biomedical electronic devices and systems have attracted a lot of attention for their potential in modern medicine. The integration of multiple functional devices to achieve complex functions has become an essential requirement for future wearable biomedical electronic devices. Chen et al. presented a flexible multifunctional device composed of a parylene-

C-based RRAM device and a photoresistor for wearable image-sensing application [4]. The RRAM device could be written and read normally with superb storage property and excellent reliability after integration with photoresistor. The multifunctional device could detect the light intensity of the incident light and simultaneously store the information in the memory devices for wearable image-sensing application. In addition, Lin et al. demonstrated an organic device based on the Al/parylene-C/W sandwich structure with functionalities of memory and temperature sensing to reduce the fabricating cost and design complexity of wearable systems [38]. The flexible device exhibits very good storage characteristics and excellent temperature-sensing capability from 25 °C to 100 °C with a temperature coefficient of resistance (TCR) of 2.138×10^{-3}/°C, indicating a great promise for a sensing application for human body temperature.

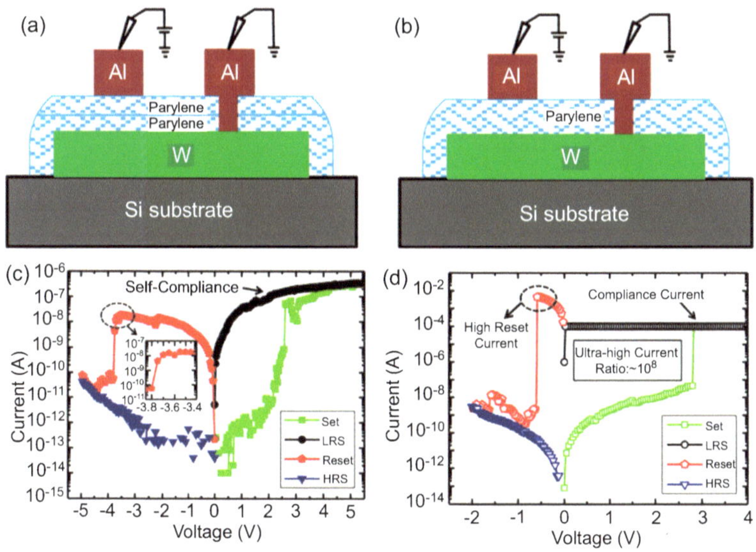

Figure 3.10 (a) Schematic views of Al/parylene-C/parylene-C/W double-layer device; (b) schematic views of Al/ parylene-C/W device; (c) typical I–V characteristics of Al/parylene-C/parylene-C/W double-layer device; (d) typical I–V characteristics of Al/parylene-C /W device. Reprinted with permission from Ref. [3], Copyright 2013, IEEE.

Parylene-C has attracted a substantial amount of attention due to good electrical and mechanical properties. In this section, we provide a brief summary of the application of parylene-C in RRAM with various structures. Parylene-C-based RRAM devices demonstrate excellent electrical performances such as a fast switching speed, a high storage window, good reliability, and remarkable mechanical stabilities, indicating great potential for next-generation nonvolatile memory devices.

3.4 Application of Parylene-C in Sensors

Parylene-C-based sensors with a wide variety of device structures and physical/chemical interaction phenomena have been used for the measurement of quantities such as protein [71], moisture [47], mechanical quantities (force sensors [19], pressure sensors [6], and shear stress sensors [24]), flow [34], temperature [25], and pH [72]. In the following section, we will describe in detail the applications of parylene-C in flow sensors, pH sensors, force sensors, and pressure sensors.

3.4.1 Flow Sensors

Flow sensors have been widely used in various applications for monitoring of gas [75] or fluid flow such as wind [55], gas chromatography [35], wall shear stress [48], and viscosity measurements [28]. Many operational principles have been introduced to realize flow sensors, which are generally divided into thermal flow sensors and non-thermal flow sensors. Due to their simple structure and implementation, thermal flow sensors have been extensively investigated. Next, a detailed introduction of parylene-C-based heat flow sensors is given.

Meng et al. demonstrated a microelectromechanical system (MEMS) thermal flow sensing array constructed of biocompatible parylene-C material [56]. The device comprised several thin-film platinum sensing elements sandwiched in a parylene-C membrane, which was integrated into the wall of a bulk-micromachined channel. Parylene-C was used as the membrane material, because it was transparent to facilitate viewing of the microchannel contents, possessed excellent barrier properties and biocompatibility. The array format allowed flexibility in sensing both flow and direction and could operate in multiple flow sensing modes using a single device, including hot-film (constant current and constant temperature biasing methods), calorimetric, and time-of-flight modes at low overheat ratios [36, 58, 59], which provides a flexible solution for accurately measuring the flow rate.

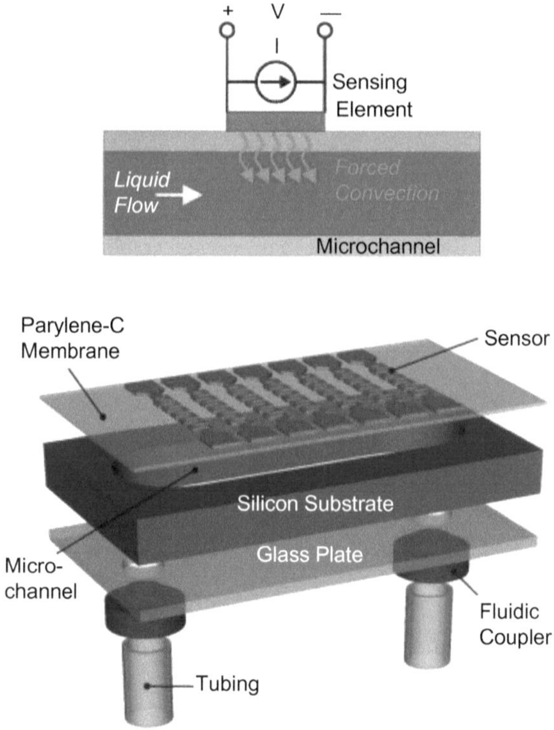

Figure 3.11 Schematic views of the sensing array layout and design. Reprinted with permission from Ref. [56], Copyright 2008, Elsevier.

Monitoring cerebral blood flow (CBF) plays an importance role for head injury patients because inadequate blood supply can disrupt neuronal functions and cause irreversible cellular damage [49, 57]. Li et al. developed a parylene-C-based micromachined blood flow sensor for continuous and quantitative measurement of intraparenchymal regional cerebral blood flow [50]. The smart catheter flow sensor consisted of a temperature sensor and a flow sensor, and parylene-C was used as substrate and insulation layer due to its biocompatibility, flexibility, and insulation (Fig. 3.11a).

At the steady state, the current applied to the sensor varied when the sensor was subjected to different flow rates so that the current (or voltage) variations could be correlated to blood flow rate. The smart catheter flow sensor achieved a sensitivity of 1.467 mV/ml/100 g-min in the range from 0 to 160 ml/100 g-min (Fig. 3.11b), which represented a promising monitoring tool for the CBF and had numerous applications for biomedical measurements.

Yu et al. also fabricated a biocompatible and flexible intravascular MEMS sensor to measure real-time shear stress in the aortas of New Zealand white rabbits [76]. Titanium and platinum layers embedded in the flexible parylene-C were used as the sensing elements. The use of parylene-C provided both insulation of electrodes and flexibility of the sensors, which allowed this sensor to conformally cover the geometry of arterial bifurcations and curvatures while maintaining its mechanical strength and operational function. The intravascular shear stress analysis in the abdominal aorta overlapping with that of the wall shear stress by computational fluid dynamic code indicated an experimental error at 9%, which provided an effective way for us to better understand between the blood flow and vascular disease.

3.4.2 pH Sensors

Due to the good chemical inertness of parylene-C, the material has been used together with ion-sensitive field-effect transistors (ISFETs) to suppress their pH sensitivity and serves as a solid-state reference electrode, namely reference field-effect transistor (REFET), for differential ISFET/REFET pH measurements. Matsuo et al. introduced a parylene-C-gated FET (PGFET), which consisted of an Si_3N_4-gated ISFET covered with a thin film of parylene-C to make it chemically inert. Although the PGFET demonstrated a sensitivity of 28 mV/pH below pH 4, the sensor was ion insensitive for larger pH. Chemical modification of parylene-C's surface with various functional groups and different cleaning procedures had been used for examining the ionic sensitivity, but the reduced ionic sensitivity response over time and large drift were still observed. In addition, contamination and surface chemical reactions could further damage the stability of parylene-C REFET and thus reduced the accuracy of the ISFET/REFET system.

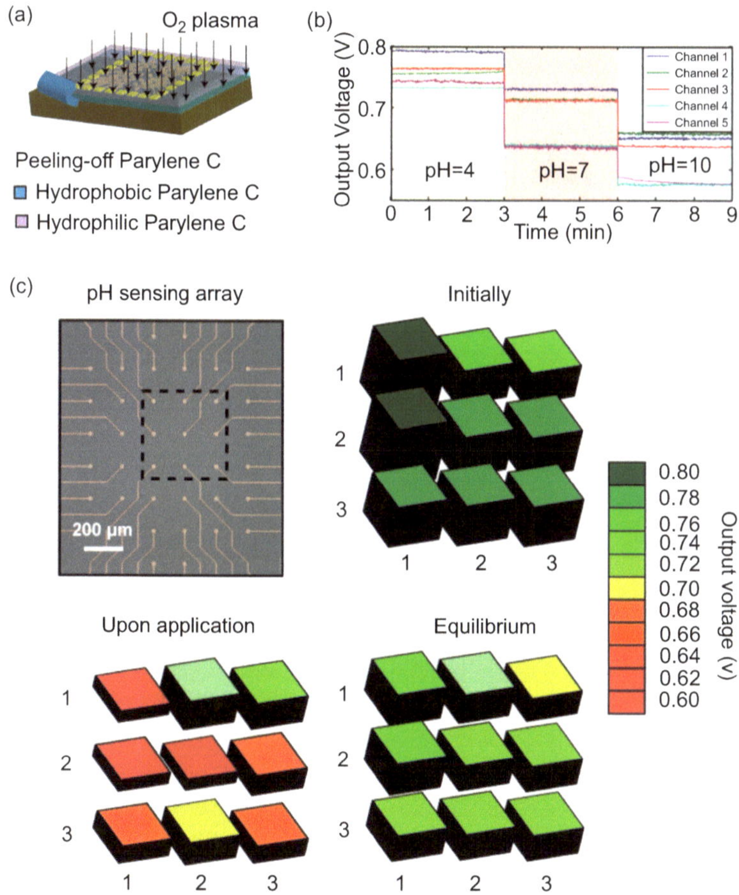

Figure 3.12 (a) Schematic views of the flexible pH-sensing arrays. (b) Filtered data indicating chemical sensitivity of the sensor at distinct pH values. (c) Spatiotemporal response of the sensing arrays. Reprinted from Ref. [73] under Creative Commons license.

Trantidou et al. fabricated flexible parylene-C-based high-density electrode arrays for functional monitoring of pH with rapid response, low chemical drifts, and excellent encapsulation integrity [73]. The sensing arrays used a simple sandwich structure of hydrophobic parylene-C-metal-hydrophilic parylene-C. The top hydrophilic parylene-C was obtained by plasma oxidation and served as the pH-sensing membrane (Fig. 3.12a). The operation principle of these sensors was identical with ISFETs, where the output (or threshold) voltage of the sensor varied in relation to the electrolyte pH. A linear relationship between the output voltage and the pH value was observed within the pH range 4–10 with a chemical sensitivity of 23 mV/pH, as shown in Fig. 3.12b. The excellent properties of the pH-sensing array were also verified by local pH detection, and a diffusion process was clearly seen in Fig. 3.12c. Initially, a microdroplet (50 μl) of pH 4 buffer solution was applied on the flexible platform. A region consisting of a 3 × 3 electrode grid was continuously monitored, while an equal volume of pH 7 buffer solution was added. The flexible parylene-C-based pH sensors highlight a novel multifunctional approach for parylene-C and can also promote various biomedical applications.

3.4.3 Force Sensors

Recent advances in the fields of material science and microfabrication technology have facilitated the development of bioimplantable devices. But the ignored mechanical consideration in the implantable-device/tissue interfaces has posed a significant threat to the device efficacy and functionality. Accurate measurement of these mechanical parameters such as contact force and pressure is important to guide the design of devices, which needs a highly sensitive sensor. Considering the environment in living organisms, the sensor should meet some basic requirements such as small footprint, simplicity, and high sensitivity.

Gutierrez et al. designed an all-parylene-C electrochemical impedance-based contact force sensor to deal with this challenge [19]. With its biocompatibility, electrical insulation, and mechanical strength, parylene-C was used as the structure material. During operation, external contact forces deformed the compliant fluid-filled structure in the normal direction and thus induced change to the cross-sectional area available to current-carrying ions in the fluid within the chamber. The change in cross-sectional area was registered as a change in the magnitude of the solution impedance. Therefore, the impedance variations could reflect the mechanical contact forces exerted by external sources on the chamber diaphragm. Instead of seeking to hermetically seal or isolate the sensor from the medium, the surrounding liquid environment as the sensing medium also greatly facilitated packaging. Kim et al. also demonstrated a parylene-C-based electrochemical-MEMS sensor array capable of integration directly onto the existing implantable ceramic electrode shanks to examine the relative force distribution of interfacial forces produced on the shank surface during insertion [37]. The sensor array consisted of a liquid-filled parylene-C-based microchannel and an array of enclosed electrodes. A mechanical analysis limited to normal insertion forces had shown that the normal insertion forces increased with the increasing insertion speeds. However, by investigating interfacial forces, Kim et al. found that a majority of the interfacial forces during insertion were within the first millimeter of the electrode shank and the interfacial forces at the shank tip decreased as the speed increased, which imposed a significant portion of the tissue displacement forces.

In brief, substrate material and large sensor footprint usually restrict the application of traditional sensors in living organisms, but the parylene-C-based electrochemical force sensors with good biocompatibility, excellent compatibility with various substrates, small footprint, and high sensitivity show great prospects for applications in bio-implantable devices.

3.4.4 Pressure Sensors

Implantable small-size pressure-sensing devices are of paramount importance for various future applications, such as intraocular pressure monitoring [18], intracranial pressure (IOP) monitoring [8], and MEMS drug-delivery system [20]. Among various materials, parylene-C has been widely used for the design of implantable pressure sensors due to its excellent physical and chemical properties.

Successful IOP monitoring is crucial in the management of glaucoma patients because it is an effective method to evaluate the progression of the chronic eye disease. Many techniques have been explored for IOP recoding, but reliable and repeatable measurements and continuous tracking are still difficult. Passive telemetric sensing is one of the viable methods for reliable and continuous IOP measurements, while delicate multiple wafer fabrication with wafer bonding greatly increases manufacturing complexity. Chen et al. investigated a monolithically microfabricated parylene-C-based wireless pressure sensor with resonant circuitry for ophthalmic applications [8]. By wirelessly interrogating the sensor implanted to the anterior chamber of the eye through an inductive coupling link, an external reader could register IOP variations [13]. This pressure sensor comprises a flexible parylene-C diaphragm chamber integrated with a capacitor and an inductor to create the resonant circuit that linked with the external reader. This wireless pressure sensor demonstrated highly sensitive pressure responses with more than 7000 ppm/mmHg sensitivity and 1 mmHg resolution, which was enough for practical IOP monitoring. The excellent device biocompatibility was also verified by in vivo testing more than 150 days without surgical complications, inflammatory responses or device encapsulation, indicating this sensor was very suitable for long-term IOP monitoring in glaucoma patients.

Noh et al. also proposed a corrugated parylene-C-diaphragm-based pressure transducer for monitoring pressure change within confined spaces. Similar to the above-mentioned pressure sensors made by Chen et al. [8], the sacrificial photoresist mold process was used to fabricate these parylene-C-based pressure sensors, which greatly limited the manufacturability of these sensors. In order to reduce the fabrication cost and improve reproducibility, Luharuka et al. developed a reusable silicon micromold based process for parylene-C pressure transducers, which greatly improved its manufacturability (Fig. 3.13) [51].

However, the conventional diaphragm-based pressure transducers generally occupy a large footprint, such as the pressure sensors introduced above. Gutierrez et al. designed an all-parylene-C microbubble pressure transducer [20], which enabled significant reduction in footprints, packaging complexity, and power consumption. Parylene-C was used as both the flexible substrate and the structural material, and an electrolytically generated microbubble was used to respond instantaneously to external pressure variations. The design of the fluidic access ports allowed for an open package instead of hermetic sealing or encapsulation techniques, which greatly reduced the complexity of packaging. The monitoring of microbubble size variations by measuring electrolyte impedance (resistance) correlated to external pressure variations. Electrolyte impedance could closely track the applied pressure, indicating excellent pressure sensing between −2 and 4 psi (−100 to 200 mmHg). The impedance was linearly related to the applied pressure with a ±0.1 psi (689.4 pa) pressure measurement resolution. This device also showed the promise of low-power applications with less than 1 nW excitation powers and more than 20 signal-to-noise ratio (SNR). By using microbubble sensing technology and open-package process, this parylene-C-based pressure sensor with small size, reduced cost, mechanical flexibility, and low power is very attractive for wet, in vivo pressure measurement.

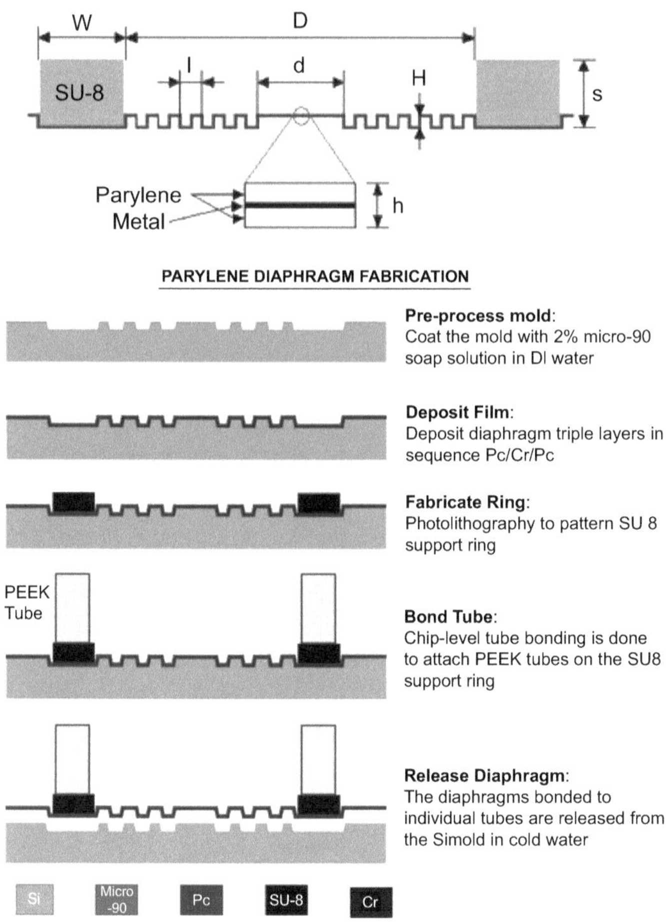

Figure 3.13 Fabrication steps of parylene-C diaphragms using Si micro-molding. Reprinted with permission from Ref. [51], Copyright 1991, IOP Publishing.

3.5 Conclusion

Parylene-C offers a lot of advantages, including relatively low cost, excellent biocompatibility and chemical stability, brilliant mechanical flexibility, and conformal coverage with various substrates, which can be used for different applications and may offer unique possibilities. The chapter gives a broad summary of parylene-C-based devices used in the field of MOSFETs, RRAMs, and sensors. The device structures, working principle, and the advantages of parylene-C-based devices are described in detail.

References

1. Aguirre, C. M., Levesque, P. L., Paillet, M., Lapointe, F., St-Antoine, B. C., Desjardins, P., and Martel, R. (2009). The role of the oxygen/water redox couple in suppressing electron conduction in field-effect transistors, *Adv. Mater.*, **21**: 3087–3091.
2. Boeck, J. D., Roy, W. V., Dsa, J., Motsnyi, V., Liu, Z., Lagae, L., Boeve, H., Desssein, K., and Borghs, G. (2002). Technology and materials issues in semiconductor-based magnetoelectronics, *Semicond. Sci. Technol.*, **17**: 342–354.
3. Bai, W., Huang, R., Cai, Y., Tang, Y., Zhang, X., and Wang, Y. (2013). Record low-power organic RRAM with sub-20-nA reset current, *IEEE Electron. Device Lett.*, **34**: 223–225.
4. Chen, Q., Lin, M., Fang, Y., Wang, Z., Yang, Y., Xu, J., Cai, Y., and Huang, R. (2018). Integration of biocompatible organic resistive memory and photoresistor for wearable image sensing application, *Sci. China Inf. Sci.* (unpublished).
5. Chiang, C. C., Wuu, D. S., Lin, H. B., Chen, Y. P., Chen, T. N., Lin, Y. C., Chen, W. C., Jaw, T. H., and Horng, R. H. (2006). Deposition and permeation properties of SiN_X/parylene multilayers on polymeric substrates, *Surf. Coat. Technol.*, **200**: 5843–5848.
6. Chen, P. J., Rodger, D. C., Agrawal, R., Saati, S., Meng, E., Varma, R., Humayun, M. S., and Tai, Y. C. (2007). Implantable micromechanical parylene-based pressure sensors for unpowered intraocular pressure sensing, *J. Micromach. Microeng.*, **17**: 1931–1938.
7. Chen, P. J., Rodger, D. C., Saati, S., Humayun, M. S., and Tai, Y. C. (2008). Microfabricated implantable parylene-based wireless passive

intraocular pressure sensors, *J. Microelectromech. Syst.*, **17**: 1342–1351.

8. Chen, P. J., Rodger, D. C., Saati, S., Humayun, M. S., and Tai, Y. C. (2008). Implantable parylene-based wireless intraocular pressure sensor, *Proceedings of the 21st IEEE International Conference on Micro Electro Mechanical Systems (MEMS)*, Wuhan, China, 13–17 January, pp. 58–61.
9. Chua, L. L., Zaumseil, J., Chang, J. F., Ou, E. C. W., Ho, P. K. H., Sirringhaus, H., and Friend, R. H. (2005). General observation of n-type field-effect behaviour in organic semiconductors, *Nature*, **434**: 194–199.
10. Chamlagain, B., Li, Q., Ghimire, N. J., Chuang, H. J., Perera, M. M., Tu, H., Xu, Y., Pan, M., Xaio, D., Yan, J., Mandrus, D., and Zhou, Z. (2014). Mobility improvement and temperature dependence in $MoSe_2$ field-effect transistors on parylene-C substrate, *ACS Nano*, **8**: 5079–5088.
11. Carchano, H., Lacoste, R., and Segui, Y. (1971). Bistable electrical switching in polymer thin films, *Appl. Phys. Lett.*, **19**: 414–415.
12. Cai, Y., Tan, J., Liu, Y., Lin, M., and Huang, R. (2016). A flexible organic resistance memory device for wearable biomedical applications, *Nanotechnology*, **27**: 275206.
13. Collins, C. C. (1967). Miniature passive pressure transensor for implanting in the eye, *Proceedings of IEEE Transactions on Biomedical Engineering*, BME-**14**: 74 –83.
14. Donhauser, Z. J., Mantooth, B. A., Kelly, K. F., Bumm, L. A., Monnell, J. D., Stapleton, J. J., Price, D. W. Jr, Rawlett, A. M., Allara, D. L., Tour, J. M., and Weiss, P. S. (2001). Conductance switching in single molecules through conformational changes, *Science*, **292**: 2303–2307.
15. Das, B. C. and Pal, A. J. (2008). Switching between different conformers of a molecule: Multilevel memory elements, *Org. Electron.*, **9**: 39–44.
16. Feng, Y., Hagiwara, K., Iguchi, Y., and Suzuki, Y. (2012). Trench-filled cellular parylene electret for piezoelectric transducer, *Appl. Phys. Lett.*, **100**: 262901.
17. Feili, D., Schuettler, M., Doerge, T., Kammer, S., Hoffmann, K. P., and Stieglitz, T. (2006). Flexible organic field effect transistors for biomedical microimplants using polyimide and parylene C as substrate and insulator layers, *J. Micromech. Microeng.*, **16**: 1555–1561.
18. Frischholz, M., Sarmento, L., Wenzel, M., Aquilina, K., Edwards, R., and Coakham, H. B. (2007). Telemetric implantable pressure sensor for short- and long-term monitoring of intracranial pressure, *Proceedings of the 29th IEEE Annual International Conference on Engineering in Medicine and Biology Society*, Lyon, France, 22–26 August, 514.

19. Gutierrez, C. A., McCarty, C., Kim, B., Pahwa, M., and Meng, E. (2010). An implantable all-parylene liquid-impedance based MEMS force sensor, *Proceedings of the 23rd IEEE International Conference on Micro Electro Mechanical Systems (MEMS)*, Hong Kong, China, 24–28 January, pp. 600–603.

20. Gutierrez, C. A. and Meng, E. (2011). A subnanowatt microbubble pressure sensor based on electrochemical impedance transduction in a flexible all-parylene package, *Proceedings of the 24th IEEE International Conference on Micro Electro Mechanical Systems (MEMS)*, Cancun, Mexico, 23–27 January, pp. 549–552.

21. Hsu, J. M., Rieth, L., Normann, R. A., Tathireddy, P., and Solzbacher, F. (2009). Encapsulation of an integrated neural interface device with parylene C, *IEEE Trans. Biomed. Eng.*, **56**: 23–29.

22. Hudgens, S. and Johnson, B. (2004). Overview of phase-change chalcogenide nonvolatile memory technology, *MRS Bull.*, **11**: 829–832.

23. Henisch, H. K. and Smith, W. R. (1974). Switching in organic polymer films, *Appl. Phys. Lett.*, **24**: 589–591.

24. Hongyu, Y., Lisong, A., Rouhanizadeh, M., Patel, D., Kim, E. S., and Hsiai, T. K. (2008). Flexible polymer sensors for in vivo intravascular shear stress analysis, *J. Microelectromech. Syst.*, **17**: 1178–1186.

25. He, S., Mench, M. M., and Tadigadapa, S. (2006). Thin film temperature sensor for real-time measurement of electrolyte temperature in a polymer electrolyte fuel cell, *Sens. Actuators A: Phys.*, **A125**: 170–177.

26. Iba, S., Sekitani, T., Kato, Y., Someya, T., Kawaguchi, H., Takamiya, M., Sakurai, T., and Takagi, S. (2005). Control of threshold voltage of organic field-effect transistors with double-gate structures, *Appl. Phys. Lett.*, **87**: 023509.

27. Iguchi, N., Kasanuki, H., Matsuda, N., Shoda, M., Ohnishi, S., and Hosoda, S. (1997). Contact sensitivity to polychloroparaxylene-coated cardiac pacemaker, *Pacing Clin. Electrophysiol.*, **20**: 372–373.

28. Ju, P. Y., Tsai, C. H., Fu, L. M., and Lin, C. H. (2011). Microfluidic flow meter and viscometer utilizing flow-induced vibration on an optic fiber cantilever, *Proceedings of the 16th IEEE International Conference on Solid-State Sensors, Actuators and Microsystems Conference (TRANSDUCERS)*, Beijing, China, 5–9 June, pp. 1428–1431.

29. Kawasaki, N., Kalb, W. L., Mathis, T., Kaji, Y., Mitsuhashi, R., Okamoto, H., Sugawara, Y., Fujiwara, A., Kubozono, Y., and Batlogg, B. (2010). Flexible picene thin film field-effect transistors with parylene gate dielectric and their physical properties, *Appl. Phys. Lett.*, **96**: 113305.

30. Kaji, Y., Kawasaki, N., Lee, X., Okamoto, H., Sugawara, Y., Oikawa, S., Ito, A., Okazaki, H., Yokoya, T., Fujiwara, A., and Kubozono, Y. (2009). Low voltage operation in picene thin film field-effect transistor and its physical characteristics, *Appl. Phys. Lett.*, **95**: 183302.

31. Kim, S., Kwon, H. J., Lee, S., Shim, H., Chun, Y., Choi, W., Kwack, J., Han, D., Song, M., Kim, S., Mohammadi, S., Kee, I., and Lee, S. Y. (2011). Low-power flexible organic light-emitting diode display device, *Adv. Mater.*, **23**: 3511–3516.

32. Kaminishi, D., Ozaki, H., Ohno, Y., Maehashi, K., Inoue, K., Matsumoto, K., Seri, Y., Masuda, A., and Matsumura, H. (2005). Air-stable n-type carbon nanotube field-effect transistors with Si_3N_4 passivation films fabricated by catalytic chemical vapor deposition, *Appl. Phys. Lett.*, **86**: 113115.

33. Kim, W., Javey, A., Vermesh, O., Wang, Q., Li, Y., and Dai, H. (2003). Hysteresis caused by water molecules in carbon nanotube field-effect transistors, *Nano Lett.*, **3**: 193–198.

34. Kuo, J. T. W., Yu, L., and Meng, E. (2012). Micromachined thermal flow sensors: A review, *Micromachines*, **3**: 550–573.

35. Kaanta, B. C., Chen, H., Lambertus, G., Steinecker, W. H., Zhdaneev, O., and Zhang, X. (2009). High sensitivity micro-thermal conductivity detector for gas chromatography, *Proceedings of the 22rd IEEE International Conference on Micro Electro Mechanical Systems (MEMS)*, Sorrento, Italy, 25–29 January, pp. 264–267.

36. Kuijk, J., Lammerink, T. S. J., Bree, H. E., Elwenspoek, M., and Fluitman, J. H. J. (1995). Multi-parameter detection in fluid fows, *Sens. Actuators A: Phys.*, **46–47**: 369–372.

37. Kim, B. J., Gutierrez, C. A., Gerhardt, G. A., and Meng, E. (2012). Parylene-based electrochemical-MEMS force sensor array for assessing neural probe insertion mechanics, *Proceedings of the 25rd IEEE International Conference on Micro Electro Mechanical Systems (MEMS)*, Paris, France, 29 January–2 February, pp. 124–127.

38. Lin, M., Chen, Q., Wang, Z., Fang, Y., Liu, J., Yang, Y., Wang, W., Cai, Y., and Huang, R. (2017). Flexible polymer device based on parylene-C with memory and temperature sensing functionalities, *Polymers*, **9**: 310.

39. Lee, X., Sugawara, Y., Ito, A., Oikawa, S., Kawasaki, N., Kaji, Y., Mitsuhashi, R., Okamato, H., Fujiwara, A., and Omote, K. (2010). Quantitative analysis of O_2 gas sensing characteristics of picene thin film field-effect transistors, *Org. Electron.*, **11**: 1394–1398.

40. Leonov, V. and Vullers, R. J. M. (2009). Wearable electronics self-powered by using human body heat: The state of the art and the perspective, *J. Renew. Sustain. Energy*, **1**: 062701.

41. Lau, C. N., Stewart, D. R., Williams, R. S., and Bockrath, M. (2004). Direct observation of nanoscale switching centers in metal/molecule/metal structures, *Nano Lett.*, **4**: 569.

42. Lim, S. L., Ling, Q. D., Teo, E. Y. H., Zhu, C. X., Chan, D. S. H., Kang, E. T., and Neoh, K. G. (2007). Conformation-induced electrical bistability in non-conjugated polymers with pendant carbazole moieties, *Chem. Mater.*, **19**: 5148–5157.

43. Laiho, A., Majumdar, H. S., Baral, J. K., Jansson, F., Österbacka, R., and Ikkala, O. (2008). Tuning the electrical switching of polymer/fullerene nanocomposite thin film devices by control of morphology, *Appl. Phys. Lett.*, **93**: 203309.

44. Li, F., Son, D. I., Seo, S. M., Cha, H. M., Kim, H. J., Kim, B. J., Jung, J. H., and Kim, T. W. (2007). Organic bistable devices based on core/shell CdSe/ZnS nanoparticles embedded in a conducting poly(*N*-vinylcarbazole) polymer layer, *Appl. Phys. Lett.*, **91**: 122111.

45. Li, F., Kim, T. W., Dong, W., and Kim, Y.-H. (2008). Formation and electrical bistability properties of ZnO nanoparticles embedded in polyimide nanocomposites sandwiched between two C60 layers, *Appl. Phys. Lett.*, **92**: 011906.

46. Liu, Y., Cai, Y., Li, Q., Wang, Z., Pan, Y., and Huang, R. (2014). Inorganic–organic hybrid resistive switching memory with high uniformity and multilevel operation, *Proceedings of International Symposium on VLSI Technology, Systems and Application (VLSI-TSA)*, Hsinchu, Taiwan, 28–30 April.

47. Lo, H. and Tai, Y. C. (2007). Characterization of parylene as a water barrier via buried-in pentacene moisture sensors for soaking tests, *Proceedings of the 2nd IEEE International Conference on Nano/Micro Engineered and Molecular Systems*, Bangkok, Thailand, 16–19 January, pp. 872–875.

48. Löfdahl, L. and Gad-el-Hak, M. (1999). MEMS-based pressure and shear-stress sensors for turbulent flows, *Meas. Sci. Technol.*, **10**: 665–686.

49. Lee, S. C., Chen, J. F., and Lee, S. T. (2005). Continuous regional cerebral blood flow monitoring in the neurosurgical intensive care unit, *J. Clin. Neurosci.*, **12**: 520–523.

50. Li, C., Wu, P. M., Hartings, J. A., Wu, Z., Ahn, C. H., and Narayan, R. K. (2012). Cerebral blood flow sensor with in situ temperature and thermal conductivity compensation, *Proceedings of the 25th IEEE International Conference on Micro Electro Mechanical Systems (MEMS)*, Paris, France, 29 January–2 February, 264–267.

51. Luharuka, R., Noh, H., Kim, S. K., Mao, H., Wong, L., and Hesketh, P. J. (2006). Improved manufacturability and characterization of a corrugated parylene diaphragm pressure transducer, *J. Micromech. Microeng.*, **16**: 1468–1474.

52. Majid, N., Dabral, S., and Mc Donald, J. F. (1989). The parylene–aluminum multilayer interconnection system for wafer scale integration and wafer scale hybrid packaging, *J. Electron. Mater.*, **18**: 301–311.

53. Mach, P., Rodriguez, S. J., Nortrup, R., Wiltzius, P., and Rogers, J. A. (2001). Monolithically integrated, flexible display of polymer-dispersed liquid crystal driven by rubber-stamped organic thin-film transistors, *Appl. Phys. Lett.*, **78**: 3592–3594.

54. Marszalek, T., Nosal, A., Pfattner, R., Jung, J., Kotarba, S., Mas-Torrent, M., Krause, B., Veciana, J., Gazicki-Lipman, M., Crickert, C., Schmidt, G., Rovira, C., and Ulanski, J. (2012). Role of geometry, substrate and atmosphere on performance of OFETs based on TTF derivatives, *Org. Electron.*, **13**: 121–128.

55. Makinwa, K. A. A. and Huijsing, J. H. (2001). A wind sensor with an integrated low-offset instrumentation amplifier, *Proceedings of the 8th IEEE International Conference on Electronics, Circuits and Systems*, Malta, 2–5 September, pp. 1505–1508.

56. Meng, E., Li, P.-Y., and Tai, Y.-C. (2008). A biocompatible parylene thermal flow sensing array, *Sens. Actuators A: Phys.*, **144**: 18–28.

57. Mette, D., Strunk, R., and Zuccarello, M. (2011). Cerebral blood flow measurement in neurosurgery, *Transl. Stroke Res.*, **2**: 152–158.

58. Nguyen, N. T. and Kiehnscherf, R. (1995). Low-cost silicon sensors for mass flow measurement of liquids and gases, *Sens. Actuators A: Phys.*, **49**: 17–20.

59. Nguyen, N. T. and Dötzel, W. (1997). Asymmetrical locations of heaters and sensors relative to each other using heater arrays: A novel method for designing multi-range electrocaloric mass-flow sensors, *Sens. Actuators A: Phys.*, **62**: 506–512.

60. Prodromakis, T., Michelakis, K., Zoumpoulidis, T., Dekker, R., and Toumazou, C. (2009). Biocompatible encapsulation of CMOS based

chemical sensors, *IEEE Sensors Conference*, Christchurch, New Zealand, 25–28 October, pp. 791–794.

61. Podzorov, V., Pudalov, V. M., and Gershenson, M. E. (2003). Field-effect transistors on rubrene single crystals with parylene gate insulator, *Appl. Phys. Lett.*, **82**: 1739–1741.
62. Park, D. W., Schendel, A. A., Mikael, S., Brodnick, S. K., Richner, T. J., Ness, J. P., Hayat, M. R., Atry, F., Frye, S. T., Pashaie, R., Thongpang, S., Ma, Z., and Williams, J. C. (2014). Graphene-based carbon-layered electrode array technology for neural imaging and optogenetic applications, *Nat. Commun.*, **5**: 5258.
63. Park, D. W., Kim, H., Bong, J., Mikael, S., Kim, T. J., Williams, J. C., and Ma, Z. (2016). Flexible bottom-gate graphene transistors on parylene C substrate and the effect of current annealing, *Appl. Phys. Lett.*, **109**: 152105.
64. Stassen, A. F., de Boer, R. W. I., Iosad, N. N., and Morpurgo, A. F. (2004). Influence of the gate dielectric on the mobility of rubrene single-crystal field-effect transistors, *Appl. Phys. Lett.*, **85**: 3899–3901.
65. Sabri, S. S., Lévesque, P. L., Aguirre, C. M., Guillemette, J., Martel, R., and Szkopek, T. (2009). Graphene field effect transistors with parylene gate dielectric, *Appl. Phys. Lett.*, **95**: 242104.
66. Stoppa, M. and Chiolerio, A. (2014). Wearable electronics and smart textiles: A critical review, *Sensors*, **14**: 11957–11992.
67. Stieglitz, T. (2001). Flexible biomedical microdevices with double-sided electrode arrangements for neural applications, *Sens. Actuators*, **A90**: 203.
68. Stieglitz, T., Schuettler, M., and Koch, K. P. (2005). Implantable biomedical microsystems for neural prostheses, *IEEE Eng. Med. Biol.*, **24**: 58–65.
69. Selvarasah, S., Li, X., Busnaina, A., and Dokmeci, M. R. (2010). Parylene-C passivated carbon nanotube flexible transistors, *Appl. Phys. Lett.*, **97**: 153120.
70. Setter, N., Damjanovic, D., Eng, L., Fox, G., Gevorgian, S., Hong, S., Kingon, A., Kohlstedt, H., Park, N. Y., Stephenson, G. B., Stolitchnov, I., Taganstev, A. K., Taylor, D. V., Yamada, T., and Streiffer, S. (2006). Ferroelectric thin films: Review of materials, properties, and applications, *J. Appl. Phys.*, **100**: 051606.
71. Takahashi, K., Ozawa, R., Oyama, H., Futagawa, M., Dasai, F., Ishida, M., and Sawada, K. (2012). A CMOS-MEMS-based label-free protein sensor for high-sensitive and compact system, *Proceedings of IEEE*

International Conference on Electron Devices Meeting (IEDM), San Francisco, CA, USA, 10–13 December.

72. Trantidou, T., Payne, D. J., Tsiligkiridis, V., Chang, Y.-C., Toumazou, C., and Prodromakis, T. (2013). The dual role of parylene C in chemical sensing: Acting as an encapsulant and as a sensing membrane for pH monitoring applications, *Sensors Actuators B*, **186**: 1–8.
73. Trantidou, T., Tariq, M., Terracciano, C., Toumazou, C., and Prodromakis, T. (2014). Parylene C-based flexible electronics for pH monitoring applications, *Sensors*, **14**: 11629–11639.
74. Waser, R. and Aono, M. (2007). Nanoionics-based resistive switching memories, *Nat. Mater.*, **6**: 833–840.
75. Yu, B., Gan, Z., Cao, S., Xu, J., and Liu, S. (2008). A micro channel integrated gas flow sensor for high sensitivity, *Proceedings of the 11th Intersociety Conference on Thermal and Thermomechanical Phenomena in Electronic Systems*, Orlando, FL, USA, 28–31 May, pp. 215–220.
76. Yu, H., Ai, L., Rouhanizadeh, M., Patel, D., Kim, E. S., and Hsiai, T. K. (2008). Flexible polymer sensors for in vivo intravascular shear stress analysis, *J. Microelectromech. Syst.*, **17**: 1178–1186.
77. Zhou, L., Wanga, A., Wu, S.-C., Sun, J., Park, S., and Jackson, T. N. (2006). All-organic active matrix flexible display, *Appl. Phys. Lett.*, **88**: 083502.
78. Zeng, W., Shu, L., Li, Q., Chen, S., Wang, F., and Tao, X. M. (2014). Fiber-based wearable electronics: A review of materials, fabrication, devices, and applications, *Adv. Mater*, **26**: 5310–5336.
79. Zhang, J., Boyd, A., Tselev, A., Paranjape, M., and Barbara, P. (2006). Mechanism of NO_2 detection in carbon nanotube field effect transistor chemical sensors, *Appl. Phys. Lett.*, **88**: 123112.

Chapter 4

Resistive Switching Phenomenon for Flexible and Stretchable Memories

Xiaohui Yi,[a,b] Shuang Gao,[a,b] Jie Shang,[a,b] Bin Chen,[a,b] Gang Liu,[a,c] and Run-Wei Li[a,b]

[a]*CAS Key Laboratory of Magnetic Materials and Devices, Ningbo Institute of Materials Technology and Engineering, Chinese Academy of Sciences, Ningbo 315201, China*

[b]*Zhejiang Province Key Laboratory of Magnetic Materials and Application Technology, Ningbo Institute of Materials Technology and Engineering, Chinese Academy of Sciences, Ningbo 315201, China*

[c]*School of Chemistry and Chemical Engineering, Shanghai Jiao Tong University, No. 800 Dongchuan Road, Shanghai 200240, China*

yixiaohui@nimte.ac.cn

With the advancement of wearable electronics and the Internet of Things, enormous effort has been put into flexible electric devices where traditional silicon-based electronics suffer from severe mechanical mismatch. Resistive switching memories, which have emerged as one of the promising candidates for the new generation of nonvolatile memory, have recently aroused great interest among the scientific community for fulfilling the function of information storage required in flexible electronic systems. In this chapter,

Flexible and Stretchable Electronics: Materials, Designs, and Devices
Edited by Run-Wei Li and Gang Liu

ISBN 978-981-4800-46-4 (Hardcover), 978-0-429-05890-5 (eBook)
www.jennystanford.com

we first briefly outline the basic concepts of resistive switching memories and their designing principle for flexible applications. Second, we summarize the recent progress in storage media used in flexible resistive switching memories. We briefly start with inorganic materials and then place particular emphasis on organic materials and inorganic–organic hybrid materials. Finally, we provide a short discussion on the current challenges and future prospects of flexible resistive switching memories.

4.1 Introduction

Continuous increase in the integration density of logic and memory on a chip with higher speed and reduced power consumption, which is known as "Moore's law," has been the principal driver of semiconductor technology in both academic and industrial communities. However, the ability to follow Moore's law has encountered a bottleneck because of the scale-down limitation of silicon-based semiconductor devices [1]. On the other hand, the ever-increasing concerns about the monitoring of human health and collecting data from everyday physical devices have greatly inspired the recent development of wearable electronic equipment and the Internet of Things, including epidermal electrode arrays, sensors, bendable displays, flexible label, etc., the most fundamental features of which are mechanical flexibility and environmental stability [2–7]. Being fabricated on soft substrates or even used as freestanding films [8], flexible electronics need to be lighter, more conformal, more portable, and less expensive for manufacturing, in comparison with their rigid traditional silicon-based counterparts. To overcome these challenges toward future high-performance electronics, efforts should be devoted both to discover new functionalities of emerging materials and to design novel data-storage device concepts.

Among the emerging memory technologies, one particular example is that of memories based on resistive switching, which generally has a simple electrode/insulator/electrode sandwich-like cell structure, as shown in Fig. 4.1a. The size of memory cells can be down-scaled to tens of nanometers, enabling them to be easily integrated into passive crossbar array and three-dimensional stacking architecture with high information storage density. Under external electrical stimuli, memory cells can be switched between a high-resistance state (HRS or OFF state) and a low-resistance state (LRS or ON state). The switching from HRS to LRS is defined as the set process, with its opposite as the reset process. According to the resistive switching behavior, the types of the memory include volatile memory, nonvolatile write-once-read-many memory (WORM memory), and nonvolatile rewritable memory (resistive random-access memory, RRAM). The set and reset processes can be realized by either the same voltage polarity or different polarities, which are defined as unipolar type and bipolar type resistive switching, respectively. Thus, binary digital data can be stored by encoding "0" and "1" as the low (OFF) and high (ON) resistance states in response to the external electric field [9–10]. Moreover, resistive switching may occur in nanoseconds, which promises fast writing, reading, and erasing operations. Besides simple cell structure, excellent miniaturization potential, and fast operating speed, they have also demonstrated low-energy consumption (<0.1 pJ) and high switching endurance (>10^{12} cycles) [11–13].

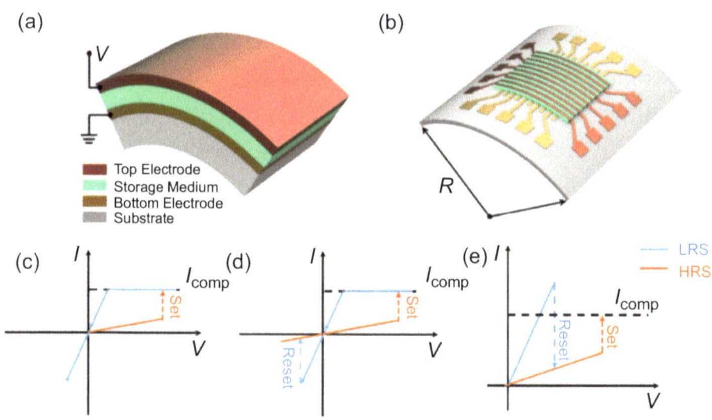

Figure 4.1 Schematic device structure and *I–V* characteristics of flexible resistive switching memories. (a) Schematic memory cell structure with measurement configuration. (b) Schematic of a flexible crossbar resistive switching memory array under bending with radius *R*. Schematic *I–V* characteristics of (c) WORM, (d) unipolar, and (e) bipolar resistive switching behaviors, where I_{Comp} denotes the compliance current during set process to prevent the device from permanent breakdown.

Another distinguished advantage of resistive switching memories is the possibility of multilevel storage and multifunction via deliberately controlling ion transport and related redox processes. For example, the phenomena of controllable quantized conductance have been observed in Nb/ZnO/Pt and ITO/ZnO/ITO resistive switching memories [14]. Uniform multilevel resistive switching is also observed in Pt/PAM–TsOH/Pt organic RRAM, where PAM–TsOH is polyazomethine doped with protonic acid [15]. More importantly, the resistive characteristics can be effectively modulated by various physical fields of electricity, temperature, magnetism, or light [16], which make resistive switching memories exceed conventional silicon-based data-storage technology and recognized as an appealing candidate for the next-generation "universal memory."

On the other hand, the various resistive switching mechanisms and abundant choices of information storage material allow the possibility of constructing flexible mechanical resistive switching memory. For example, the switching mechanisms can be generally sorted into three classes, i.e., ionic, electronic, trapping/detrapping, and thermal-effect-induced resistive switching. The ionic-effect-induced resistive switching relies on the transport of mobile ions and their electrochemical reactions in confined regions to form filamentary conductive paths. The formation and rupture of conductive filaments determine the device resistance states [17–20]. In electronic resistive switching memories, the trapping/detrapping of electrons at the intrinsic crystalline defects alters the intrinsic charge transport behaviors of the thin film, while that occurring at the electrode/media interface may modify the Schottky barrier height and influence the charge injection process [16, 21]. Resistive switching in a thermal-effect-controlled memory is similar to that of an ionic-effect-induced device, except for the fact that the rupture of filaments is mainly induced by thermal fusing rather than electrochemical dissolution [22]. Usually, all these phenomena may coexist during the entire course of the resistive switching process, while only one or two of them play a dominant role [23]. Meanwhile, plenty of materials have been reported for their application in resistive switching memories, including oxides [24–26], chalcogenides [27], nitrides [28–29], small molecules [30], polymers [31–32], and organic–inorganic composites [33–35], providing resistive switching more opportunity for flexible applications.

This chapter focuses on designing strategy and potential materials used for developing flexible RRAMs. In the following section, the design principles of flexible resistive switching memory are briefly introduced first. Then we discuss the recent progress in the storage media used in flexible resistive switching memories. We briefly introduce inorganic materials and then place particular emphasis on organic materials and inorganic–organic materials. Finally, a short discussion on the current challenges and future prospects of flexible resistive switching memories is provided.

4.2 Design Principle of Flexible Resistive Switching Memory

Although the phenomenon of resistive switching has been reported since the 1960s [36], RRAM gradually attracted intensive interest after the observation of resistive switching in $Pr_{0.7}Ca_{0.3}MnO_3$ thin films by A. Ignatieva at the University of Houston in 2000. Because of their excellent advantages, developing flexible resistive switching memories has recently aroused great interest among the scientific community for fulfilling the function of information storage required in this flexible electric system [37–40].

The structure of a generic resistive switching memory is composed of (1) a substrate, (2) bottom electrodes and top electrodes, and (3) an information storage medium. The mechanical stability of flexible RRAMs is usually estimated by bending tests as a function of bending radius and the number of bending cycles. Thus, the key performance parameters of flexible resistive switching memories include minimum bending radius, maximum bending strain, bending endurance, switching endurance under bending, and data retention time under bending. Under a given bending radius R (Fig. 4.1b), the real bending strain on memory cells can be roughly calculated to be $d/2R$, where d represents substrate thickness [41]. This is because the thickness of memory cells (mostly <1 μm) is usually much smaller than that of the substrate (mostly >10 μm) and, thus, can be reasonably ignored for strain calculation [20, 26]. To make the structure flexible, all components must comply with bending to some degree without losing their function. Three types of substrate materials are available for flexible applications: metals, organic polymers (plastics), and flexible glass; while for the electrodes, metallic materials, metal oxide (such as ITO), organic conducting polymers, and carbon-based materials are commonly employed due to their flexibility under certain bending condition. On the other hand, the insulator sandwiched between the two electrodes plays a key role in RRAM since the resistive switching behavior for information storage takes place in the insulator or at the insulator–electrode interface. In this section, we only focus on the design consideration of flexible RRAM from the point of view of storage media.

Two basic strategies have been employed in designing information storage materials in flexible resistive random access memories: One approach is to explore the potential flexibility of the reported resistive switching systems, in which materials with reported resistive switching phenomenon are employed as information media on rigid substrate. For instance, RRAM employing inorganic materials features excellent stability, endurance, and fast read/write speed with fabrication compatible with the complementary metal oxide semiconductor (CMOS) process flow. Resistive switching in inorganic materials has been extensively investigated to understand their switching mechanism. With these knowledge in mind, inorganic materials are deposited on the flexible substrate and their switching memory behavior upon bending condition are examined in the as-fabricated resistive switching memories. Two basic approaches have been employed to make flexible resistive switching memories based on inorganic materials: (1) fabricating the resistive switching memory device on a rigid substrate and then transfer the complete device to a flexible substrate and (2) fabricating the device directly on a flexible substrate. However, the performance of these resistive switching memories is still limited due to the rigid nature of inorganic materials.

Another approach to constructing flexible resistive switching memories is selecting flexible media as the function layer. Organic materials are intrinsically flexible and have easy solution processability and roll-to-roll processing, which highly satisfies the requirements of flexible electronics. These materials are, thus, considered one of the most promising information storage media in flexible resistive switching memories. For example, polymer materials can be employed either as matrices or functional unit for resistive switching. Organic–inorganic hybrid materials are beneficial for their organic and inorganic components featuring reliable resistive switching behavior and potential mechanical flexibility. Thus, these materials have also attracted tremendous attention in the field of flexible electronics very recently.

4.3 Flexible Resistive Switching Storage Media Materials

4.3.1 Inorganic Materials

Generally, inorganic materials for resistive switching include metal oxides and chalcogenide, deposited in the form of thin films through conventional CMOS-compatible physical or chemical deposition method. The type of resistance-switching mechanism in most inorganic materials is ionic. With the aid of an electric field, oxygen ions or metal ions are created and electromigrated, which in turn induce the redox reaction of the materials. Ion migration and redox reaction take place in very limited region, giving rise to a filament conducting path. The rupture of the filament can be induced by opposite bias voltage, current-induced joule heating, or the synergistic effect of both. This mechanism has been evidenced by direct observation of the conduction filament in several oxide-based RRAMs through in situ current–voltage (I–V) measurements in high-resolution transmission electron microscopy (HRTEM) [42–45].

Based on the understanding of switching behavior and effective control of switching characteristics, many flexible RRAMs based on inorganic materials have been developed. The first paper on flexible resistive switching memory was published in 2008, in which an $Al/Al_xO_y/Al$ structure was fabricated on a flexible polyether sulfone (PES) substrate, as shown in Fig. 4.2 [40]. Aluminum oxide employed as storage media is formed by plasma oxidation, which exhibits good endurance under substrate bending (bending radius of 27.5 mm, 10^5 times). Since then, AlO_x [24, 46], TiO_x [47, 48], NiO_x [25, 49], ZrO_x [50, 51], HfO_x [52, 53], WO_{3-x} [26, 54], $BaTiCoO_3$ [55], SiO_x [56, 57], and CuO_x [58, 59] have been directly deposited on a flexible substrate and employed as information storage media in flexible resistive switching systems. For example, an $IZO/Al_2O_3/IZO$ fabricated on PET substrate shows a minimum bending radius of 10 mm [44], while an $Ag/TiO_2/Pt$ device demonstrates a maximum bending radius of 8.4 mm [45]. Pd-interlayer SiO_x junction fabricated on a flexible substrate maintains low electroforming voltage and demonstrates a minimum bending radius of 5 mm [54].

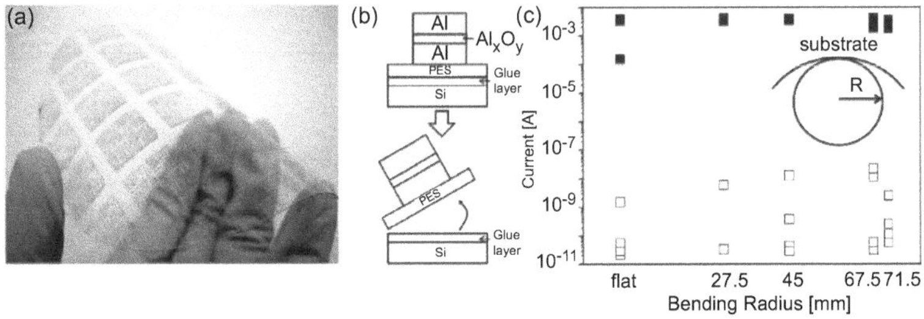

Figure 4.2 (a) Photograph of flexible RRAM with $Al/Al_xO_y/Al$ structure, (b) structure and process flow of fabricated flexible RRAM, and (c) current ratio between the ON and OFF states as a function of the bending radius. Reprinted with permission from Ref. [40], Copyright 2008, AIP Publishing.

Except for the direct deposition of inorganic materials on a flexible substrate, another approach to fabricating flexible resistive switching memory devices is to transfer entire devices fabricated on rigid substrates at high temperature to flexible substrates. Lee et al. reported a conceptual strategy for the fabrication of flexible memory employing a one selector–one resistor (1S–1R) crossbar structure with $Ni/TiO_2/Ni$ selector (1S) and $Ni/NiO_x/Pt$ resistor (1R) on a plastic substrate via an inorganic-based laser lift-off (ILLO) process, as shown in Fig. 4.3 [25]. An XeCl excimer laser with a wavelength of 308 nm was irradiated through the back side of the glass substrate to reduce the adhesion between the exfoliation layer and the substrate by localized melting and vaporizing to detach the upper inorganic layer from the substrate. Finally, the exfoliated layers were transferred on a receiver plastic substrate. The developed flexible device demonstrates reliable and reproducible resistive switching with excellent mechanical stability on a plastic substrate.

Efforts have also been devoted to optimize the microstructure of inorganic storage layer to further enhance the mechanical performance of RRAM devices. For instance, a non-stoichiometric hafnium oxide layer consisting of an amorphous nanocrystalline mixed structure has been obtained through carefully adjusting the RF power, sputtering time, deposition temperature, and atmosphere, as shown in Fig. 4.4 [20]. It is found that amorphous nanocrystalline mixed structure of the hafnium oxide layer promises reliable formation and evolution of the pseudo-straight metallic hafnium conductive filaments (CFs), wherein the suppression of random annihilation and regeneration of branch-like CFs under bending strains can lead to superior mechanical performance of the device.

With the amorphous nanocrystalline structure of hafnium oxide as the switching layer, the transparent ITO/HfO_x/ITO device shows good optical transmittance, thermal stability, and, more importantly, uniform and reproducible resistive switching behavior that can withstand the mechanical tensile stress of the bending radius of 5 mm (up to 2.12%). By replacing the electrode with platinum metal, the mechanical failure threshold of the Pt/HfO_x/Pt device can be further extended to the bending level of 2 mm (strain of 3.18%). Nevertheless, the brittle nature of the inorganic materials still makes the memory devices suffer severely from the limited adaptivity to large strains upon repeated bending or stretching operations.

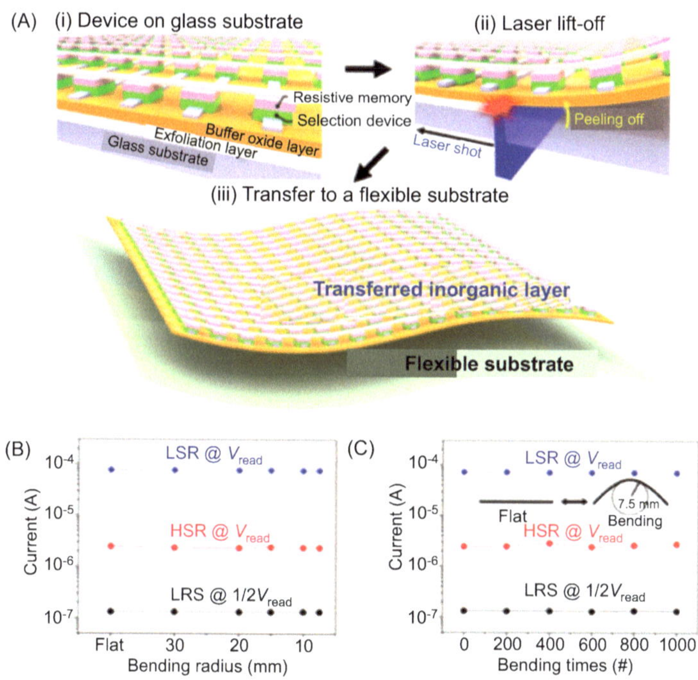

Figure 4.3 (A) Schematic illustrations of the process for fabricating flexible crossbar-structured memory with Ni/TiO_2/Ni selector (1S) and NiO_x-based resistive switching materials on a plastic substrate via the laser lift-off method. (B) The resistance values of HRS and LRS as a function of bending radius. (C) Continuous bending fatigue test results for the 1S–1R cell up to 1000 iterations of bending and unbending. The inset shows a schematic illustration for repeated bending deformations (bending and release states) at the fixed bending radius of 7.5 mm. Reprinted with permission from Ref. [25], Copyright 2014, John Wiley & Sons.

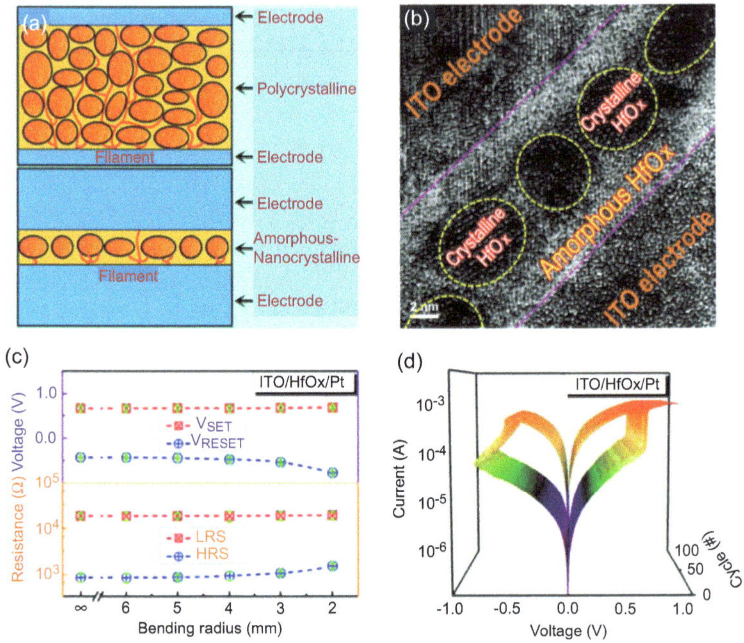

Figure 4.4 (a) Schematic illustration of the formation of conductive filaments in polycrystalline (upper panel) and pseudo-amorphous nanocrystalline (lower panel) hafnium oxide layers. (b) Cross-sectional high-resolution transmission electron microscopy image of the as-fabricated ITO/HfOx/ITO structure. (c) Evolution of SET/RESET voltages and ON/OFF state resistances of the ITO/HfOx/Pt device as a function of the bending radius. (d) Current–voltage (I–V) characteristic of the ITO/HfOx/Pt device over 100 consecutive cycles with a bending radius of 2 mm. Reproduced from Ref. [20] with permission from The Royal Society of Chemistry.

4.3.2 Organic Materials

Compared to metal oxides and other inorganic counterparts, organic materials are considered one of the most promising information storage media in flexible and even stretchable resistive switching memories, because of their intrinsic advantages such as low-temperature processing, printability, and inherent mechanical flexibility. Organic materials with electrical switching behavior, including small molecules, graphene, and polymers, have attracted tremendous scientific and industrial interests and are well studied.

The mechanism of the resistive switching behavior in organic systems has been well summarized in the reviews [60, 61]. The electrical conduction mechanism in organic materials is much more complex than in ordered inorganic materials. The common conduction mechanisms are listed as follows:

Filament conduction mechanism: Resistive switching in devices of this type is due to the formation of filamentary conduction path. Two types of filamentary conduction are widely reported in polymer resistive switching memory devices. One type is related to the carbon-rich filaments formed by the local degradation of polymer films. The other is related to the metallic filaments resulting from ion migration and redox reaction of the metal ions from the active metallic electrodes.

Charges trap and de-trap mechanism: This type of resistive switching usually takes place in the insulating polymer doped with electroactive component as the charge trap center or polymer with donor–acceptor structure. Initially the device is at its OFF state with high resistance. When the voltage reaches a certain value, the traps are completely filled and induce an ON state with lower resistance compared to the initial state.

Charge transfer mechanism: Charge transfer mechanism describes the resistive switching that occurs in a charge transfer (CT) complex, which is defined by electronic transition to an excited state in which there is a partial transfer of electronic charges from the donor (D) to the acceptor (A) moieties.

Tunneling effects: If the electrode contact is ohmic and the applied field across an insulating thin film (e.g., less than 10 nm) is very high, the probability of finding an electron on the other side of the potential barrier by quantum mechanical tunneling will increase to give rise to a low-resistance state.

Conformation change effects: Electrically induced conformational changes in molecules or molecular bundles may influence the effective charge transfer through π–π stacking system and modulate the material's conductance.

4.3.2.1 Organic resistive switching memory with small molecules

Molecule-based resistive switching memory has been of interest to the research community because of its solution processability and chemically tunable functionalities. For instance, the molecular skeleton or the substituent of molecule can be modified to adjust the storage performance. Besides, the well-defined structures and molecular packing of small molecules not only allow them to have good reproducibility in processing, but also favor better understanding of the switching mechanisms with established electronic structure getting from the experiment-simulation matching. In order to design a molecule for the resistive switching memory effect, the factors that influence resistive switching performance, including the arrangement of donor and acceptor moieties, molecular packing, and crystallite orientation of the molecular thin film, are well investigated.

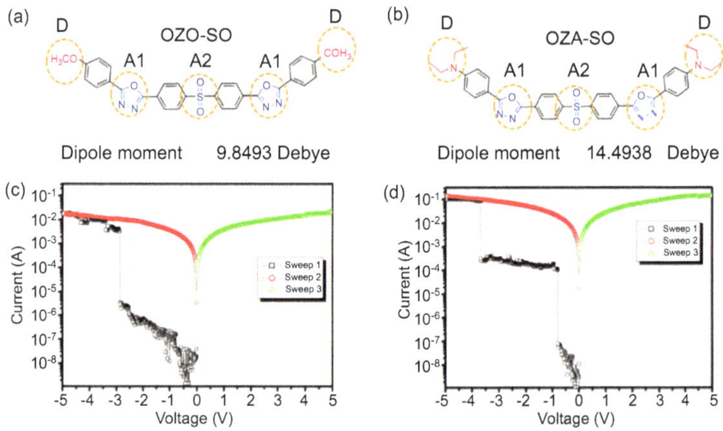

Figure 4.5 (a,b) Two compounds OZO-SO and OZA-SO with D–A1–A2–A1–D symmetric structure and their dipole moments. (c,d) I–V curves of the memory device fabricated by OZO-SO and OZA-SO under negative bias. Reprinted with permission from Ref. [62], Copyright 2015, John Wiley & Sons.

Lu and Xu et al. synthesized two compounds, p,p′-bis (2-methoxyl-1,3,4-oxadiazol-5-yl)diphenyl sulfone (OZO-SO) and (p,p′-bis(2-aryl-1,3,4-oxadiazol-5-yl)diphenyl sulfone (OZA-SO), with symmetric Donor-Aceptor1-Aceptor2-Aceptor1-Donor structure, as shown in Fig. 4.5 [62]. The 1,3,4-oxadiazole group was selected because of a classic nitroxide heterocycle with good charge transport property and strong electron affinity, which could increase the degree of conjugation and molecular rigidity of the two molecules, improving thermal stability and facilitating carrier transport. Moreover, in these molecules, two terminal electron-donating groups (methoxy and diethylamino groups) with distinct electron-donating abilities were introduced into the terminal position of the backbone. OZA-SO has a higher dipole moment because of the strong electron-donating ability of the diethylamino group. The strong electron donor moiety of OZA-SO results in strong intermolecular interaction and ordered molecular aggregation, enhancing the field-induced intra-charge transfer effect, which is beneficial for ternary memory behavior. In contrast, OZO-SO employed weak electron donor moiety and showed an amorphous aggregation and very weak intra-charge transfer effect, with only binary memory performance observed.

The same group also investigated the effect of backbone planarity on the crystallite orientation and the resulted resistive switching performance. They have synthesized two diketopyrrolopyrrole-based small molecules (NI2PDPP and NI2FDPP) with increasing planarity by furan substitution for phenyl rings (Fig. 4.6) [63]. The highly planar NI2FDPP in film prefers out-of-plane crystallite orientation with respect to the substrate normal, while the nonplanar NI2PDPP displays less ordered packing with a broad orientation distribution relative to the substrate. Both devices exhibit ternary WORM memory behavior with different switching threshold voltages. In particular, the highly preferred crystallite orientation is essential to obtain efficient multilevel memory performance and favors good stability and reproducibility of memory effects.

The length of molecular conjugation and alkyl chain plays an important role in molecular stacking, which could strongly affect the migration of charge carriers through the films and, thus, the electronic performance. Lu et al. designed a small molecule, 1,5-bis((E)-(4-(di-butylamino)phenyl)-diazenyl)anthracene-9,10-dione (BAzoAN), by mounting the conjugated backbone with alkyl chains, as shown in Fig. 4.7 [64]. 1,5-bis((E)-(4-(dimthyl-amino)phenyl)diazenyl)anthracene-9,10-dione (MAzoAN) and 1,5-bis ((E)-(4-(diethylamino)phenyl) diaze-nyl) anthracene-9,10-dione (EAzoAN) were synthesized as the control to consider the influence of side chain length in film forming. These molecules are deposited on the PET substrates, and flexible ternary memory is realized for the first time. In the flat state, the memory device fabricated with BAzoAN shows a typical ternary memory behavior. The memory device fabricated with MAzoAN shows no conductive switching effect when a voltage from 0 to 5 V was applied. The memory device Al/EAzoAN/Al/PET shows a typical binary WORM memory property. The electronic properties of the flexible device with BAzoAN were measured under various bending conditions, which exhibited stable ternary data-storage properties. Meanwhile, the threshold voltages and current values of the flexible device were almost the same within the maximum bending conditions. Moreover, the memory characteristics were stable for up to 500 bending cycles in the highest bending condition with a radius of 10 mm. When the bending radius reaches 5 mm, an obvious crack is observed in the film, indicating the film is damaged, which may influence the migration of the charges and the memory effects.

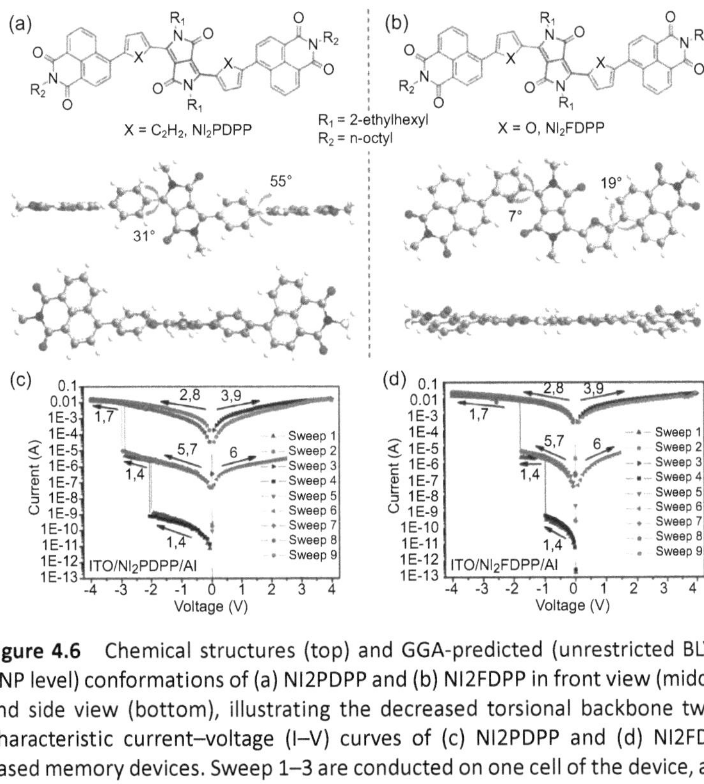

Figure 4.6 Chemical structures (top) and GGA-predicted (unrestricted BLYP/DNP level) conformations of (a) NI2PDPP and (b) NI2FDPP in front view (middle) and side view (bottom), illustrating the decreased torsional backbone twist. Characteristic current–voltage (I–V) curves of (c) NI2PDPP and (d) NI2FDPP based memory devices. Sweep 1–3 are conducted on one cell of the device, and sweep 4–9 are conducted on another cell. Reprinted with permission from Ref. [63], Copyright 2015, John Wiley & Sons.

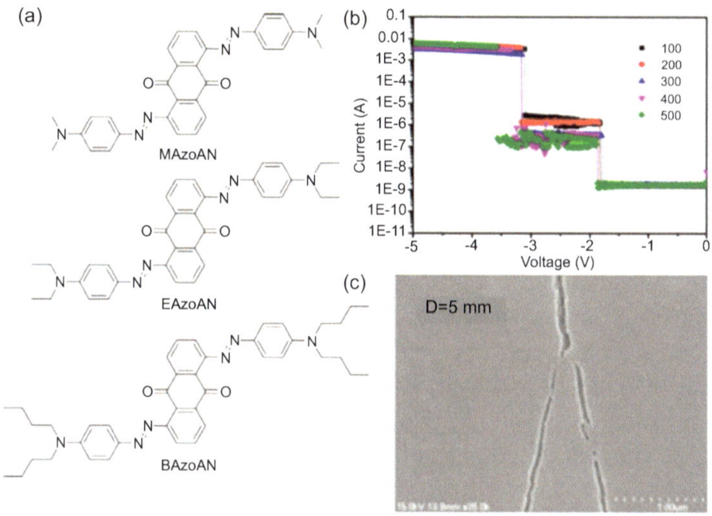

Figure 4.7 (a) Chemical structures of MAzoAN, EAzoAN, and BAzoAN. (b) The memory device can maintain ternary memoryper formance within 500 bending cycles in the maximum bending condition. (c) SEM surface images of BAzoAN-based films under the bending conditions with *D* of 5 mm. Reprinted with permission from Ref. [64], Copyright 2016, John Wiley & Sons.

It is noteworthy that although multilevel resistive switching phenomena and great improvements have been reported for several molecular devices, they also suffer from relatively poor endurance performance. Besides, the intrinsic crystallization nature of the small molecules during film formation limits their application in flexible memory.

4.3.2.2 Blends or mixtures of memory polymer materials

In comparison to small organic molecules, polymers demonstrate improved intrinsic flexibility with their flexible chain structures. In the state-of-the-art studies of the polymer memories, blends of polymers or mixtures of polymers with small molecules or carbon nanomaterials, in which both or all the constituting components are electroactive, are often used as the switching media. For instance, Forrest et al. spun a layer of polyethylene dioxythiophene: polystyrene sulfonic acid (PEDOT:PSS) onto the surface of an Si-based thin-film diode that was pre-deposited onto a stainless steel substrate for the construction of memory devices, wherein PEDOT:PSS is a well-known organic conductive polymer. With a sufficiently high current applied through the electrode/PEDOT:PSS/Si/electrode structure, the device can be irreversibly switched from the initial low-resistance state (ON state) to the high-resistance state (OFF state) with a large ON/OFF ratio, which permits the polymer/semiconductor junction to function as a WORM memory. The underlying mechanism is revealed by X-ray photoelectron spectroscopy and temperature-induced conductivity measurements. It is found that the aggregation of the insulating PSS chains near the electrode/polymer interface upon high current density reduces the conductivity.

Shang et al. also reported a reduced GO poly(*N*-vinylcarbazole) (rGO-PVK) composite based memory device, in which the composite was prepared by simply mixing PVK with rGO in solution [65]. It was found that the electrical switching behavior is strongly dependent on the weight content of rGO in the composite. Specially, when the weight content of rGO increased to 4%, the obtained device presented a rewritable switching behavior with an ON/OFF current of ratio more than 10^3, in which case the rGO serves as the electron charge trapping center because of its electron withdrawing while PVK polymer serves as the electron donor.

Nevertheless, the operation parameters of polymer memory devices usually suffer from severe fluctuations, which lead to consequently false programming and error readout limitations that hinder their practical applications. Therefore, obtaining reproducible and uniformly distributed switching parameters is always a big challenge for the polymer memory communities. Hu et al. proposed a strategy to overcome this problem by controlling the resistance of the polymer device gradually and exactly via manipulating the doping level of a toluenesulfonic acid (TsOH): polyazomethine (PAM) system (Fig. 4.8) [15]. It was found that the Pt/PAM–TsOH/Pt devices exhibited smooth and self-rectifying I–V characteristics during the resistive switching process with a rectifying ratio of 10–100, an ON/OFF ratio of 600, and a small dispersive index of less than 5% for both the switching voltage and the ON/OFF ratio in over 1200 switching cycles. The promising uniformity of the PAM–TsOH system is superior to those reported for resistive switching memory devices, suggesting that the controllable protonic doping is an effective way to optimize the uniformity of resistive switching behaviors. Using protonic acids with larger molecular weights and sizes can further increase the self-rectifying property of poly(schiff base), as well as enhance the device retention capability of the doped system. Moreover, by using different reset voltages and compliance current levels, multilevel storage was established during the consecutive modulation of the device resistance, which is important for increasing the data-storage capacity.

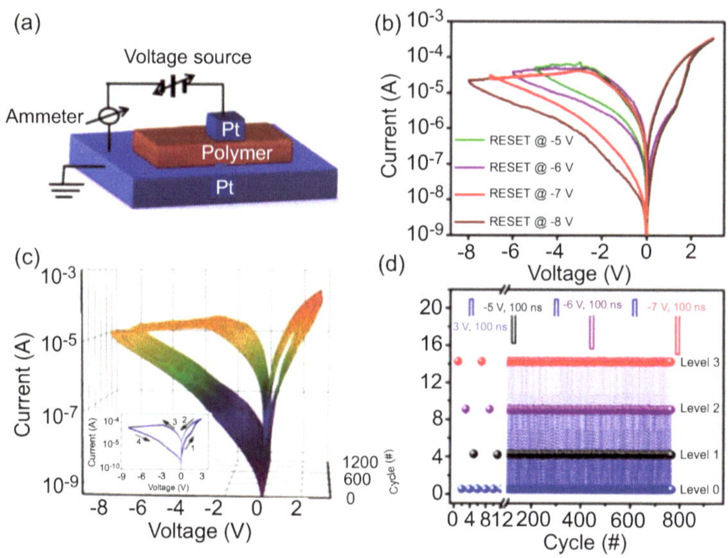

Figure 4.8 (a) Schematic illustration of the poly(Schiff base)-based Pt/PAM–TsOH/Pt devices. (b) Consecutive I–V cycles of a Pt/PA-TsOH/Pt device. The inset shows the first I–V cycle. Multilevel operation of the Pt/PA-TsOH/Pt devices in a dc sweeping mode (c) and a pulse mode (d). The inset of (d) illustrates the voltage pulse utilized for device programming. The reading voltage is −0.5 V. Reprinted with permission from Ref. [15], Copyright 2012, American Chemical Society.

4.3.2.3 Polymer matrices for electroactive components

Another facile and effective strategy to fabricate flexible organic RRAM device is employing polymers mixed with electroactive components such as small organic molecules, inorganic nanoparticles, nanosheets, and carbon nanomaterials, in which polymers are employed as a non-active flexible matrix.

Lee et al. developed an organic, nonvolatile memory device on flexible substrates using polyimide (PI): 6-phenyl-C61 butyric acid methylester (PCBM) composite as the active layer, as shown in Fig. 4.9 [38]. PI is selected because of its thermal stability, good chemical resistance, and excellent mechanical properties, while PCBM is selected because it is commonly used as electron acceptor in organic heterojunction solar cells. Thus, PI and PCBM composite materials may exhibit mechanical robustness and expected electronic properties for resistive switching. The flexible memory devices integrating PI and PCBM showed ON/OFF ratio of 10^4 and no overlapping of SET/RESET voltage under the bending radius of 9 mm, indicating stable and reliable resistive switching properties under bending condition.

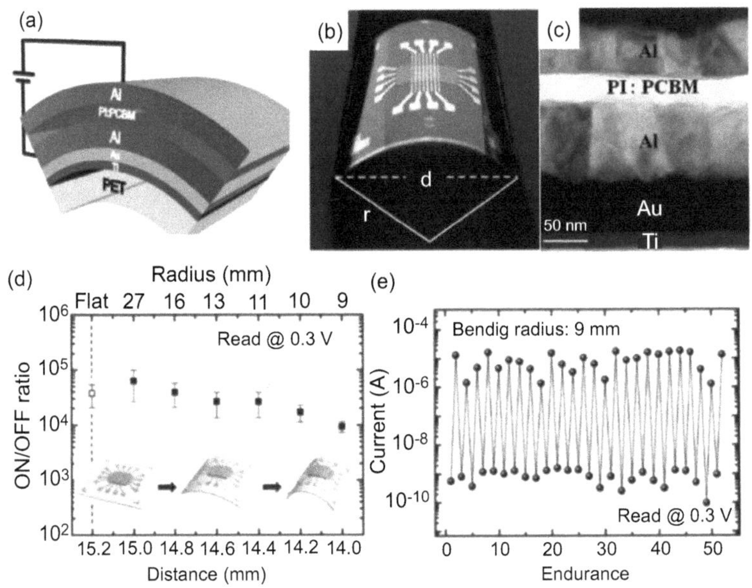

Figure 4.9 (a) Schematic view of a Ti/Au/Al/PI:PCBM/Al memory cell. (b) Optical images of the 8 × 8 array-type Ti/Au/Al/PI:PCBM/Al flexible organic memory devices. (c) Cross-sectional TEM image of the layers in a memory cell. (d) ON/OFF ratio as a function of the bending condition (the substrate distance and the bending radius). (e) Endurance cycles when the memory device was bent. Reprinted with permission from Ref. [38], Copyright 2010, John Wiley & Sons.

In order to prevent the PCBM molecules from severely aggregating, Kim et al. reported an organic nonvolatile resistive switching memory in which PCBM is incorporated into the nanostructure of self-assembled poly(styrene-b-methyl methacrylate) (PS_{10}-b-$PMMA_{130}$) diblock copolymer (Fig. 4.10) [66]. The favorable miscibility of PCBM molecules with the PS component allows the zero-dimensional carbon nanospheres to be confined in PS without aggregation. The PS-b-PMMA/PCBM nanocomposite spontaneously formed nanostructures with 10 nm PS nanosphere domains surrounded by a PMMA matrix. The as-fabricated flexible RRAM demonstrates bipolar switching behavior without significant degradation before and after bending ratio of 4 mm.

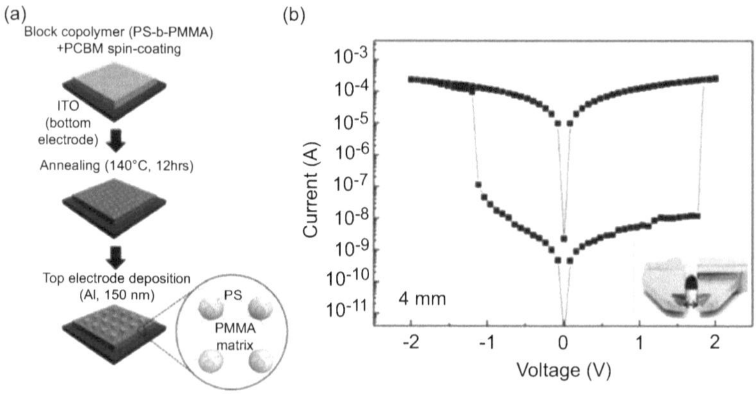

Figure 4.10 (a) Schematic procedure for the fabrication of PS_{10}-b-$PMMA_{130}$ /PCBM nanocomposite memory device with a hypothetical diagram of the active layer. (b) The current–voltage curves of flexible ReRAM (Al/PS-b-PMMA: PCBM (0.05 wt%)/ITO PET) with the bending radius of 4 mm. Reprinted with permission from Ref. [66], Copyright 2013, John Wiley & Sons.

On the other hand, two-dimensional materials such as GO and MoS_2 have shown attractive physical and electronic properties for future electronics and optoelectronics [67]. Integrating these functional 2D materials into flexible polymer matrix has also been investigated and employed as the information storage media.

Choi et al. demonstrated the fabrication of a memory device with the configuration of PET/Ag/MoS_2-PVA/Ag via an all-printed fabrication approach, as shown in Fig. 4.11, wherein the MoS_2-PVA ink is prepared by mixing PVA with MoS_2 in solution. Bendability test at various bending diameters (50 mm to 2 mm) suggested that the as-fabricated device exhibited mechanical robustness over 1500 deformation cycles [68]. Gao et al. reported a flexible memory device based on MoS_2-polyvinylpyrrolidone nanocomposites films, which showed rewritable resistive switching under bending condition [69].

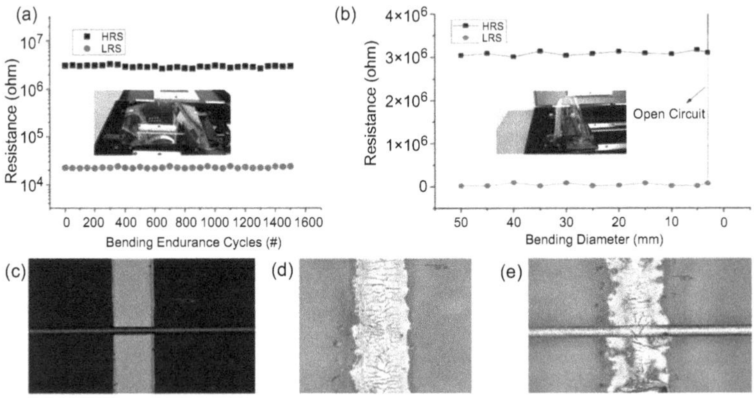

Figure 4.11 (a) Endurance curve of device PET/Ag/MoS_2-PVA/Ag showing HRS and LRS values for 1500 bending cycles at a fixed diameter value of 10 mm. (b) Endurance plot clearly showing very less deviation in HRS and LRS with changing values of bending diameters in the range of 30 mm to 2 mm. The device behaved like an open circuit at a bending diameter of 2 mm. Optical images of top and bottom electrodes at different values of bending diameters. (c) Before bending. (d) Appearance of micro-cracks in bottom electrode. (e) Development of a major crack in the bottom electrode and completely broken top electrode at a bending diameter of 2 mm, resulting in the open circuit of memory device [68].

Other carbon nanomaterials, such as carbon nanotubes, also possess unique electric properties and have been selected for the fabrication of resistive switching device. Liu et al. presented an indium-tin oxide/PVK-CNT/ aluminum (ITO/PVK–CNT/Al) sandwich structure memory, which can be tuned in a controlled manner by varying the carbon nanotube (CNT) content in PVK composite thin films (Fig. 4.12) [70]. PVK serves as the matrix for CNT in the presented devices. CNT has a much higher workfunction (−5.1 eV) compared to the LUMO level of PVK (−2.0 eV) and is more likely to serve as the electron-trapping center and electron transporter. Distinctly different electrical conductance behaviors, such as (i) insulator behavior, (ii) WORM and rewritable memory effect, and (iii) conductor behavior, are observed from the current density–voltage characteristics of the composite films. The turn-on voltage of the two bistable conductance switching devices decreases, and the ON/OFF state current ratio of the WORM device increases with the increase in CNT content of the composite film. Both the WORM and rewritable devices are stable under a constant voltage stress or a continuous pulse voltage stress, with an ON/OFF state current ratio in excess of 10^3.

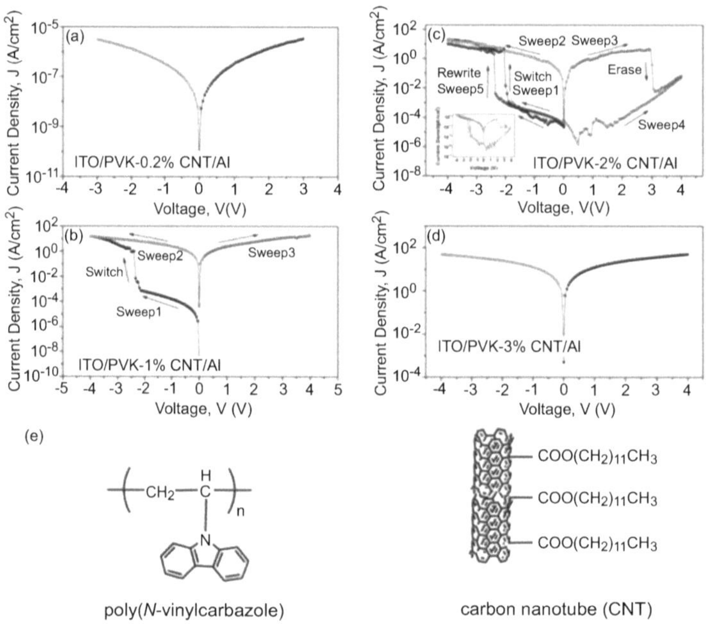

Figure 4.12 J–V characteristics of the ITO/PVK-CNT/Al devices containing (a) 0.2%, (b) 1%, (c) 2%, and (d) 3% carbon nanotubes. (e) Structural formulas of poly (*N*-vinylcarbazole) and surfaced-functionalized carbon nanotube. Reprinted with permission from Ref. [70], Copyright 2009, American Chemical Society.

Another example of polymer matrix employed in resistive switching memories is the solid-state polymer electrolytes (SPE). Lee et al. suggested a flexible transient RRAM device utilizing Ag^+-ions-doped chitosan-based biopolymer on plastic substrate. The device demonstrated an ON/OFF current ratio $>10^2$, good retention capability of $\sim10^4$ s, and low operation voltages. The resistive switching is reliable when bending the devices with a 5 mm radius of curvature [31]. Mohapatra et al. reported an Ag/SPE/Pt cell fabricated on a plastic substrate using a mixture of poly(ethylene oxide) (PEO) and silver salt ($AgClO_4$). The device showed stable bipolar resistive switching under bias voltage sweeping due to the transport of Ag^+ ions and redox reactions at the electrode interfaces.

Moreover, the cell demonstrated very stable switching behavior without any deterioration when bended to the radius of 15.2 mm [71].

4.3.2.4 Single-component polymer active materials

Doping or mixing of polymers with guest components indeed provides a facile approach to fine-tune electrical properties and, therefore, the resistive switching memory performance in flexible devices. However, it may not always give rise to uniformly dispersed and compatible components and, thus, may result in phase separation and ion aggregation, which are unfavorable to the device performance. To avoid such problems, much attention has been paid to uniform memory polymer materials, which can provide the required electronic properties within a single macromolecule, while still possessing good chemical, mechanical, and morphological characteristics.

Designing a polymer with donor–acceptor (D–A) structure is an effective approach to fabricating single-component polymer-based resistive switching memory since the D–A structure favors the electron charge transfer process resulting in resistance switching. The versatile memory behaviors—including volatile or nonvolatile digital memory behavior such as WORM and rewritable resistive switching—of polymer-based devices could be manipulated through the donor/acceptor characteristics or the nanostructures of the polymer active layer.

Polyazothmine (PAM) materials are a family of conjugated poly(Schiff base) polymers containing imine groups (C=N) in the backbone and exhibiting excellent thermal stability, good mechanical properties, metal-chelating ability, and molecular-doping-controlled electrical properties. Pan et al. have successfully synthesized two D–A type poly(Schiff base) (polyazothine) derivatives based on triphenylamine, oxadiazole, and 3,3′-dinitro-diphenylsulfone moieties (PAM-1 and PAM-2), as shown in Fig. 4.13 [72]. By varying the chemical structure of the electroactive polymers, as well as the top electrode metal, the four kinds of memory devices—Pt/PAM-1/Pt, Pt/PAM-2/Pt, Al/PAM-1/Pt, and Al/PAM-2/Pt—are fabricated and found to exhibit different resistive switching properties. In the case of Pt/PAM/Pt devices, PAM-1 shows rewriteable memory behavior with poor endurance of less than 20 cycles, while PAM-

2 exhibits WORM memory characteristics. The different memory properties of the Pt-electrode devices originate from the different HOMO–LUMO band gaps of the PAMs, which influence the degree of intra- and intermolecular charge transfer interaction between the electron donor–acceptor pairs. By using Al to replace the Pt top electrodes, the Al/PAM-2/Pt devices exhibit a repeatable resistive switching effect.

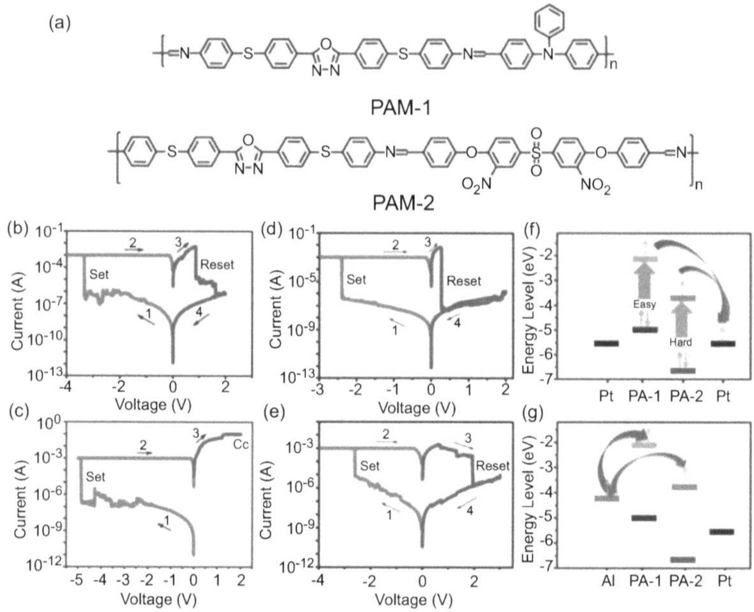

Figure 4.13 (a) Two donor–acceptor (D–A) polyazothines (PAM-1 and PAM-2); current–voltage (I–V) characteristics of (b) Pt/PAM-1/Pt, (c) Pt/PAM-2/Pt, (d) Al/PAM-1/Pt, and (e) Al/PAM-2/Pt devices. Energy diagram of M/PAM/M sandwich systems, and the different RS mechanisms for (f) Pt/PAM/Pt and (g) Al/PAM/Pt devices. Reproduced from Ref. [72] with permission from The Royal Society of Chemistry.

In order to further improve the reliability of the resistive switching behavior, a hyperbranched poly(azothine)s based on triphenylamine moieties are synthesized and used as the functioning layers in Ta/PAM/Pt resistive switching memory devices by the same group (Fig. 4.14) [73]. The hyperbranched polymers usually characterized unique three-dimensional structures, good solubility, low melting temperature, and solution viscosity, and excellent physiochemical properties. The as-synthesized hyperbranched

PAM with isotropic architecture and semi-crystalline nature shows enhanced memory behaviors with more uniform distribution of the HRS and LRS resistances.

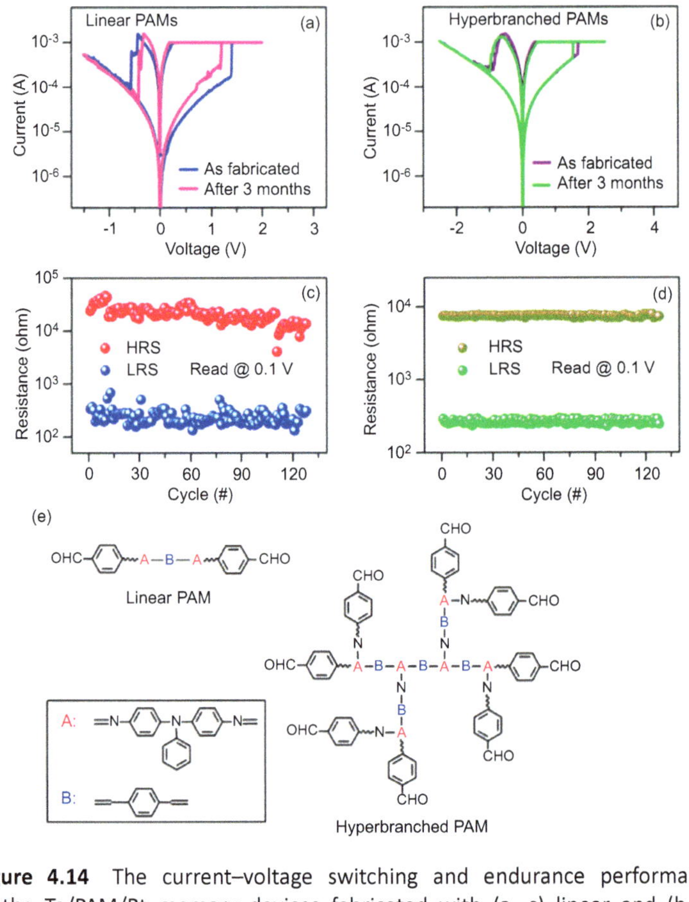

Figure 4.14 The current–voltage switching and endurance performance of the Ta/PAM/Pt memory devices fabricated with (a, c) linear and (b, d) hyperbranched PAMs, respectively. (e) Molecular structures of the linear and hyperbranched PAMs. Reproduced from Ref. [73] with permission from The Royal Society of Chemistry.

The large ON/OFF ratio favors the practical application for reading the stored information. In order to enhance the ON/OFF ratio of polymer-based memory, thermally stable poly(triphenylamine) (Fig. 4.15) has been synthesized as the switching media. In this PTPA, the unpaired electrons on the nitrogen atoms of triphenylamine unit can be easily removed from the polymer to give a stable cationic conducting pathway [74]. The as-fabricated device exhibits excellent nonvolatile bistable resistive switching behavior with a long retention time over 8×10^3 s, superior ON/OFF ratio of 5×10^8, and a wide working temperature range of 30–390 K.

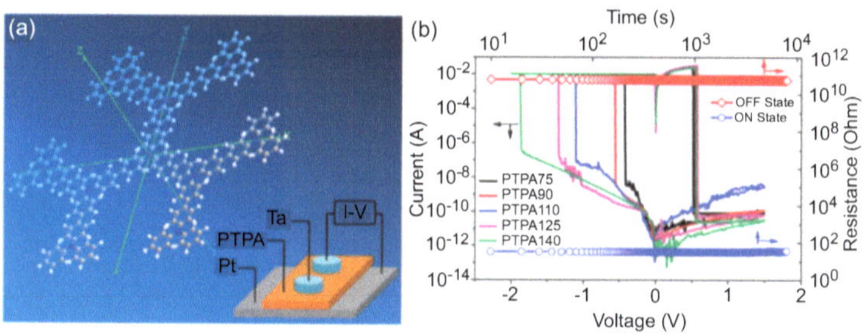

Figure 4.15 (a) Possible molecular structure of PTPA and the schematic illustration of the Ta/PTPA/Pt memory device. (b) Room-temperature current–voltage characteristics and retention capability of Ta/PTPA/Pt structured devices. Reproduced from Ref. [74] with permission from The Royal Society of Chemistry.

Taking advantage of the electron-donating feature of the triphenylamine unit, Liou et al. also reported a flexible dual functional device based on triphenylamine-containing polymers with multi-colored electrochromic and volatile memory characteristics, as shown in Fig. 4.16 [75]. The incorporation of the electron-donating, packing-disruptive starburst triphenylamine moiety with the methoxyl group at the para-position of phenyl groups not only enhances the thermal stability and solubility of the polymer, but also helps to lower the oxidation potential associated with increasing electrochemical stability and multi-stage electrochromic characteristic. Consequently, the device shows electrochromism with high contrast ratio, high coloration efficiency, and low switching time on the bending ratio of 5 mm.

Carbon nanomaterials are also covalent bonded to the polymer backbone or pendant groups for the construction of resistive switching memories. As a typical example, Liu et al. reported for the first time a GO-polymer-based memory by employing poly(N-vinylcarbazole) (PVK) covalently grafted GO sheets as the active material, as shown in Fig. 4.17 [76]. The GO-PVK composite can be well dissolved in a number of organic solvents, while the as-fabricated Al/GO-PVK/ITO structure with 100-nm-thick GO layer showed bistable electrical switching behavior and rewritable memory effects, with a turn-on voltage of about 2 V and an ON/OFF state current ratio in excess of 10^3. The device can be switching for 10^8 times, and the stored data can be maintained for more than 10 years, thus fulfilling the practical requirements for nonvolatile information storage applications.

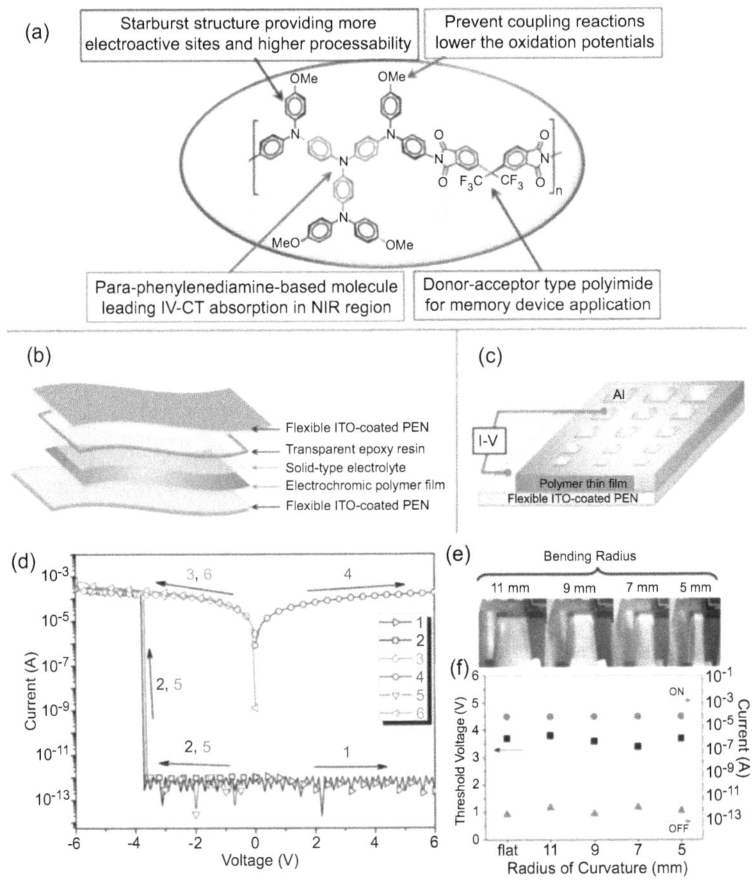

Figure 4.16 (a) Design strategy of starburst triphneylamine-based polyimide 9Ph-6FDA, (b) schematic diagram of flexible EC device based on single-layer EC polyimide, and (c) configuration of the flexible memory device consisting of the polyimide thin film sandwiched between an ITO bottom electrode and an Al top electrode, (d) current–voltage (I–V) characteristics, (e) appearance in various bent states, and (f) variation of current and threshold voltage with different bending radii of ITO/ 9Ph-6FDA (50 ± 3 nm)/Al flexible memory device. Reprinted with permission from Ref. [75], Copyright 2013, John Wiley & Sons.

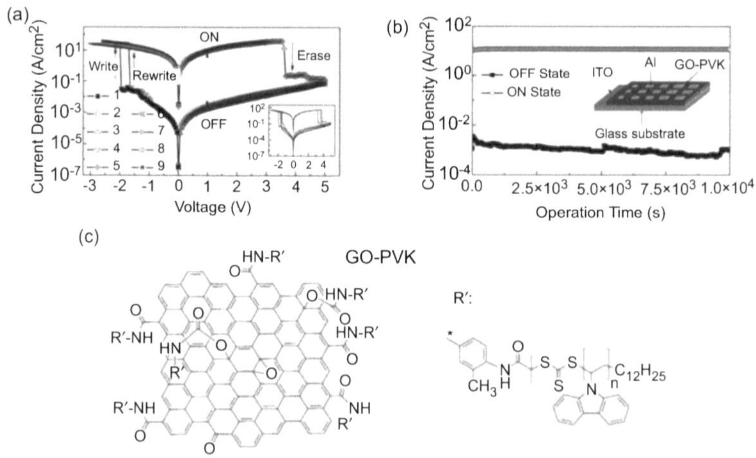

Figure 4.17 (a) J–V characteristics of the Al/GO-PVK/ITO device. Stability of the Al/GO-PVK/ITO device in the ON and OFF states (b) under a constant stress of 1 V. (c) Schematic illustration of GO-PVK composite. Reprinted with permission from Ref. [76], Copyright 2009, AIP Publishing.

Instead of designing flexible electric active polymers with donor–acceptor structure, another approach to designing high-performance flexible RRAM is employing flexible insulating polymer as the information storage media in which carbon filament is formed. Choi et al. reported a poly(ethylene glycol dimethacrylate) (pEGDMA) based RRAM showing stable memory operations up to a bending radius of 4 mm (Fig. 4.18) [77]. The polymer applied in this work with a highly cross-linked structure has a proper band structure as an insulator and is deposited by use of an initiated chemical vapor deposition process (i-CVD process). The reliability is verified by a robust bending endurance of more than 10^3 cycles. Direct observation of a carbon filament by means of cross-sectional transmission electron microscopy (TEM) is reported for the first time, in which the switching operation of the RRAM is attributed to the thermal pyrolysis of the pEGDMA resistive switching layer. The RRAM shows robust flexibility and stable operation, even under water, which originates from the high water stability of the pEGDMA layer, as demonstrated through a bending endurance test (4 mm) and strong durability against water.

Choi et al. reported another iCVD-deposited poly(1,3,5-trimethyl-1,3,5-trivinyl cyclotrisiloxane) (pV3D3) based RRAM, in which the device can be switched to the low-resistance state at a bending radius of 3 mm (Fig. 4.19). The pV3D3 is selected because of its excellent chemical stability. With its high purity and electrically robust polymer thin film, the pV3D3-RRAM exhibits unipolar resistive switching behavior with a high ON/OFF ratio >10^7, good retention over 10^5 s, excellent cycling endurance of 10^5 cycles, and robust immunity to mechanical stress [32].

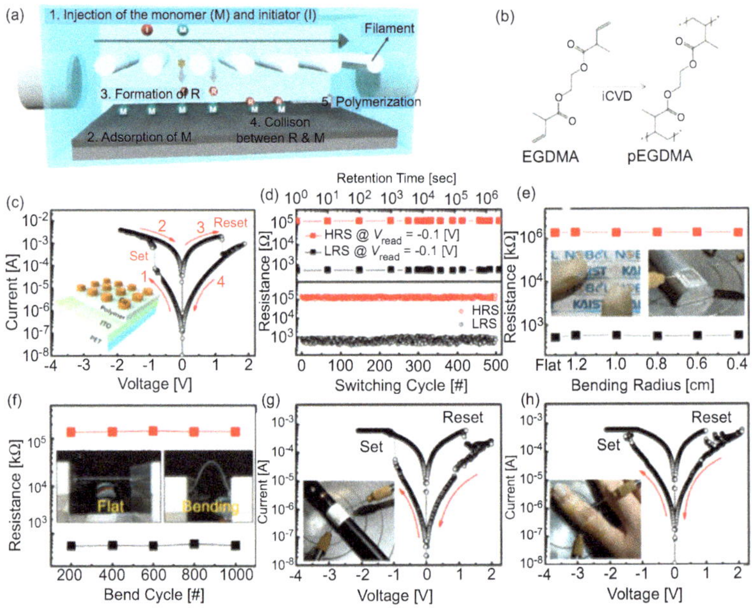

Figure 4.18 (a) Schematic to explain the iCVD process. (b) Schematic of EGDMA and pEGDMA. (c) Bistable memory operation with the bipolar switching mode. The arrow signifies the sweep of the applied voltage. The inset image shows a schematic of the fabricated i-RRAM. (d) Variation of the HRS and LRS as a function of the retention time and the switching cycle. (e) Variation of the HRS and LRS as a function of the bending radius. (f) Variation of HRS and LRS as a function of the bending cycle. A custom-made bending machine was used to evaluate the bending endurance at a radius of 0.4 cm. (e) I–V characteristics for memory operation on a laser pointer. (f) I–V characteristics for memory operation on a finger. The RRAM, with remarkable flexibility, is easily attachable to various curved structures. Reprinted with permission from Ref. [77], Copyright 2015, American Chemical Society.

Despite considerable advances in fundamental research on polymer-based RRAMs during the last decade, it is still in the infancy and obstacles persist against the commercially viable devices. For instance, the severe fluctuations during operation and the poor resistance to the changing environments such as oxygen, water in the air make them improper for working under harsh conditions. The underlying switching and conduction mechanisms in polymer-based RRAMs have yet to be clearly understood in order to effectively control their resistive switching. Despite the great challenges faced by it at the moment, the organic flexible device proves to be one of the most promising solutions toward flexible electronic information memory.

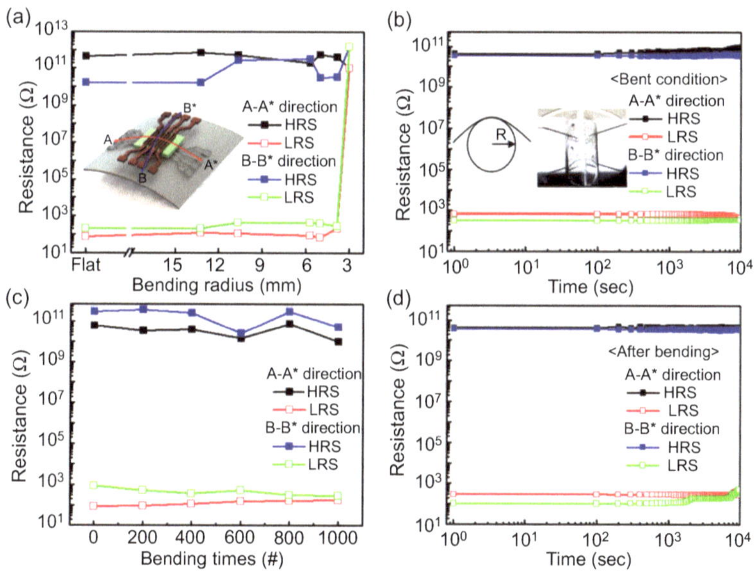

Figure 4.19 (a) The resistance ratio between LRS and HRS as a function of bending radius and (inset) a photograph of Cu/pV3D3/Al/PES structure and an I–V measurement being performed under a fixed condition. (b) Retention times of LRS and HRS under the bent condition. (c) Continuous bending fatigue test results for pV3D3-RRAM during 1000 bending cycles. (d) After repeated 1000 bending cycles, retention characteristics of LRS and HRS. Reprinted with permission from Ref. [32], Copyright 2016, American Chemical Society.

4.3.3 Inorganic–Organic Hybrid Materials

4.3.3.1 Metal-organic frameworks

Although the memories based on brittle inorganic materials, which usually demonstrate reliable storage performance on rigid substrates, their limited adaptivity to large strains upon repeated bending or stretching operations hinders their application in flexible electronics. On the contrary, though the soft organic electronic materials carry inherent mechanical flexibility, their poor resistance to the changing environments makes them improper for working under harsh conditions. Thus, more efforts should be devoted by material scientists toward exploring and designing novel materials, which may combine the advantages of both inorganic and organic materials, that is, memory devices should remain stable and reliable under various deformation scenarios to avoid any possible mis-operation on the stored data.

Metal-organic frameworks (MOFs) are a new class of crystalline microporous materials that exhibit tunable functionalities, large surface areas, and framework flexibility. Their frameworks can be taken as the rigid metal oxide matrix extended by soft organic linkers and possess the features of both inorganic and organic materials, providing alternative strategy for the construction of flexible resistive memories [78, 79]. Recently, it has been documented that the conductivity states of a metal/macroscopic MOF single crystal/metal structure can be tuned bistably by the self-limiting oxidative reaction of the metal anode involving liquid species [80]. Pan et al. also observed stable and low-power resistance-switching effect in MOF single crystals, which is probably related to the ferroelectric transition of N···H-O···H-N bridge-structured dipoles of guest water molecules inside the MOF nanochannels [81].

Inspired by these, efforts are further devoted to the direct deposition of high-quality MOF nanofilm on flexible substrate and the investigation of its promising memory performance under bending conditions (Fig. 4.20). It has been found that uniform and reproducible resistive switching effect, which can be sustained under the strain of as high as 2.8%, has been observed for the first time in all solid-state Au/HKUST-1/Au/PET thin-film structures. The mechanical flexibility and environmental stability well meet the requirements for wearable application under earth surface conditions. Through conductive atomic force microscopic (C-AFM) and depth-profiling X-ray photoelectron spectroscopic (XPS) analysis, it has been proposed that the electric-field-induced migration of the Cu^{2+} ions, which may lead to subsequent pyrolysis of the trimesicacid linkers and thus the formation of highly conducting filaments, could be the possible origin for the observed uniform resistance switching in HKUST-1 nanofilms. With the intrinsic properties of the MOF material, namely the 3D periodical nature of the framework and the synergistic interplay between the metal species and the organic linkers, to produce the resistive switching effect, the mechanically flexible HKUST-1 nanofilm may provide a more appropriate and controllable platform for the construction of wearable information storage devices [82].

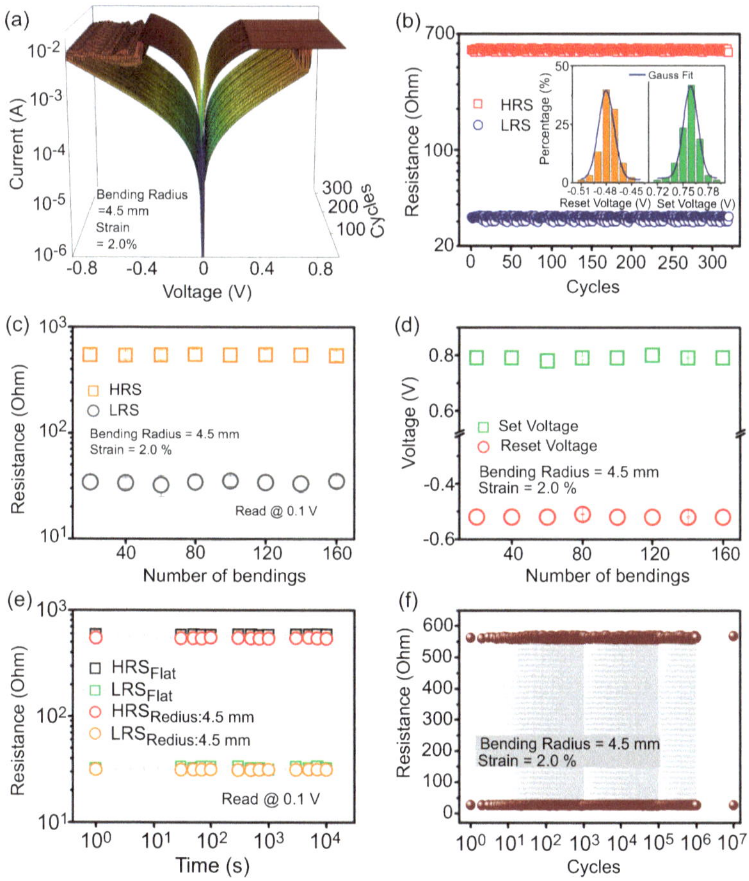

Figure 4.20 Current–voltage characteristics of the Au/HKUST-1/Au/PET flexible device (a) at the flat state and (b) during the dynamic bending test with the bending rates ranging from 0.25 to 2 mm/s. The inset of Fig. (a) shows the schematic configuration and the forming process of the memory device. Evolution of the (c) HRS and LRS resistances and (d) the set and reset voltages as a function of the bending radius or strain level of the Au/HKUST-1/Au/PET flexible device. (e) Schematic illustration of the switching mechanism in HKUST-1 nanofilms. Reprinted with permission from Ref. [82], Copyright 2015, John Wiley & Sons.

Another example by Chen et al. showed that ZIF-8 consisting of Zn^{2+} metallic ion nodes and 2-mIm organic linkers could be used as the resistive switching layer in RRAM devices under a bending radius of 1.35 cm [83]. They reported MOF-films-based memory devices with alcohol-mediated resistance-switching property, where the resistance state is controlled by applying alcohol vapors to achieve multilevel information storage. The ordered alcohol molecules absorbed in ZIF-8 pores with extensive hydrogen bonded network would work as electron transfer path to decrease the HRS resistance of ZIF-8-based memory device and finally form a new high-resistance state. The mechanism of resistance switching relies on the combined action of the formation of Ag nanoparticles in the insulator layer through electro-migration and the electron hopping effect in between the nanoparticles. The alcohol-mediated resistance-switching property can still perform well in the bending state with minimal bending radius of 14.2 mm, making this chemically mediated memory device a candidate in constructing flexible and wearable electronics.

4.3.3.2 Perovskite

Perovskite materials are another example of organic–inorganic hybrid materials with the chemical formula ABX3 (A is an organic cation, B is Pb, and X is a halide anion). They have strong optical absorption, tunable band gap, ambipolar charge transport, and long electron–hole diffusion length [85]. They also show a wide range of light absorption and high extinction coefficients. These characteristics make them suitable for various applications, such as thin-film transistors [86], solar cells [87], light-emitting diodes (LEDs) [88], and ambipolar transistors [89, 90]. Perovskite layers can be prepared on flexible substrates with solution processing at low temperatures. Lee et al. reported the fabrication and characterization of flexible RRAM based on the $CH_3NH_3PbI_3$ perovskite as the resistive element on the plastic substrate, as illustrated in Fig. 4.21 [84]. Use of a smaller bending radius of 7.5 mm also did not cause degradation. Repeated bending cycles up to 100 cycles were measured for a mechanical flexibility test. A ratio of HRS current to LRS current was maintained over 100 bending cycles. These results indicate that $CH_3NH_3PbI_3$ shows bipolar resistive switching behavior with good electrical reliability as well as mechanical stability.

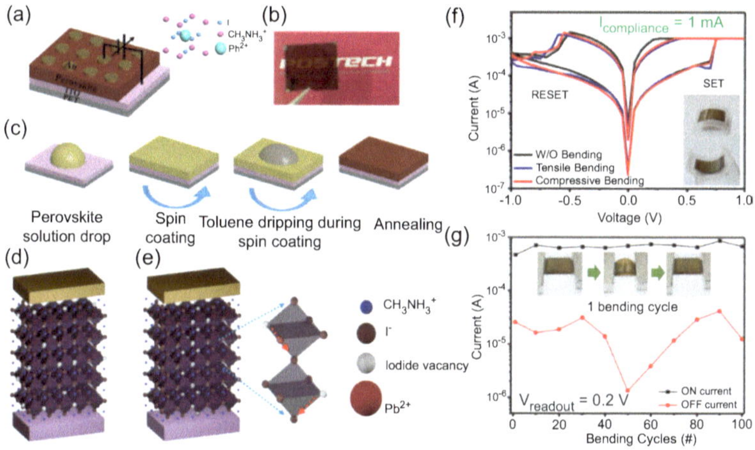

Figure 4.21 (a) Schematic diagram of the memory device with a structure of Au (top electrode)/$CH_3NH_3PbI_3$ perovskite layer/ITO (bottom electrode)/PET substrate. The inset shows the perovskite structure. (b) Photograph of a flexible perovskite resistive switching memory. (c) Schematic procedures to synthesize the perovskite layer using a solvent dripping process. (d) Proposed resistive switching mechanism of hybrid inorganic–organic perovskite resistive switching memory devices. Initial state showing randomly distributed iodide vacancies. (e) Alignment of iodide vacancies under the applied bias voltage. Inset: movement of iodide vacancies along the octahedral edge of perovskite structure. (f) I–V characteristics without and with bending stress (a bending radius of 1.5 cm). The inset photographs show tensile and compressive bending states of devices. (g) Bending stability of perovskite RRAM with repetitive bending cycles. Inset photographs show procedure of one bending cycle. Reprinted with permission from Ref. [84], Copyright 2016, American Chemical Society.

Kim et al. also reported resistive switching behavior with structural flexibility in organolead-halide-perovskites-based Ag/$CH_3NH_3PbI_3$/Pt devices (Fig. 4.22) [91]. Morphology-improved smooth thin films with a small crystal grain size were obtained by adding a suitable amount of hydrogen iodide (HI) as the additive. Therefore, the use of this smooth thin film as an insulating layer significantly increased the endurance from 350 cycles to more than 1300 cycles with an on/off ratio of 10^6, and an operation time at the level of a few hundreds of microseconds is achieved. It has been demonstrated that the organolead halide perovskites-based resistive switching devices can be operated at an extremely low bending radius of 5 mm. In addition, transparent thin films formed by the perovskite quantum dots (PeQDs) and PMMA nanocomposite have been demonstrated for resistive switching with flexibility as well. Li et al. have synthesized $CH_3NH_3PbBr_3$ PeQDs, and the resistance-switching phenomenon was observed when the nanocomposite was sandwiched between the memory device structures [92]. The PeQDs-based memory shows excellent electric characteristics, including the ON/OFF current ratio of larger than 10^3, retention time longer than 4000 s as well as good reproducibility and flexibility with a bending radius of 7 mm.

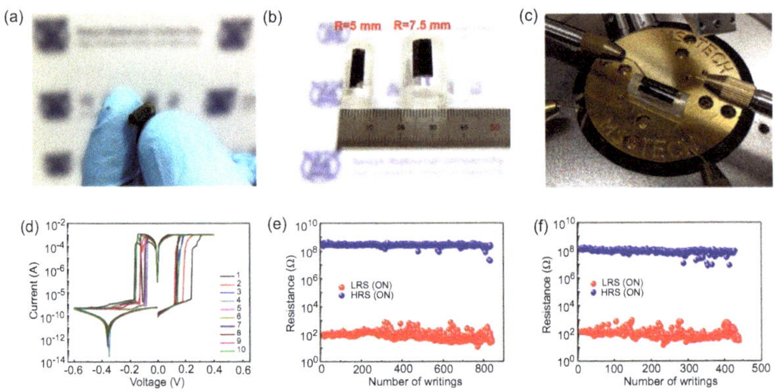

Figure 4.22 (a) Photograph of a fabricated resistive switching device Ag/$CH_3NH_3PbI_3$/Pt on a Pt-deposited cyclo-olefin polymer substrate. (b) Photograph showing bended states of the device. Bending radii were 5 mm (left) and 7.5 mm (right). (c) Photograph of a measuring resistive switching device with a bent state. (d) Typical current–voltage (I–V) characteristics of a flexible device with bending stress (a bending radius of 5 mm). (e, f) Endurance properties measured under a set voltage of +0.4 V and a reset voltage of −0.6 V. The pulse widths of the writing voltage were 50 ms (set) and 100 ms (reset). Bending radii were (e) 7.5 and (f) 5 mm. Reprinted with permission from Ref. [91], Copyright 2017, American Chemical Society.

4.4 Conclusion and Outlook

In this chapter, we have introduced the common strategy toward flexible resistive switching device and summarized the reported materials employed to construct the memory unit from inorganic, organic to inorganic–organic hybrid materials. On one hand, it has been indicated that the polycrystalline inorganic-materials-based memories, which usually demonstrate reliable storage performance on rigid substrates, suffer severely from limited adaptivity (the maximum bending radius being 5 mm or strain below 1%) to large strains upon repeated bending or stretching operations. On the other hand, though the soft organic electronic materials carry inherent mechanical flexibility, the poor resistance to the changing environments makes them improper for working under harsh conditions, high temperature, or strong UV irradiation. The MOFs, which can be taken as the rigid metal oxide matrix extended by soft organic linkers, possess the features of both inorganic and organic materials. Employing MOF materials as memory active material presented stable resistive switching behavior even under bending condition, demonstrating their potential and promise in flexible resistive switching memory. Organic–inorganic hybrid perovskites are also promising candidates for flexible memory applications.

Despite that considerable advances have been achieved in resistive switching memory from various materials, there are still challenges and obstacles in employing commercially available devices. First, a thorough understanding of the dependence of device bendability on switching mechanisms (for example, ion migration versus charge trapping/de-trapping) and device structures (for instance, vertical versus planer) is urgently needed. Such knowledge will be highly beneficial to the performance optimization of flexible resistive switching memories in the future. In addition, studies on materials appropriate for flexible resistive switching memories under twisting and stretching conditions deserve much attention since real application environments of flexible electronic devices are usually more complex than being only bended. It is also necessary to mention that integration of flexible electrodes such as polymer or liquid metal into flexible resistive switching memories is urgent as well, since the commonly used metallic or carbon-based electrodes have been demonstrated as the main limit for device performance when working under stretching or twisting conditions. Besides, research on flexible selector devices for suppressing the sneaking path and misreading problem of high-density crossbar memory arrays is still rare. Much devotion needs to be engaged since the sneaking current is one of the key obstacles impeding the progress of arraying the memory unit. Finally, since the analog-type resistive switching behavior is possible and can be tuned by the external field (light, electric field, and magnetic field), the potential of the resistive switching toward the flexible neuromorphic information computing and multifunctional device is highly valued.

In spite of these challenges at the moment, the flexible resistive memory proves to be one of the most promising electronic technologies of the 21st century.

Acknowledgments

This work is supported by the National Natural Science Foundation of China (11474295, 51525103, 61504154, 61674153, 61722407, 61774161, 61704178), National Key R&D Program of China (2017YFB0405604), K. C. Wong Education Foundation (rczx0800), Natural Science Foundation of Zhejiang Province (LR17E020001), and Ningbo Science and Technology Innovation Team (2015B11001).

References

1. Pan, F., Gao, S., Chen, C., Song, C., and Zeng, F. (2014). Recent progress in resistive random access memories: Materials, switching mechanisms, and performance. *Materials Science and Engineering: R: Reports*, **83**, 1–59.
2. Han, S.-T., Peng, H., Sun, Q., Venkatesh, S., Chung, K.-S., Lau, S. C., Zhou, Y., and Roy, V. A. L. (2017). An overview of the development of flexible sensors. *Advanced Materials*, **29**(33), 1700375-n/a.
3. Liu, Y., Pharr, M., and Salvatore, G. A. (2017). Lab-on-skin: A review of flexible and stretchable electronics for wearable health monitoring. *ACS Nano*, **11**(10), 9614–9635.
4. Kim, D. H., Lu, N. S., Ma, R., Kim, Y. S., Kim, R. H., Wang, S. D., Wu, J., Won, S. M., Tao, H., Islam, A., Yu, K. J., Kim, T. I., Chowdhury, R., Ying, M., Xu, L. Z., Li, M., Chung, H. J., Keum, H., McCormick, M., Liu, P., Zhang, Y. W., Omenetto, F. G., Huang, Y. G., Coleman, T., and Rogers, J. A. (2011). Epidermal electronics. *Science*, **333**(6044), 838–843.
5. Noda, M., Kobayashi, N., Katsuhara, M., Yumoto, A., Ushikura, S., Yasuda, R., Hirai, N., Yukawa, G., Yagi, I., Nomoto, K., and Urabe, T. (2011). An OTFT-driven rollable OLED display. *Journal of the Society for Information Display*, **19**(4), 316–322.
6. Rosset, S. and Shea, H. R. (2013). Flexible and stretchable electrodes for dielectric elastomer actuators. *Applied Physics A: Materials Science & Processing*, **110**(2), 281–307.

7. Meng, Y. N., Zhao, Y., Hu, C. G., Cheng, H. H., Hu, Y., Zhang, Z. P., Shi, G. Q., and Qu, L. T. (2013). All-graphene core-sheath microfibers for all-solid-state, stretchable fibriform supercapacitors and wearable electronic textiles. *Advanced Materials*, **25**(16), 2326–2331.

8. Shen, L. K., Wu, L., Sheng, Q., Ma, C. R., Zhang, Y., Lu, L., Ma, J., Ma, J., Bian, J. H., Yang, Y. D., Chen, A. P., Lu, X. L., Liu, M., Wang, H., Jia, C. L. (2017). Epitaxial lift-off of centimeter-scaled spinel ferrite oxide thin films for flexible electronics. *Advanced Materials*, **29**(33), 7.

9. Waser, R., Dittmann, R., Staikov, G., and Szot, K. (2009). Redox-based resistive switching memories: Nanoionic mechanisms, prospects, and challenges. *Advanced Materials*, **21**(25–26), 2632.

10. Waser, R. and Aono, M. (2007). Nanoionics-based resistive switching memories. *Nature Materials*, **6**(11), 833–840.

11. Tsai, C.-L., Xiong, F., Pop, E., and Shim, M. (2013). Resistive random access memory enabled by carbon nanotube crossbar electrodes. *ACS Nano*, **7**(6), 5360–5366.

12. Torrezan, A. C., Strachan, J. P., Medeiros-Ribeiro, G., and Williams, R. S. (2011). Sub-nanosecond switching of a tantalum oxide memristor. *Nanotechnology*, **22**(48), 485203.

13. Lee, M.-J., Lee, C. B., Lee, D., Lee, S. R., Chang, M., Hur, J. H., Kim, Y.-B., Kim, C.-J., Seo, D. H., Seo, S., Chung, U. I., Yoo, I.-K., and Kim, K. (2011). A fast, high-endurance and scalable non-volatile memory device made from asymmetric Ta_2O_{5-x}/TaO_{2-x} bilayer structures. *Nature Materials*, **10**, 625.

14. Zhu, X., Su, W., Liu, Y., Hu, B., Pan, L., Lu, W., Zhang, J., and Li, R.-W. (2012). Observation of conductance quantization in oxide-based resistive switching memory. *Advanced Materials*, **24**(29), 3941–3946.

15. Hu, B., Zhu, X., Chen, X., Pan, L., Peng, S., Wu, Y., Shang, J., Liu, G., Yan, Q., and Li, R.-W. (2012). A multilevel memory based on proton-doped polyazomethine with an excellent uniformity in resistive switching. *Journal of the American Chemical Society*, **134**(42), 17408–17411.

16. Tan, H., Liu, G., Zhu, X., Yang, H., Chen, B., Chen, X., Shang, J., Lu, W. D., Wu, Y., and Li, R.-W. (2015). An optoelectronic resistive switching memory with integrated demodulating and arithmetic functions. *Advanced Materials*, **27**(17), 2797–2803.

17. Yang, Y., Gao, P., Gaba, S., Chang, T., Pan, X., and Lu, W. (2012). Observation of conducting filament growth in nanoscale resistive memories. *Nature Communications*, **3**, 732.

18. Gao, S., Song, C., Chen, C., Zeng, F., and Pan, F. (2012). Dynamic processes of resistive switching in metallic filament-based organic memory devices. *The Journal of Physical Chemistry C*, **116**(33), 17955–17959.

19. Gao, S., Zeng, F., Li, F., Wang, M., Mao, H., Wang, G., Song, C., and Pan, F. (2015). Forming-free and self-rectifying resistive switching of the simple Pt/TaOx/n-Si structure for access device-free high-density memory application. *Nanoscale*, **7**(14), 6031–6038.

20. Shang, J., Xue, W., Ji, Z., Liu, G., Niu, X., Yi, X., Pan, L., Zhan, Q., Xu, X.-H., and Li, R.-W. (2017). Highly flexible resistive switching memory based on amorphous-nanocrystalline hafnium oxide films. *Nanoscale*, **9**(21), 7037–7046.

21. Hwang, S. K., Lee, J. M., Kim, S., Park, J. S., Park, H. I., Ahn, C. W., Lee, K. J., Lee, T., and Kim, S. O. (2012). Flexible multilevel resistive memory with controlled charge trap B- and N-doped carbon nanotubes. *Nano Letters*, **12**(5), 2217–2221.

22. Chen, J.-Y., Hsin, C.-L., Huang, C.-W., Chiu, C.-H., Huang, Y.-T., Lin, S.-J., Wu, W.-W., and Chen, L.-J. (2013). Dynamic evolution of conducting nanofilament in resistive switching memories. *Nano Letters*, **13**(8), 3671–3677.

23. Kim, S., Kim, S. J., Kim, K. M., Lee, S. R., Chang, M., Cho, E., Kim, Y. B., Kim, C. J., Chung, U. I., and Yoo, I. K. (2013). Physical electro-thermal model of resistive switching in bi-layered resistance-change memory. *Scientific Reports*, **3**, 6.

24. Yeom, S.-W., Shin, S.-C., Kim, T.-Y., Ha, H. J., Lee, Y.-H., Shim, J. W., and Ju, B.-K. (2016). Transparent resistive switching memory using aluminum oxide on a flexible substrate. *Nanotechnology*, **27**(7), 07LT01.

25. Kim, S., Son, J. H., Lee, S. H., You, B. K., Park, K.-I., Lee, H. K., Byun, M., and Lee, K. J. (2014). Flexible crossbar-structured resistive memory arrays on plastic substrates via inorganic-based laser lift-off. *Advanced Materials*, **26**(44), 7480–7487.

26. Ji, Y., Yang, Y., Lee, S.-K., Ruan, G., Kim, T.-W., Fei, H., Lee, S.-H., Kim, D.-Y., Yoon, J., and Tour, J. M. (2016). Flexible nanoporous WO_{3-x} nonvolatile memory device. *ACS Nano*, **10**(8), 7598–7603.

27. Jang, J., Pan, F., Braam, K., and Subramanian, V. (2012). Resistance switching characteristics of solid electrolyte chalcogenide Ag_2Se nanoparticles for flexible nonvolatile memory applications. *Advanced Materials*, **24**(26), 3573–3576.

28. Qian, K., Tay, R. Y., Nguyen, V. C., Wang, J., Cai, G., Chen, T., Teo, E. H. T., and Lee, P. S. (2016). Hexagonal boron nitride thin film for flexible

resistive memory applications. *Advanced Functional Materials*, **26**(13), 2176–2184.

29. Zhao, F., Cheng, H., Hu, Y., Song, L., Zhang, Z., Jiang, L., and Qu, L. (2014). Functionalized graphitic carbon nitride for metal-free, flexible and rewritable nonvolatile memory device via direct laser-writing. *Scientific Reports*, **4**(1).
30. Nau, S., Wolf, C., Sax, S., and List-Kratochvil, E. J. W. (2015). Organic non-volatile resistive photo-switches for flexible image detector arrays. *Advanced Materials*, **27**(6), 1048–1052.
31. Hosseini, N. R. and Lee, J.-S. (2015). Biocompatible and flexible chitosan-based resistive switching memory with magnesium electrodes. *Advanced Functional Materials*, **25**(35), 5586–5592.
32. Jang, B. C., Seong, H., Kim, S. K., Kim, J. Y., Koo, B. J., Choi, J., Yang, S. Y., Im, S. G., and Choi, S.-Y. (2016). Flexible nonvolatile polymer memory array on plastic substrate via initiated chemical vapor deposition. *ACS Applied Materials & Interfaces*, **8**(20), 12951–12958.
33. Son, D. I., Kim, T. W., Shim, J. H., Jung, J. H., Lee, D. U., Lee, J. M., Park, W. I., and Choi, W. K. (2010). Flexible organic bistable devices based on graphene embedded in an insulating poly(methyl methacrylate) polymer layer. *Nano Letters*, **10**(7), 2441–2447.
34. Nagashima, K., Koga, H., Celano, U., Zhuge, F., Kanai, M., Rahong, S., Meng, G., He, Y., De Boeck, J., Jurczak, M., Vandervorst, W., Kitaoka, T., Nogi, M., and Yanagida, T. (2014). Cellulose nanofiber paper as an ultra flexible nonvolatile memory. *Scientific Reports*, **4**(1).
35. Siddiqui, G. U., Rehman, M. M., Yang, Y.-J., and Choi, K. H. (2017). A two-dimensional hexagonal boron nitride/polymer nanocomposite for flexible resistive switching devices. *Journal of Materials Chemistry C*, **5**(4), 862–871.
36. Hickmott, T. W. (1962). Low-frequency negative resistance in thin anodic oxide films. *Journal of Applied Physics*, **33**(9), 2669.
37. Hwang, S. K., Lee, J. M., Kim, S., Park, J. S., Park, H. I., Ahn, C. W., Lee, K. J., Lee, T., and Kim, S. O. (2012). Flexible multilevel resistive memory with controlled charge trap band N-doped carbon nanotubes. *Nano Letters*, **12**(5), 2217–2221.
38. Ji, Y., Cho, B., Song, S., Kim, T.-W., Choe, M., Kahng, Y. H., and Lee, T. (2010). Stable switching characteristics of organic nonvolatile memory on a bent flexible substrate. *Advanced Materials*, **22**(28), 3071–3075.
39. Ji, Y., Zeigler, D. F., Lee, D. S., Choi, H., Jen, A. K. Y., Ko, H. C., and Kim, T. W. (2013). Flexible and twistable non-volatile memory cell array with all-

organic one diode-one resistor architecture. *Nature Communications*, **4**, 7.

40. Kim, S. and Choi, Y.-K. (2008). Resistive switching of aluminum oxide for flexible memory. *Applied Physics Letters*, **92**(22).
41. Han, S.-T., Zhou, Y., and Roy, V. A. L. (2013). Towards the development of flexible non-volatile memories. *Advanced Materials*, **25**(38), 5425–5449.
42. Kwon, D.-H., Kim, K. M., Jang, J. H., Jeon, J. M., Lee, M. H., Kim, G. H., Li, X.-S., Park, G.-S., Lee, B., Han, S., Kim, M., and Hwang, C. S. (2010). Atomic structure of conducting nanofilaments in TiO_2 resistive switching memory. *Nature Nanotechnology*, **5**, 148.
43. Park, G.-S., Kim, Y. B., Park, S. Y., Li, X. S., Heo, S., Lee, M.-J., Chang, M., Kwon, J. H., Kim, M., Chung, U. I., Dittmann, R., Waser, R., and Kim, K. (2013). In situ observation of filamentary conducting channels in an asymmetric Ta_2O_{5-x}/TaO_{2-x} bilayer structure. *Nature Communications*, **4**, 2382.
44. Yao, Y., Li, C., Huo, Z. L., Liu, M., Zhu, C. X., Gu, C. Z., Duan, X. F., Wang, Y. G., Gu, L., and Yu, R. C. (2013). In situ electron holography study of charge distribution in high-κ charge-trapping memory. *Nature Communications*, **4**, 2764.
45. Yang, Y., Zhang, X., Qin, L., Zeng, Q., Qiu, X., and Huang, R. (2017). Probing nanoscale oxygen ion motion in memristive systems. *Nature Communications*, **8**, 15173.
46. Kim, S. and Choi, Y.-K. (2008). Resistive switching of aluminum oxide for flexible memory. *Applied Physics Letters*, **92**(22), 223508.
47. Lien, D.-H., Kao, Z.-K., Huang, T.-H., Liao, Y.-C., Lee, S.-C., and He, J.-H. (2014). All-printed paper memory. *ACS Nano*, **8**(8), 7613–7619.
48. Kim, S., Jeong, H. Y., Kim, S. K., Choi, S.-Y., and Lee, K. J. (2011). Flexible memristive memory array on plastic substrates. *Nano Letters*, **11**(12), 5438–5442.
49. Wang, H., Zou, C., Zhou, L., Tian, C., and Fu, D. (2012). Resistive switching characteristics of thin NiO film based flexible nonvolatile memory devices. *Microelectronic Engineering*, **91**, 144–146.
50. Lee, S., Kim, H., Yun, D.-J., Rhee, S.-W., and Yong, K. (2009). Resistive switching characteristics of ZnO thin film grown on stainless steel for flexible nonvolatile memory devices. *Applied Physics Letters*, **95**(26), 262113.
51. Park, S., Lee, J. H., Kim, H.-D., Hong, S. M., An, H.-M., and Kim, T. G. (2013). Resistive switching characteristics of sol-gel based ZnO

nanorods fabricated on flexible substrates. *Physica Status Solidi (RRL): Rapid Research Letters*, **7**(7), 493–496.

52. Cheng, C.-H., Yeh, F.-S., and Chin, A. (2011). Low-power high-performance non-volatile memory on a flexible substrate with excellent endurance. *Advanced Materials*, **23**(7), 902.

53. Zheng, Z.-W., Cheng, C.-H., Chou, K.-I., Liu, M., and Chin, A. (2012). Improved current distribution in resistive memory on flexible substrate using nitrogen-rich TaN electrode. *Applied Physics Letters*, **101**(24), 243507.

54. Liang, L., Li, K., Xiao, C., Fan, S., Liu, J., Zhang, W., Xu, W., Tong, W., Liao, J., Zhou, Y., Ye, B., and Xie, Y. (2015). Vacancy associates-rich ultrathin nanosheets for high performance and flexible nonvolatile memory device. *Journal of the American Chemical Society*, **137**(8), 3102–3108.

55. Yang, Y., Yuan, G., Yan, Z., Wang, Y., Lu, X., and Liu, J.-M. (2017). Flexible, semitransparent, and inorganic resistive memory based on $BaTi_{0.95}Co_{0.05}O_3$ Film. *Advanced Materials*, **29**(26).

56. Huber, B., Popp, P. B., Kaiser, M., Ruediger, A., and Schindler, C. (2017). Fully inkjet printed flexible resistive memory. *Applied Physics Letters*, **110**(14), 143503.

57. Wang, G., Raji, A.-R. O., Lee, J.-H., and Tour, J. M. (2014). Conducting-interlayer SiO_x memory devices on rigid and flexible substrates. *ACS Nano*, **8**(2), 1410–1418.

58. Yoo, H. G., Kim, S., and Lee, K. J. (2014). Flexible one diode–one resistor resistive switching memory arrays on plastic substrates. *RSC Advances*, **4**(38), 20017–20023.

59. Park, K. and Lee, J.-S. (2016). Flexible resistive switching memory with a $Ni/CuO_x/Ni$ structure using an electrochemical deposition process. *Nanotechnology*, **27**(12), 125203.

60. Chen, Y., Zhang, B., Liu, G., Zhuang, X., and Kang, E.-T. (2012). Graphene and its derivatives: switching ON and OFF. *Chemical Society Reviews*, **41**(13), 4688–4707.

61. Chen, Y., Liu, G., Wang, C., Zhang, W., Li, R.-W., and Wang, L. (2014). Polymer memristor for information storage and neuromorphic applications. *Materials Horizons*, **1**(5), 489–506.

62. Gu, Q. F., He, J. H., Chen, D. Y., Dong, H. L., Li, Y. Y., Li, H., Xu, Q. F., and Lu, J. M. (2015). Multilevel conductance switching of a memory device induced by enhanced intermolecular charge transfer. *Advanced Materials*, **27**(39), 5968–5973.

63. Yang, L., Hua, L., Hongfei, C., Yu, W., Najun, L., Qingfeng, X., Jinghui, H., Dongyun, C., Lihua, W., and Jianmei, L. (2015). Controlling crystallite orientation of diketopyrrolopyrrole-based small molecules in thin films for highly reproducible multilevel memory device: Role of furan substitution. *Advanced Functional Materials*, **25**(27), 4246–4254.

64. Zhang, Q.-J., He, J.-H., Zhuang, H., Li, H., Li, N.-J., Xu, Q.-F., Chen, D.-Y., and Lu, J.-M. (2016). Ternary flexible electro-resistive memory device based on small molecules. *Chemistry: An Asian Journal*, **11**(10), 1624–1630.

65. Zhang, Q., Pan, J., Yi, X., Li, L., and Shang, S. (2012). Nonvolatile memory devices based on electrical conductance tuning in poly(N-vinylcarbazole)–graphene composites. *Organic Electronics*, **13**(8), 1289–1295.

66. Jo, H., Ko, J., Lim, J. A., Chang, H. J., and Kim, Y. S. (2013). Organic nonvolatile resistive switching memory based on molecularly entrapped fullerene derivative within a diblock copolymer nanostructure. *Macromolecular Rapid Communications*, **34**(4), 355–361.

67. Chen, Y., Zhang, B., Liu, G., Zhuang, X. D., and Kang, E. T. (2012). Graphene and its derivatives: Switching ON and OFF. *Chemical Society Reviews*, **41**(13), 4688–4707.

68. Rehman, M. M., Siddiqui, G. U., Gul, J. Z., Kim, S.-W., Lim, J. H., and Choi, K. H. (2016). Resistive switching in all-printed, flexible and hybrid MoS2-PVA nanocomposite based memristive device fabricated by reverse offset. *Scientific Reports*, **6**.

69. Zhang, P., Gao, C., Xu, B., Qi, L., Jiang, C., Gao, M., and Xue, D. (2016). Structural phase transition effect on resistive switching behavior of MoS2-polyvinylpyrrolidone nanocomposites films for flexible memory devices. *Small*, **12**(15), 2077–2084.

70. Liu, G., Ling, Q.-D., Teo, E. Y. H., Zhu, C.-X., Chan, D. S.-H., Neoh, K.-G., and Kang, E.-T. (2009). Electrical conductance tuning and bistable switching in poly(*N*-vinylcarbazole)-carbon nanotube composite films. *Acs Nano*, **3**(7), 1929–1937.

71. Mohapatra, S. R., Tsuruoka, T., Hasegawa, T., Terabe, K., and Aono, M. (2012). Flexible resistive switching memory using inkjet printing of a solid polymer electrolyte. *Aip Advances*, **2**(2).

72. Pan, L., Hu, B., Zhu, X., Chen, X., Shang, J., Tan, H., Xue, W., Zhu, Y., Liu, G., and Li, R.-W. (2013). Role of oxadiazole moiety in different D-A polyazothines and related resistive switching properties. *Journal of Materials Chemistry C*, 1(30), 4556–4564.

73. Zhang, W., Wang, C., Liu, G., Wang, J., Chen, Y., and Li, R.-W. (2014). Structural effect on the resistive switching behavior of triphenylamine-based poly(azomethine)s. *Chemical Communications*, **50**(78), 11496–11499.

74. Zhang, W., Wang, C., Liu, G., Zhu, X., Chen, X., Pan, L., Tan, H., Xue, W., Ji, Z., Wang, J., Chen, Y., and Li, R.-W. (2014). Thermally-stable resistive switching with a large ON/OFF ratio achieved in poly(triphenylamine). *Chemical Communications*, **50**(80), 11856–11858.

75. Yen, H. J., Chen, C. J., and Liou, G. S. (2013). Flexible multi-colored electrochromic and volatile polymer memory devices derived from starburst triarylamine-based electroactive polyimide. *Advanced Functional Materials*, **23**(42), 5307–5316.

76. Liu, G., Zhuang, X., Chen, Y., Zhang, B., Zhu, J., Zhu, C.-X., Neoh, K.-G., and Kang, E.-T. (2009). Bistable electrical switching and electronic memory effect in a solution-processable graphene oxide-donor polymer complex. *Applied Physics Letters*, **95**(25).

77. Lee, B.-H., Bae, H., Seong, H., Lee, D.-I., Park, H., Choi, Y. J., Im, S.-G., Kim, S. O., and Choi, Y.-K. (2015). Direct observation of a carbon filament in water-resistant organic memory. *ACS Nano*, **9**(7), 7306–7313.

78. Zhou, H. C., Long, J. R., and Yaghi, O. M. (2012). Introduction to metal-organic frameworks. *Chemical Reviews*, **112**(2), 673–674.

79. Yaghi, O. M., O'Keeffe, M., Ockwig, N. W., Chae, H. K., Eddaoudi, M., and Kim, J. (2003). Reticular synthesis and the design of new materials. *Nature*, **423**(6941), 705–714.

80. Yoon, S. M., Warren, S. C., and Grzybowski, B. A. (2014). Storage of electrical information in metal-organic-framework memristors. *Angewandte Chemie-International Edition*, **53**(17), 4437–4441.

81. Pan, L., Liu, G., Li, H., Meng, S., Han, L., Shang, J., Chen, B., Platero-Prats, A. E., Lu, W., Zou, X., and Li, R.-W. (2014). A resistance-switchable and ferroelectric metal–organic framework. *Journal of the American Chemical Society*, **136**(50), 17477–17483.

82. Pan, L., Ji, Z., Yi, X., Zhu, X., Chen, X., Shang, J., Liu, G., and Li, R.-W. (2015). Metal-organic framework nanofilm for mechanically flexible information storage applications. *Advanced Functional Materials*, **25**(18), 2677–2685.

83. Liu, Y., Wang, H., Shi, W., Zhang, W., Yu, J., Chandran, B. K., Cui, C., Zhu, B., Liu, Z., Li, B., Xu, C., Xu, Z., Li, S., Huang, W., Huo, F., and Chen, X. (2016). Alcohol-mediated resistance-switching behavior in metal-organic framework-based electronic devices. *Angewandte Chemie-International Edition*, **55**(31), 8884–8888.

84. Gu, C. and Lee, J.-S. (2016). Flexible hybrid organic-inorganic perovskite memory. *ACS Nano*, **10**(5), 5413–5418.

85. Burschka, J., Pellet, N., Moon, S.-J., Humphry-Baker, R., Gao, P., Nazeeruddin, M. K., and Grätzel, M. (2013). Sequential deposition as a route to high-performance perovskite-sensitized solar cells. *Nature*, **499**, 316.

86. Kagan, C. R., Mitzi, D. B., and Dimitrakopoulos, C. D. (1999). Organic-inorganic hybrid materials as semiconducting channels in thin-film field-effect transistors. *Science*, **286**(5441), 945–947.

87. Jeon, N. J., Noh, J. H., Kim, Y. C., Yang, W. S., Ryu, S., and Seok, S. I. (2014). Solvent engineering for high-performance inorganic–organic hybrid perovskite solar cells. *Nature Materials*, **13**, 897.

88. Kim, Y.-H., Cho, H., Heo, J. H., Kim, T.-S., Myoung, N., Lee, C.-L., Im, S. H., and Lee, T.-W. (2015). Multicolored organic/inorganic hybrid perovskite light-emitting diodes. *Advanced Materials*, **27**(7), 1248–1254.

89. Chin, X. Y., Cortecchia, D., Yin, J., Bruno, A., and Soci, C. (2015). Lead iodide perovskite light-emitting field-effect transistor. *Nature Communications*, **6**, 7383.

90. Li, F., Ma, C., Wang, H., Hu, W., Yu, W., Sheikh, A. D., and Wu, T. (2015). Ambipolar solution-processed hybrid perovskite phototransistors. *Nature Communications*, **6**, 8238.

91. Choi, J., Quyet Van, L., Hong, K., Moon, C. W., Han, J. S., Kwon, K. C., Cha, P.-R., Kwon, Y., Kim, S. Y., and Jang, H. W. (2017). Enhanced endurance organolead halide perovskite resistive switching memories operable under an extremely low bending radius. *ACS Applied Materials & Interfaces*, **9**(36), 30764–30771.

92. Yang, K., Li, F., Veeramalai, C. P., and Guo, T. (2017). A facile synthesis of $CH_3NH_3PbBr_3$ perovskite quantum dots and their application in flexible nonvolatile memory. *Applied Physics Letters*, **110**(8).

Chapter 5

Two-Dimensional Materials for Flexible In-Plane Micro-Supercapacitors

Kaiyue Jiang, Chongqing Yang, and Xiaodong Zhuang
The Soft2D Lab, School of Chemistry and Chemical Engineering, Shanghai Jiao Tong University, 800 Dongchuan Road, Shanghai 200240, China
zhuang@sjtu.edu.cn

5.1 Introduction

The development of flexible electronic devices has made intelligent living a reality. Nowadays, the ever-increasing demand for flexible wearable electronic devices has driven research on flexible energy storage and lightweight power sources [1, 2]. To obtain flexibility, electronic circuits have been assembled by mounting electronic devices on flexible substrates, such as polyimide, PEEK, or transparent conductive polyester films in order to achieve the requirements of thin, transparent, flexible, and stretchable electronics [3].

The continued miniaturization of wearable/portable electronics has made micropower sources with compact, lightweight, and

Flexible and Stretchable Electronics: Materials, Designs, and Devices
Edited by Run-Wei Li and Gang Liu

ISBN 978-981-4800-46-4 (Hardcover), 978-0-429-05890-5 (eBook)
www.jennystanford.com

flexible features a critical prerequisite [4, 5]. Among various energy storage devices, supercapacitors, known as electrochemical capacitors, have attracted tremendous attention due to their long cycle life, simple structure, and fast rates of charge/discharge. And thin-film supercapacitors are considered highly competitive candidates that can potentially be integrated on a chip along with the necessary electronic components [6]. Although significant advances in ultrahigh power density have been realized for carbon/polymer-based electrode materials, film-type supercapacitors normally suffer from low energy density. Therefore, the development of energy storage materials that combine the power density of supercapacitors with the energy density of batteries is significantly crucial to advance portable and distributed power technology. On-chip or in-plane micro-supercapacitors (MSCs) have emerged as potential candidates to complement micro-batteries in various applications due to their safety, fast charge–discharge, and silicon-compatible features [7].

Conductive two-dimensional (2D) materials, which have ultrathin thickness, high specific surface areas, and high conductivity, are one of the most suitable materials for high double-layer capacitance [8]. Inorganic 2D materials have reactive basal planes and edges. In addition, 2D materials can intercalate ions and thus can provide intercalation pseudo-capacitance. Moreover, 2D materials can store energy in the 2D channels between nanosheets by a fast ion adsorption mechanism enabled by the pre-intercalation of water molecules and a rapid expansion and contraction of the multilayered and flexible 2D nanomaterials [9]. Unlike rigid porous 3D materials, such as hard carbons undergoing slow diffusion processes, the nature of 2D nanomaterials favors a fast ionic transport through 2D channels that are free to expand and contract that can provide pseudo-capacitance. The future development of portable and wearable micro-supercapacitors promotes a great number of researches on 2D materials.

5.2 In-Plane Micro-Supercapacitors

The electrochemical performances of the flexible all-solid-state MSCs can be characterized by CV and galvanostatic charge–discharge experiments using an electrochemical station. In order to give better overview of all characters of MSCs, the calculation methods, equations, and parameters are briefly illustrated in this part [10].

The top view (left) and side view (right) of a cartoon in-plane MSC are shown in Chart 5.1. In Chart 5.1, w is the finger width, l is the finger length, s is the interspatial distance, b is the edge breadth, h is the finger height, and N is the number of interdigital fingers. Therefore, the area (A_{Device}) and volume (V_{Device}) of finger electrodes can be calculated as below:

$$A_{\text{Device}} = 2\,N \times l \times w + ((w + s) \times (2N - 1) + w) \times b \tag{5.1}$$

$$V_{\text{Device}} = A_{\text{Device}} \times h \tag{5.2}$$

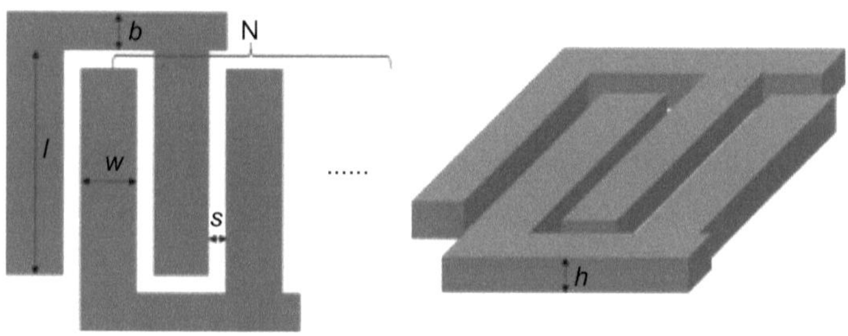

Chart 5.1 The parameters for in-plane micro-supercapacitors.

The areal specific capacitance (C_A) and volumetric specific capacitance (C_V) of electrode materials can be calculated from galvanostatic charge–discharge curves according to Eqs. (5.3) and (5.4), respectively:

$$C_A = I/(A_{\text{Device}} \times (\mathrm{d}V/\mathrm{d}t)) \tag{5.3}$$

$$C_V = I/(V_{\text{Device}} \times (\mathrm{d}V/\mathrm{d}t)) \tag{5.4}$$

where I is the applied current and $\mathrm{d}V/\mathrm{d}t$ is the slope of the discharge curve.

The capacitance values can also be calculated from the CV curves according to the following equation (5.3′):

$$C_{\text{Device}} = \frac{1}{v(V_f - V_i)} \int_{V_i}^{V_f} I(V)dV \tag{5.3′}$$

where C_{Device} denotes the capacitance, v is the scan rate (in V/s), V_f and V_i are the integration potential limits of the voltammetric curve, and $I(V)$ is the voltammetric discharge current (in amperes).

The areal capacitance ($C_{\text{Device, A}}$) and volumetric capacitance ($C_{\text{Device, V}}$) of MSCs can be calculated by using Eqs. (5.5) and (5.6), respectively:

$$C_{\text{Device, A}} = C_{\text{Device}}/A_{\text{Device}} \tag{5.5}$$

$$C_{\text{Device, V}} = C_{\text{Device}}/V_{\text{Device}} \tag{5.6}$$

The energy density and the power density of the device was obtained from the formula given in Eqs. (5.7) and (5.8):

$$E_{\text{Device, V}} = 1/2 \times C_{\text{Device, V}} \times (\Delta V)^2/3600 \tag{5.7}$$

$$P_{\text{Device, V}} = E_{\text{Device, V}} \times 3600/t \tag{5.8}$$

where ΔV is the discharge voltage range and t is the discharge time.

5.3 Graphene

Graphene is the most popular material applied to supercapacitors/MSCs, which aroused widespread interest due to its unique properties with a single layer of carbon atoms arranged in a hexagonal lattice [11]. As one of the most promising carbon materials with high specific surface area (up to ~2620 m^2/g), graphene offers ultrahigh intrinsic double-layer capacitance (~21 $\mu F/cm^2$) and high theoretical capacitance (~550 F/g) [12]. Conventional supercapacitors based on curved graphene [13], activated graphene [14], vertically oriented graphene [15], doped graphene [16], and laser-scribed graphene [17], and graphene/metal oxide hybrids [18] as bulk electrodes have demonstrated excellent performance in terms of specific capacitance and energy density [8, 10, 19, 20]. More importantly, graphene-based thin films hold great promise for developing new types of flexible, transparent, and miniaturized ultrathin MSCs. In this section, we will focus on graphene-based in-plane MSCs.

5.3.1 Reduced Graphene Oxide

Graphene oxide (GO) is one of the most popular derivatives of graphene [21]. However, GO suffers from low conductivity due to the destruction of conjugation by the large amount of oxygen-containing groups. Therefore, GO was reduced and functionalized by many kinds of methods for increasing conductivity [22]. Wu et al. [23] reported a novel class of all-solid-state graphene-based in-plane interdigital MSCs on both rigid and flexible substrates through micropatterning of graphene films reduced by methane (CH_4)-plasma treatment over a short time at 700°C (Fig. 5.1). The reduced graphene films (called MPG films) were subsequently deposited on gold current collectors by lithography techniques, followed by oxidative etching and drop-casting H_2SO_4/polyvinyl alcohol (H_2SO_4/PVA) as gel electrolyte (Fig. 5.1a). To construct flexible MSCs, they utilized a PMMA film as a bridge to transfer the MPG film onto the PET substrate (PMMA/MPG/PET) via etching of Cu foil, which first spin-coated the GO film in

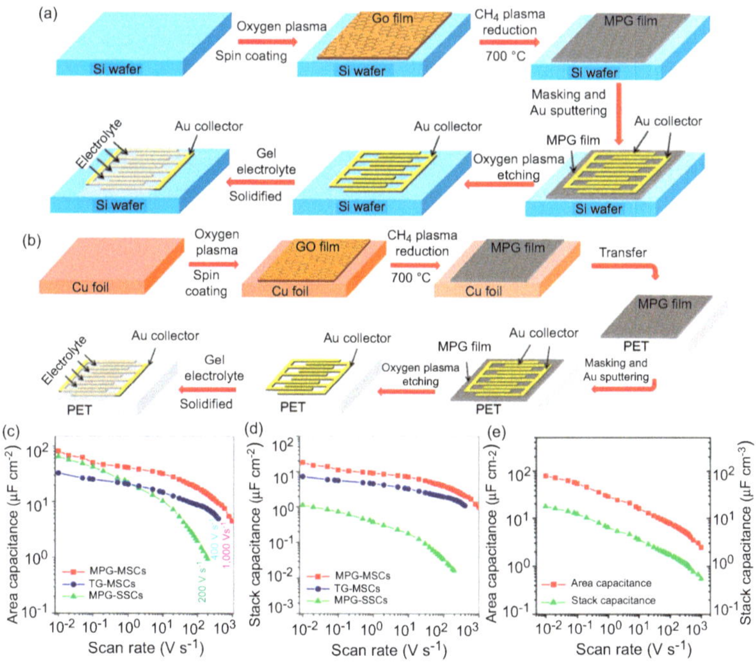

Figure 5.1 (a) Schematic illustration of the fabrication of all-solid-state interdigital graphene-based MSCs integrated onto a silicon wafer. (b) Schematic illustration of the fabrication of flexible MPG-MSCs-PET. (c,d) Evolution of the (c) area capacitance and (d) stack capacitance versus scan rate of MPG-MSCs, TG-MSCs, and MPG-SSCs. (e) Area capacitance and stack capacitance of the MPG-MSCs-PET. Reprinted with permission from Ref. [23], Copyright 2013, Springer Nature.

aqueous $Fe(NO_3)_3$ solution. The PMMA film was removed by acetone as well (Fig. 5.1b). Compared with MPG-based MSCs (MPG-MSCs) on a silicon wafer, MPG-MSCs on a PET substrate (MPG-MSCs-PET) performed equivalent electrochemical properties, such as an area capacitance of 78.9 mF/cm^2 and a stack capacitance of 17.5 F/cm^3 (at 10 mV/s, Fig. 5.1e), both of which were much higher than those of MPG-SSCs (64 mF/cm^2 and 1.1 F/cm^3, Fig. 5.1c,d). In addition, the MPG-MSCs-PET maintained 99.1% capacitance after 100,000 times at a scan rate of 200 V/s, and no capacity degradation was observed when the devices were bent 100 times, which demonstrated its excellent cycling and mechanical stability.

5.3.2 Electrochemically Exfoliated Graphene

To avoid introduction of too many defects, the electrochemical exfoliation method was developed for fast fabrication of single and multilayered graphene with high quality and in large quantity [24]. Li et al. [25] developed a full inkjet printing technique to fabricate and integrate MSCs in a scalable and massive manner (Fig. 5.2). High-concentration electrochemically exfoliated graphene (EEG) inks formulated through a solvent exchange technique can efficiently print graphene films (with enough thickness up to ~0.7 μm) that can serve not only as electrodes but also as current collectors. Moreover, a solid-state electrolyte was printed as well using a polyelectrolyte ink based on poly (4-styrenesulfonic acid) (PSSH) (Fig. 5.2a). The resultant fully printed MSCs were distinguished from the graphene-based MSCs-printed graphene electrodes with drop-cast polymer electrolytes with the highest value of ~0.1 mF/cm^2, exhibiting an areal capacitance of around 0.7 mF/cm^2. Different substrates (Fig. 5.2b–d) were adopted, showing the interdigitated electrode structures reliably printed with the EEG inks. Notably, more than 100 devices were manufactured through this method on Kapton (Fig. 5.2e,f). The excellent capacitive behavior of the MSC array was confirmed by rectangular CV profiles (Fig. 5.2g) and triangular GCD curves (Fig. 5.2h). As shown in Fig. 5.2g, the CV profiles of the 12S × 12P MSC array exhibit excellent rectangularity even at 1000 mV/s, which highlights that the large-scale MSC arrays deliver higher power density, as confirmed in the areal Ragone plot (Fig. 5.2i). In addition, the remarkable capacitive stability of these MSCs was illustrated (Fig. 5.2j).

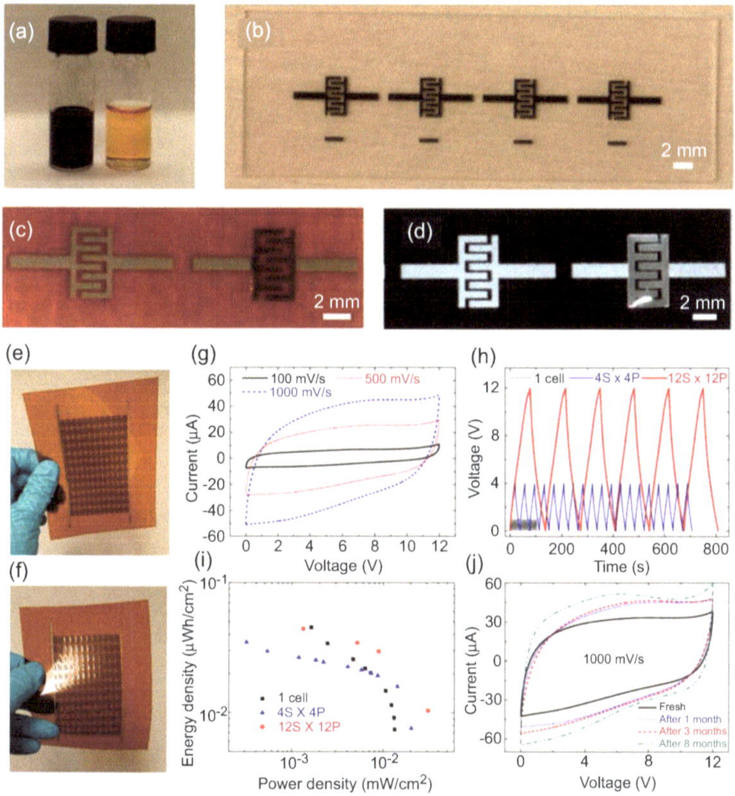

Figure 5.2 (a) Photograph of the EEG ink (left) and PSSH ink (right) used to print the electrodes and electrolyte, respectively. (b) Photograph of four fully printed planar (interdigitated) MSCs on a glass slide. (c,d) Photographs of the printed EEG electrodes (left) and complete MSCs with printed electrolytes (right) on Kapton (c) and silicon wafer (d). (e,f) Photographs of a 12S × 12P MSC array on Kapton taken in different perspective. (g) CV profiles of the 12S × 12P MSC array at different scan rates with a voltage window of 12 V (measured at one month after the fabrication). (h) GCD curves of the MSC array at the current of 10 μA. For comparison, the corresponding GCD curves at the same current (10 μA) of one single cell (10L on Kapton), and the 4S × 4P array are also presented in (h). (i) Areal energy density against power density (areal Ragone plot) for a single cell and the 4S × 4P and 12S × 12P arrays on Kapton. (j) CV profiles at the scan rate of 1000 mV/s for the 12S × 12P array measured at different periods after the fabrication. Reprinted with permission from Ref. [25], Copyright 2017, American Chemical Society.

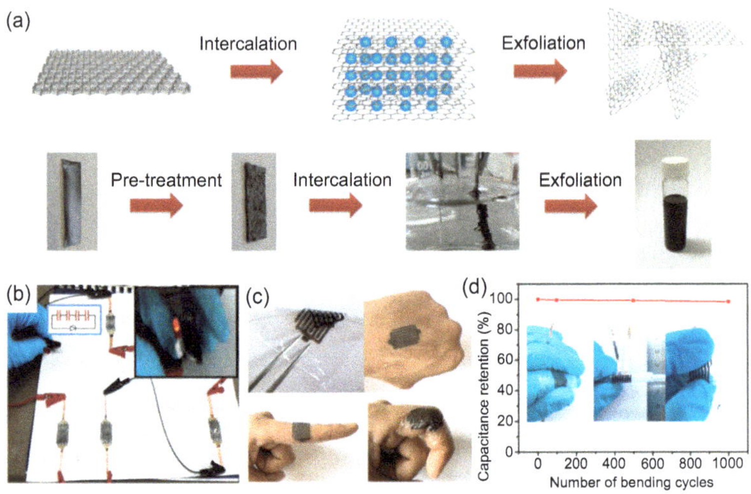

Figure 5.3 (a) Schematic illustration of the electrochemical exfoliation of graphene and photographs of the graphite foil volume change during exfoliation. (b) A charged MSC array on a paper substrate containing four single devices can power an LED. (c) Ultrathin and flexibility demonstrations of the fabricated MSC device indicating that the device is thinner than a human hair, wearable on the human hand, and capable of conforming to bent fingers. (d) The MSC capacitance retention after 1000 cycles of the bending test using ultrathin substrates. Reprinted with permission from Ref. [26], Copyright 2016, John Wiley & Sons.

Liu et al. [26] demonstrated directly printable in-plane MSCs both on paper and ultrathin polyethylene terephthalate (PET) substrates. Graphene ink was obtained by electrochemically exfoliating graphite produced with a two-electrode system (Fig. 5.3a) under mild conditions. Combined with an electrochemically active poly(3,4 ethylenedioxythiophene):poly(styrenesulfonate) (PEDOT:PSS) formulation (Clevios PH1000), the hybrid ink (EG/PH1000) was used to fabricate MSC arrays with connecting four separately printed MSC devices on paper (Fig. 5.3b). As the charged MSC arrays could power an LED, a great potential of MSCs for portable and flexible micropower supply was demonstrated. Even the expectation was confirmed by easily printing MSC arrays on an A4 paper with the stable pristine EG ink loaded into the cartridge of a commercial inkjet printer. Although the MSC device on ultrathin PET substrates delivered a maximum areal capacitance of 2 mF/cm^2 at

5 mV/s, which is lower than the device fabricated on the paper substrate (5.4 mF/cm^2 at 1 mV/s) because of the limited thickness, the ultrathin MSC device can be wrapped around a finger or other body parts (Fig. 5.3c), which demonstrates tremendous flexibility and compatibility with artificial intelligent devices, such as wearable electronics that undergo bending and stretching. To explore the mechanical stability of such an MSC device for flexible and ultrathin energy storage, the areal capacitance retention of a resultant device was examined with 1000 cycles of a bending test (Fig. 5.3d, bending radius of 5 mm).

5.3.3 Laser-Scribed Graphene

As what is reported that the intrinsic capacitance of single-layer graphene is ~21 mF/cm^2, the upper limit for EDL capacitance sets for all carbon-based materials is ~550 F/g, in principle, in case that their entire surface area could be used. El-Kady et al. [17] published a strategy transfer graphite oxide to laser-scribed graphene (LSG) through an inexpensive commercially available LightScribe CD/DVD optical drive, in which process the re-stacking of graphene sheets was avoided (Fig. 5.4a). A thin film of graphite oxide drop-casted onto the flexible substrate was irradiated by an infrared laser in the optical drive as the film color changed from golden brown to black. Due to the excellent mechanical flexibility shown by LSG with only ~1% change in the electrical resistance of the film after 1000 bending cycles, a device used as an electrochemical capacitor (EC) without any additional binders or conductive additives was readily made by sandwiching with gelled polymer electrolyte (Fig. 5.4b, device structure). The CV performance of the device under different bending conditions confirmed again that the bending had almost no effect on the capacitive behavior (Fig. 5.4c). The overall performance of the LSG-ECs using various electrolytes is shown in a Ragone plot (Fig. 5.4d) comparing the performance of LSG-ECs with different energy storage devices designed for high-power microelectronics. Besides the high power density (~20 W/cm^3) and energy densities (1.36 mWh/cm^3) of LSG-ECs, their potential for commercial applications because of the simplicity of the device architecture and the availability of the graphite oxide precursor, which is already manufactured on the ton scale, should be taken into account.

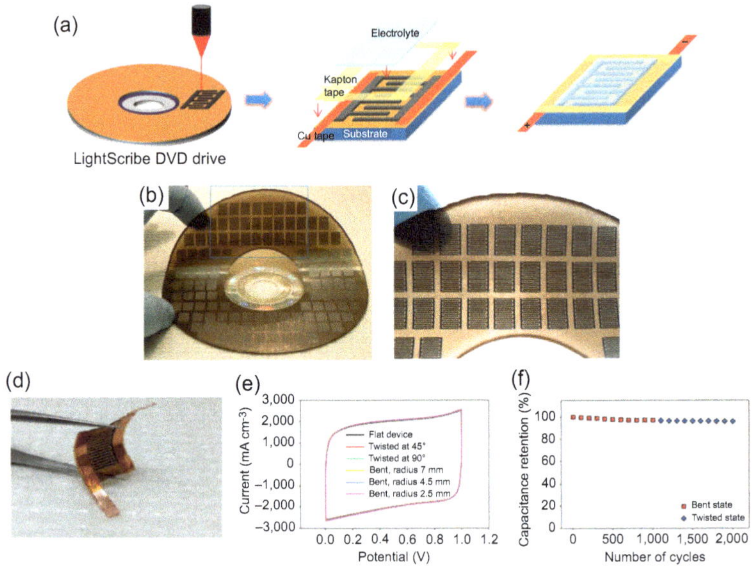

Figure 5.4 (a) Schematic diagram showing the fabrication process for an LSG micro-supercapacitor. A graphite oxide film supported on a PET sheet is placed on a DVD media disc. The disc is inserted into a LightScribe DVD drive, and a computer-designed circuit is etched onto the film. The laser inside the drive converts the golden-brown graphite oxide into black LSG at precise locations to produce interdigitated graphene circuits. (b,c) This technique has the potential for the direct writing of micro-devices with high areal density. More than 100 micro-devices can be produced on a single run. The micro-devices are completely flexible and can be produced on virtually any substrate. (d) A photograph of LSG-MSC(16) bent with tweezers demonstrates the flexibility of the micro-device. (e) Bending/twisting the device has almost no effect on its performance, as can be seen from these CVs collected under different bending and twisting conditions at 1000 mV/s. (f) Performance durability of the micro-device when tested under bending and twisting conditions. The device retains ~97% of its initial capacitance after 1000 cycles under the bent state, followed by another 1000 cycles under the twisted state. Reprinted with permission from Ref. [27], Copyright 2013, Springer Nature.

In consideration of the cost-effective problems of MSCs that apply lithographic techniques or employ masks for the definition of patterns of electrodes, El-Kady et al. [27] transferred the laser-induced production from electrochemical capacitors (ECs) to MSCs (Fig. 5.4a). The direct fabrication of interdigitated graphene MSCs used a consumer-grade LightScribe DVD burner to make graphene circuits designed on a computer converted from graphite oxide substrate. It is worth mentioning that 112 MSCs can be produced on a single piece of graphite oxide (Fig. 5.4b,c), demonstrating cost-effective and readily scalable characteristics of this technique. To explore the durability of LSG-MSC for flexible energy storage, the electrochemical performance under constant strain was tested (Fig. 5.4e). The CV performance in different bending and twisting conditions at 1000 mV/s shows exceptional electrochemical stability of as-fabricated MSCs regardless of the degree of bending. Also, the flexibility endurance of the device was tested (Fig. 5.4f), indicating the LSG-MSC as a promising device for flexible microelectronics with 97% retention of the initial capacitance after 2000 cycles.

Peng et al. [28] reported flexible MSCs using a series of boron-doped LIG (B-LIG). The synthesis and patterning process of B-LIG materials for MSCs was through a standard commercial laser writing tool as found in common machine shops. A boron precursor was synthesized by dissolving H_3BO_3 into poly-(pyromellitic dianhydride-co-4,40-oxydianiline amic acid) (or poly (amic acid), PAA) solution, followed by condensation of the PAA to produce a boric-acid-containing PI sheet (Fig. 5.5a). The reaction that PAA and H_3BO_3 dehydrated and transformed into PI and Box was shown (Fig. 5.5b). To investigate the electrochemical properties of B-LIG, a series of in-plane MSCs, which contains different percentages of B, is fabricated (Fig. 5.5c). It is shown that 5B-LIG-MSC (xB-LIG, x = 0, 1, 2, 5, and 8, which denotes the initial H_3BO_3 loading weight percentages) exhibited the largest areal capacitance (C_A) in Fig. 5.5d, demonstrating the influence of boron content on C_A. In addition to high C_A, the assembled MSC from 5B-LIG also shows excellent durability under mechanical stress. Although the device was bent and fixed (Fig. 5.5e), after 8000 bending cycles at a radius of 10 mm, the C_A of the device was unchanged (Fig. 5.5f), and CV curves during different bending cycles, as shown in Fig. 5.5g, are identical to each other, suggesting that bending had little effect on the electrochemical performance of 5B-LIG-MSC. Furthermore, a Ragone plot of volumetric power density (P_V) versus energy density (E_V) was compared to further demonstrate the high capability of 5B-LIG-MSC over non-doped devices (Fig. 5.5h).

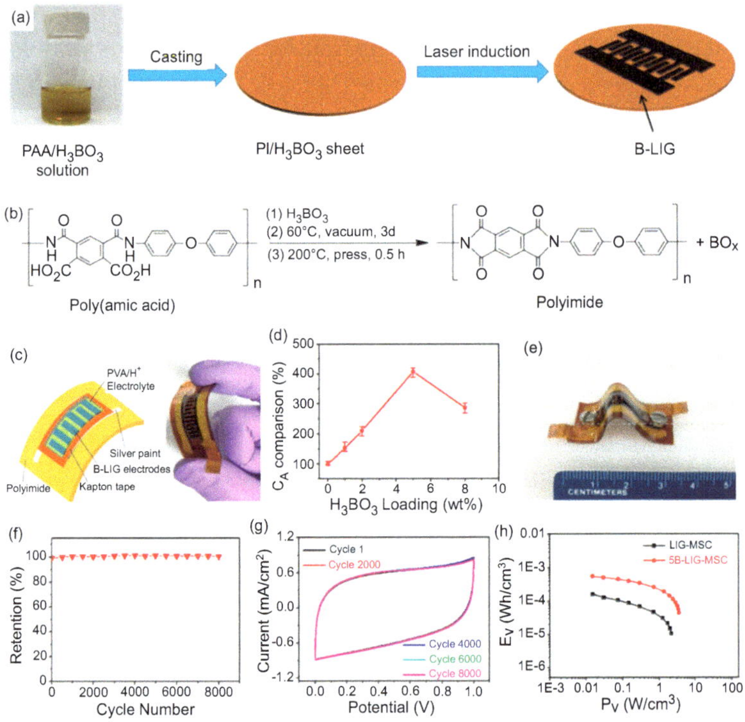

Figure 5.5 (a) Synthetic scheme for the preparation of B-LIG and fabrication of B-LIG-MSC. (b) Scheme of the dehydration reaction from PAA to a PI film during the curing process. (c) Schematic of a B-LIG-MSC device and the digital photograph of a fully fabricated device under bending. (d) Comparison of calculated C_A from LIG-MSC and B-LIG-MSC with different H_3BO_3 loadings. Current density is 1 mA/cm^2. (e) Digital photograph of a bent 5B-LIG-MSC at a bending radius of 10 mm. (f) Bent cyclability testing of flexible 5B-LIG-MSC at a fixed bending radius of ~10 mm. The C_p is calculated from discharge runtime at a current density of 1.0 mA/cm^2. (g) CV curves of the 5B-LIG-MSC at different bending cycles (f) at a scan rate of 50 mV/s^1. (h) Volumetric Ragone plot of 5B-LIG-MSC and LIG-MSC. Reprinted with permission from Ref. [28], Copyright 2015, American Chemical Society.

5.3.4 Graphene Composites

Although graphene has high theoretical capacitance, GO, reduced graphene oxide (rGO), EG, and other graphene derivatives still can only approach such value and can never exceed it. In order to overcome this problem, fabrication of graphene-based composites becomes an effective solution. Among graphene composites, graphene can contribute high conductivity and EDLC capacitance, which the other components can contribute high pseudo-capacitance [29]. Along this line, Zhang et al. [30] reported the first stimulus-responsive MSC (SR-MSC) with a reversible electrochromic window. The hybrid nanopaper prepared by 1D V_2O_5 nanoribbons and 2D EG nanosheets was used as electrode for MSCs, which was fabricated by patterning Au collector and oxygen plasma etching followed by gel electrolyte coating (Fig. 5.6a). Notably, the MSCs reached a volumetric energy density with 20 mWh/cm^3 (power density = 235 W/cm^3), which is markedly higher than the energy densities of the state-of-the-art MSCs. The electrically driven equilibrium (Fig. 5.6b) between the viologen (MV^{2+}, colorless) and the radical ion derivative of viologen (MV^{+}, purple) led to a reversible electrochromic phenomenon during the charge–discharge cycles (Fig. 5.6c) of the fabricated EG/V_2O_5-based flexible MSC (EG/V_2O_5-fMSC). Good flexibility and electrochemical stability were illustrated in the CV curves (Fig. 5.6e) at four bending angles (0°, 30°, 60°, and 90° in Fig. 5.6d), showing no obvious changes were observed under different bending states.

The layer-by-layer method is another method of fabricating composite materials, especially composite films. Wu et al. [31] demonstrated fabrication of MSCs prepared by alternating deposition of two different 2D nanosheets composed of EG nanosheets and sandwich-like mesoporous graphene-based conducting polymer nanosheets in sequence on a filtration membrane (Fig. 5.7a). The thickness can be adjusted by the number (x/y) related to "PxGy," abbreviation of the 2D nanohybrid films, where x and y represent the number of PANI-G layers and EG layers (Fig. 5.7b,c). After their group investigated areal and volumetric capacitances of different film thickness, a 4.5 μm thick P2G3 film (using 16 nm thick PANI-G) with an optimum value was chosen to be transferred to a PET substrate. The P2G3-MSCs showing high flexibility maintained ≈92.6% of the initial performance after 1000 cycles under flat and constant bending state (Fig. 5.7d) and ≈86.6% of capacitance retention after 5000 cycles. Because energy density is a critical parameter for different types of energy storage systems, these MSCs are distinguished from most reported carbon-based MSCs <5 mWh/cm^3, e.g., about 2.0 mWh/cm^3 for LSG [27], 2.5 mWh/cm^3 for MPG [23], 3.2 mWh/cm^3 for CDC

[32], 4.8 mWh/cm^3 for PDPC [33], 1.9 mWh/cm^3 for NPG [34], IPC (2.3 mWh/cm^3) [35], 1.6 mWh/cm^3 for OLC [36], and 4.0 mWh/cm^3 for LWG [37] (Fig. 5.7e,f). The landmark performance of the alternating stacked graphene-conducting polymer hybrid films for MSCs is attributed to the unique alternating stacked structure, with a strong coupling effect from the 2D pseudocapacitive nanosheets.

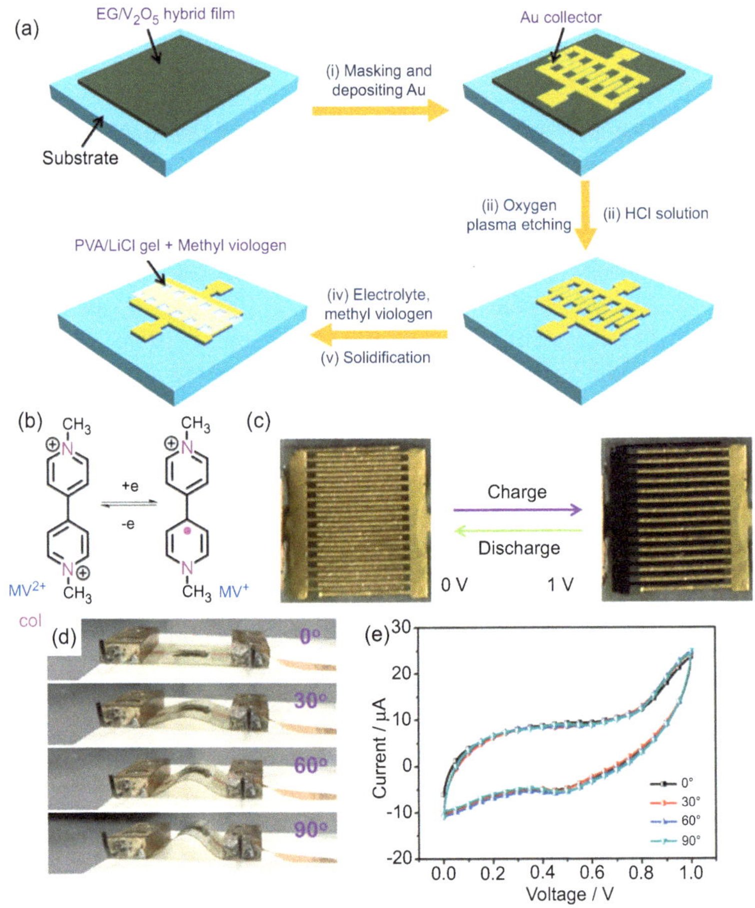

Figure 5.6 (a) Schematic fabrication of SR-MSCs. (b) The electrochromic mechanism of viologen (complex ions were not shown for clarity). (c) Photographs of the reversible electrochromic effect of EG/V_2O_5-fMSC during charge–discharge cycles (scale bar: 2000 μm). (d) Optical photos of EG/V_2O_5-fMSC under different bending angles. (e) CV curves of EG/V_2O_5-fMSC under different bending angles at a scan rate of 10 mV/s. Reprinted with permission from Ref. [30], Copyright 2016, John Wiley & Sons.

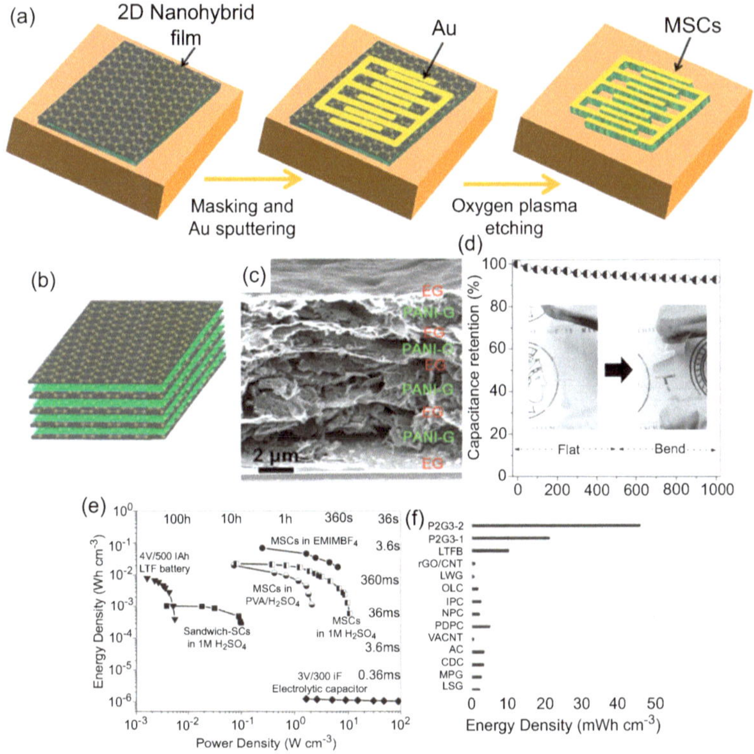

Figure 5.7 (a) Illustration of the fabrication procedure for in-plane MSCs with interdigital fingers. (b) Schematic of micrometer-thick 2D nanohybrid film, showing layer-stacked structure. (c) Cross-sectional SEM images of 2D nanohybrid film. (d) Cycling stability of P2G3-MSCs-C obtained at 50 mV/s for 1000 cycles under flat and bend states. Inset shows the optical images of the device in flat and bend states. (e) Ragone plot of P2G3-MSCs in different electrolytes, sandwich supercapacitors (sandwich SCs), LTF battery (LTFB, 4 V/500 μAh), and electrolytic capacitor (3 V/300 μF). (f) Comparison of energy density of P2G3-MSCs and various carbon-based MSCs. Reprinted with permission from Ref. [31], Copyright 2015, John Wiley & Sons.

Li et al. [38] combined the laser induction process with subsequent electrodeposition of pseudocapacitive materials for the fabrication of all-solid-state, flexible symmetric and asymmetric MSC devices that show greatly improved electrochemical performance. It was the first time that a CO_2 laser was used to convert the PI into porous LIG with an interdigitated architecture, which works not only as EDLC electrodes but also as a flexible and conductive matrix for the electrodeposition of pseudocapacitive materials. Two types of pseudocapacitive materials, manganese dioxide (MnO_2) or ferric oxyhydroxide (FeOOH), and polyaniline (PANI), representing characteristic transition metal oxides and conductive polymers, are electrodeposited onto the LIG forming LIG-MnO_2, LIG-FeOOH, and LIG-PANI composites. Then, symmetric LIG-MnO_2-MSCs and LIG-PANI-MSCs, and asymmetric MSCs using LIG-FeOOH as a negative electrode and LIG-MnO_2 as a positive electrode (LIG-FeOOH// LIG-MnO_2) were assembled (Fig. 5.8a). The flexibility of all these devices was also studied through capacitive performance with a bending angle (α_B) (Fig. 5.8b, $\alpha_B \approx 135°$). Figure 5.8c shows the good mechanical flexibility demonstrated by LIG-MnO_2-2.5 h, LIG-PANI-15, and LIG-FeOOH//LIG-MnO_2 (LIG-MnO_2-X and LIG-FeOOH-X, where X represents the deposition time; LIG-PANI-Y, where Y represents the number of deposition cycles) after bending tests.

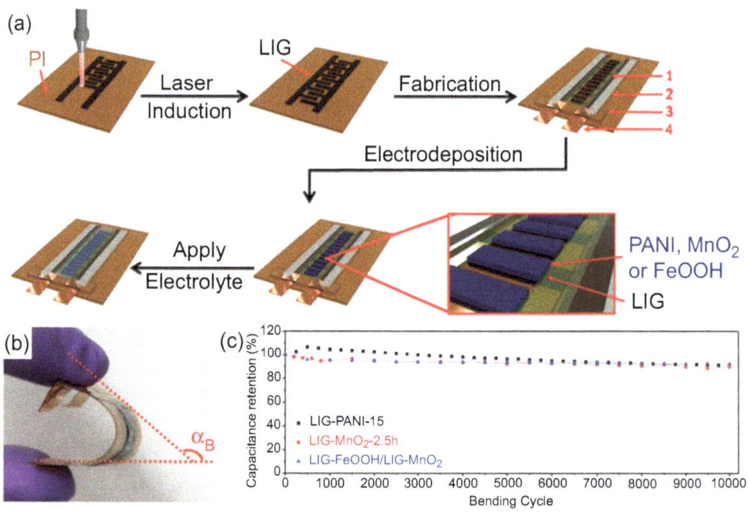

Figure 5.8 (a) Scheme of the fabrication of MSCs with LIG-MnO_2, which is similar to the formation of LIG-FeOOH, or LIG-PANI. 1, 2, 3, and 4 are epoxy adhesive, silver paste, Kapton tape, and copper tape, respectively. (b) Digital photograph of a device under bending. The angle labeled as α_B in the image is defined as the bending angle. (c) Capacitance retention of LIG-MnO_2-2.5 h, LIG-PANI-15, and LIG-FeOOH//LIG-MnO_2 at different bending cycles with an α_B of ≈90°. Reprinted with permission from Ref. [31], Copyright 2015, John Wiley & Sons.

5.4 MXenes

MXenes are a new family of 2D-layered transition metal carbides and nitrides [39], which have shown great promise as potential electrode materials for electrochemical energy storage devices [6, 40]. Ti_3C_2 is the most studied member of MXenes.

Kurra et al. [41] formed thick MXene films by Meyer rod coating on commercially available printing paper as MXene coplanar electrodes for MSCs, followed by laser patterning. Although MXene is very sensitive to temperature changes and can be readily oxidized, the major portion of MXene electrode finger remains unaffected. In addition, the separator region becomes material free, ensuring that no electrical shorts occur between adjacent interdigitated fingers. As mentioned in their essay, the clay-like Ti_3C_2 MXene performed much better than the HF-etched MXene in device electrochemical performance, such as the CVs. In addition to the simple, scalable, and cost-effective coating fabrication, these MXene-based MSCs are found to have superior energy and power densities over reported state-of-the-art paper-based MSCs.

Li et al. [42] reported an EG/MXene-composite-based supercapacitor through solution processing fabricating EGMX*x*:*y* (where *x*:*y* = 1:3, 1:9, 3:1, and 9:1) films with EGMX*x*:*y* hybrid inks (Fig. 5.9a–e). In this work, a graphite foil was first electrochemically exfoliated to produce the EG powder by using a home-made exfoliation system. Notably, the EGMX1:3 dispersion can be stable for at least 1 week without visible agglomeration, which qualified it as a reliable ink for solution processing and thin-film electrode fabrication. Thus, they also applied EGMX1:3 films to MSCs (Fig. 5.9f). The resultant MSC was fabricated by spray coating the hybrid ink onto a PET substrate through a mask. Remarkably, an ultrahigh areal capacitance of 3.26 mF/cm^2 and a volumetric capacitance of 33 F/cm^3 at 5 mV/s were obtained by the fabricated EGMX1:3-based MSC (Fig. 5.9g). In addition to the excellent capacitive performance superior to the state-of-the-art graphene-based MSCs, a long-term electrochemical stability under flat and bent states was measured with 82% of capacitance retention after 2500 cycles (Fig. 5.9h), which promised applications in flexible devices.

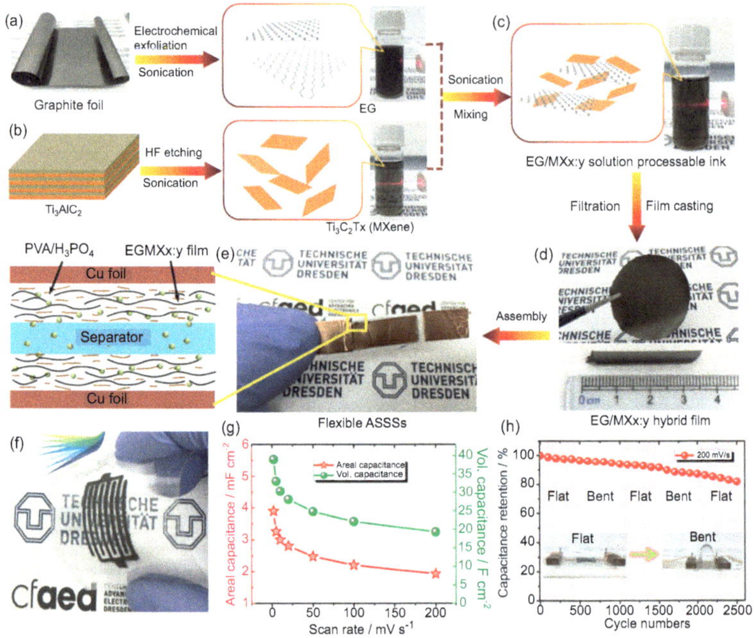

Figure 5.9 (a) Schematic description of electrochemical exfoliation graphene and stable graphene-IPA dispersion. (b) Schematic description of MXene ($Ti_3C_2T_x$) nanosheets and the stable MXene–water dispersion. (c) Illustration of the fabrication process of EGMX1:3 nanohybrid dispersion. *Note*: The digital photographs of (a), (b), and (c) show the Tyndall scattering effect in a colloidal solution of EG, MXene, and EGMX1:3 nanohybrid composite, respectively. (d,e) The digital photographs of flexible and freestanding EGMX1:3 film and corresponding assembly of ASSS. (f) Photograph of the flexible in-plane MSC based on EGMX1:3 hybrid ink on a PET substrate. (g) Areal capacitance and volumetric capacitance of MSC based on EGMX1:3 nanohybrid at different scan rates. (h) The capacitance retention in the alternative flat and bent configurations after 2500 cycles at a scan rate of 200 mV/s. The inset shows the photographs of the microdevice in the flat and bent configurations. Reprinted with permission from Ref. [42], Copyright 2016, John Wiley & Sons.

5.5 Two-Dimensional Metal Oxides

Metal oxides, which can perform fast reversible faradaic reactions, represent an outstanding candidate as pseudocapacitive materials [43]. However, metal oxides generally show poor electrical conductivity. Hybrid electrodes, in order to take advantage of both carbon-based materials and metal oxides such as RuO_2, Co_3O_4, NiO, and MnO_2 [44–47], were reported to increase the capacity of supercapacitors. In particular, MnOx has been widely used because of its natural abundance, low cost, and eco-friendly properties, as well as the high theoretical capacity of 1370 F/g [48, 49]. Vanadium oxide (V_2O_5) is also suitable as active materials for MSCs because of multiple oxidation states and similar reasons listed ahead [50–54]. What is more, the synergy of metal oxides shows a better outcome. For example, a larger drop in workfunction between MnO_2 and V_2O_5 can gain a wider voltage window for fabricated asymmetric MSCs. Layered double hydroxides (LDHs), which are composed of positively charged brucite-like host layers and charge-compensating interlayer anions, have been used as promising candidates in electrochemical energy storage/conversion owing to their tunable chemical composition, cost effectiveness, and environmentally friendly nature [55–58].

5.5.1 Layered Double Hydroxides

Energy density is one of the biggest challenges for MSCs in meeting the need for advanced portable and flexible devices, which combine the power density of SCs with the energy density of batteries. Although the hybrid systems integrate two or more individual constituents with complementary properties [59–61], most of the phase interface blocked pseudocapacitive phase from direct contact with conductive backbones leading to irreversible charge-transfer loss. Considering the crucial role of interfaces within an organized hybrid system, hybridizing ultrathin LDH structures with conducting polymers at a molecular level would be a candidate with individual components hybridized on a molecular scale. This system shows improved overall energy storage performance due to the homogeneous interface and stable charge-carrier transporting feature.

Zhao et al. [62] developed a superlattice heterostructure composed of CoNi-LDH monolayers and a commercial polymer PEDOT:PSS via the electrostatic self-assembly route (Fig. 5.10a). The resulting CoNi-LDH/PEDOT:PSS thin films are uniform, translucent, and mechanically flexible (Fig. 5.10b). To meet the demands of high-performance energy storage, it is essential to increase the energy density of SCs, without sacrificing the device power density and cycle life. An all-solid-state SC based on the CoNi-LDH/PEDOT:PSS film (positive) and rGO film (negative) with an in-plane interdigital

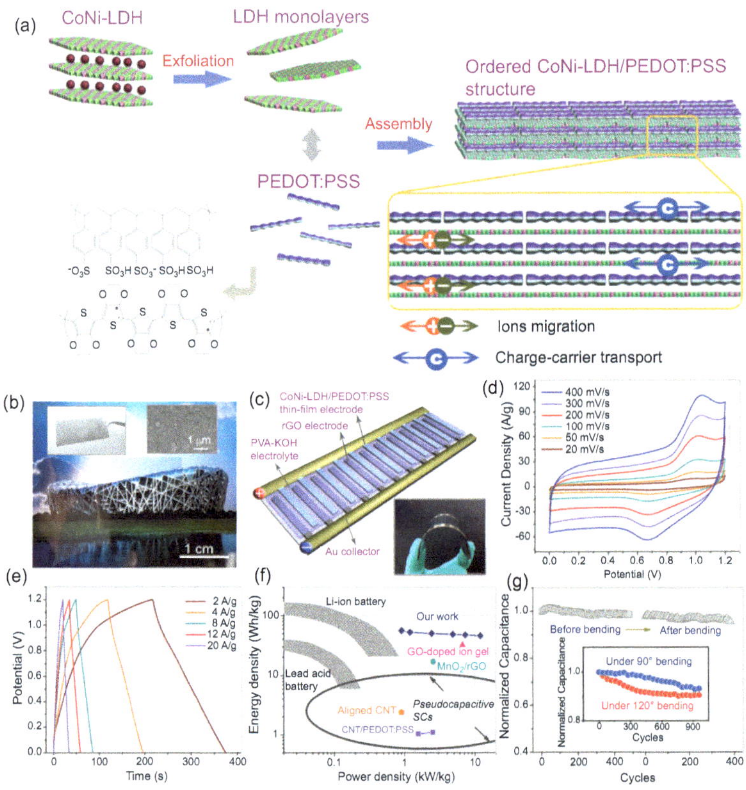

Figure 5.10 (a) Schematic illustration for the assembly of CoNi-LDH monolayers and PEDOT:PSS. (b) Optical photographs of the CoNi-LDH/PEDOT:PSS thin film on a transparent and flexible polyethylene terephthalate (PET) substrate (inset: corresponding SEM image). (c) Schematic illustration of the flexible in-plane SC device based on interdigital CoNi-LDH/PEDOT:PSS thin film and rGO thin film (inset: demonstration of the flexibility). (d) CV curves of the device at various scan rates; (e) CD curves of the device at various current densities. (f) Ragone plots of the in-plane SC device. (g) Cycling performance at a current density of 12 A/g before and after bending to 90° for 400 times (inset: capacitance retention under 90° and 120° bending angles during 1000 cycles). Reprinted with permission from Ref. [62], Copyright 2015, John Wiley & Sons.

design was fabricated to explore its application in the miniaturized energy storage (Fig. 5.10c). Compared with a symmetric design, the asymmetric SC configuration takes advantage of largely increased device operating voltage and hence significantly enhanced energy density [63]. Taking advantage of the efficient in-plane charge transport provided by an intimate and homogeneous interfacial bonding between the LDH monolayers and the interlayer PEDOT:PSS, the as-fabricated asymmetric SC showed an improved capacitance (Fig. 5.10d) and rate capability (Fig. 5.10e), excellent energy and power density (Fig. 5.10f), and good capacitance retention (Fig. 5.10g) under the certain bending angles.

5.5.2 V_2O_5/MWNT

To obtain high energy density for MSCs, carbon-based materials such as activated carbon (AC), carbon nanotube (CNT), and graphene with high surface area and high electrical conductivity are used, while a combination of MnO_2, RuO_2, and V_2O_5 with such carbon-based electrodes could be good candidates. Especially, V_2O_5 is one of the most promising electro-active materials due to its low cost, abundance in nature, and production of high specific capacitance.

In this regard, Kim et al. [64] reported a fabrication of flexible all-solid-state planar-type MSC arrays with hybrid electrodes of multi-walled carbon nanotube (MWNT)/V_2O_5 nanowire (NW) composites (Fig. 5.11a). The patterned MSCs showed an excellent energy density of 6.8 mWh/cm^3, comparable to that of an Li thin-film battery (1–10 mWh/cm^3) [17], and a power density of 80.8 W/cm^3 comparable to that of state-of-the-art MSCs with electrodes of nanoscale thickness [23]. A photographic image of a flexible MSC array is shown in Fig. 5.11b. And the flexibility of the as-fabricated MSC array in both convex and concave bending states down to the bending radius (*r*) of 1.5 mm is highlighted (Fig. 5.11c). Regardless of the bending radius, CV curves showed an ideal rectangular shape without any noticeable deterioration (Fig. 5.11d). Furthermore, the flexible MSC array exhibited excellent stability with 94% of the capacitance maintained after 1000 bending cycles from r =16 to 7 mm (Fig. 5.11e). As a result, this flexible MSC array was a potential device for flexible portable devices.

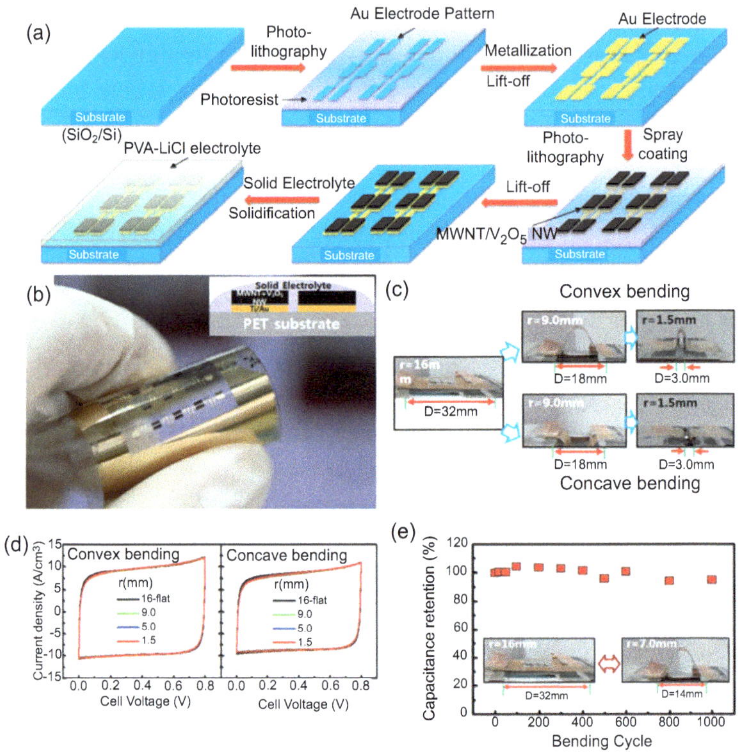

Figure 5.11 Schematics of the fabrication of the planar-type all-solid-state micro-supercapacitor (MSC) with MWNT/V_2O_5 NW composite electrodes. Here, the curved serpentine interconnections are represented as solid lines for simplicity. (b) Photograph of the fabricated flexible MSC array on a PET substrate with a cross-sectional scheme (inset). (c) Photographs showing both the convex and concave bending of the flexible MSC array with variation of the bending radius (r). (d) CV curves taken under convex (left) and concave (right) bending with variations of the bending radius (r). (e) Change in capacitance with the repetition of bending deformation with a bending radius of 7.0 mm. Reproduced from Ref. [64] with permission from The Royal Society of Chemistry.

5.6 Two-Dimensional Soft Materials

Organic 2D materials [65], mainly conducting polymers such as polyaniline (PANI) and polypyrrole (PPy), are often combined with carbon materials to increase the electrode conductivity or metal oxides to gain an extra pseudo-capacitance. The conventional activation protocol, which generates abundant micropores and ultrahigh specific surface area in graphene, also results in limited capacitance, because of the difficulty for the electrolyte ions to efficiently access the micropore surfaces [66]. In this regard, some mesoporous conducting materials were fabricated to tackle the current rate capability challenge of MSCs.

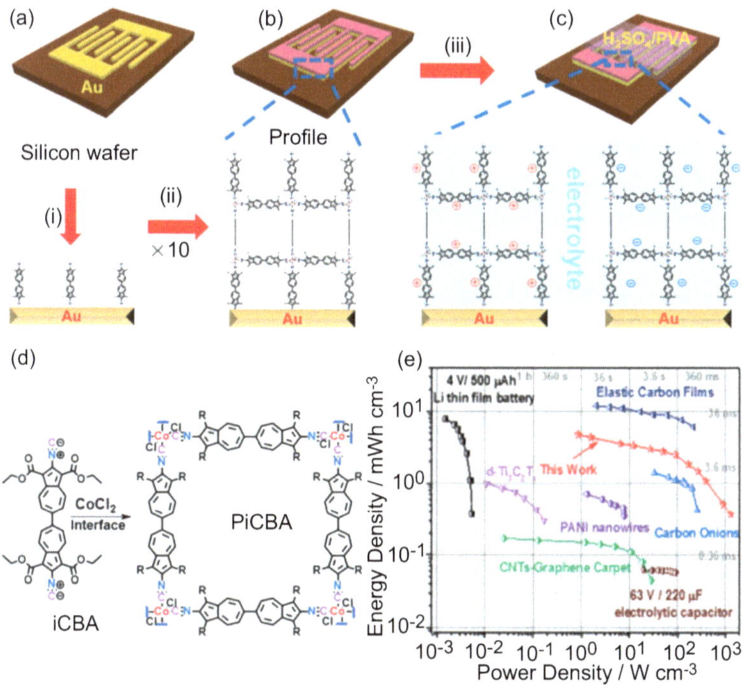

Figure 5.12 (a–c) Schematic of LBL fabrication of a PiCBA film on Au interdigital electrodes. (i) The Au interdigital electrode was immersed in iCBA solution for 24 h, followed by rinsing in fresh $CHCl_3$ and then (ii) immersed in $CoCl_2$ solution for another 60 s, followed by rinsing in DI water, iCBA solution for 60 s, and fresh $CHCl_3$, respectively, to form the one-layer PiCBA film. This process was repeated nine times to obtain a 10-layer PiCBA-Au electrode. (d) Synthesis of PiCBA through the coordination reaction between isocyanide and cobalt ions (R=–$COOC_2H_5$) (e) Ragone plots for PiCBA. This demonstrates that the PiCBA exhibits outstanding electrochemical energy storage with high power density compared with commercially applied Li thin-film batteries, electrolytic capacitors, polyaniline (PANI) nanowires, CNT-graphene carpets, $Ti_3C_2T_x$, and carbon onions. Reprinted with permission from Ref. [67], Copyright 2017, John Wiley & Sons.

5.6.1 Two-Dimensional Coordination Polymer Framework

In order to produce an ideal alternative for traditional aluminum electrolytic capacitors (AECs) under the premise of preserving good AC line-filtering performance, Yang et al. [67] developed an on-chip MSC exhibiting a high energy density. The as-fabricated MSCs were obtained by the layer-by-layer (LBL) growth of an azulene-bridged coordination polymer framework (PiCBA) on an Si-wafer-supported Au surface (Fig. 5.12a–c). The rigid substrate could also be replaced by a flexible substrate, such as PET. The PiCBA films with rich aromatic skeleton, ultra-low band gap, good carrier mobility, and the permanent dipole moment of the azulene unit were assembled by in situ formation through the coordination reaction between isocyanide and cobalt ions on the Au electrode (Fig. 5.12d). After being rinsed in a $CoCl_2$ aqueous solution, deionized water, iCBA monomer solution, and fresh $CHCl_3$ sequentially 10 times, the Au substrate deposited as 10-layer PiCBA films with enough thickness to be electrode of MSCs. As the plot demonstrates in Fig. 5.12e, the PiCBA-based MSCs exhibited a high energy density of up to 4.7 mWh/cm^3 at 50 mV/s, an order of magnitude higher than that obtained for electrolytic capacitors [17] and MSCs based on graphene-CNT carpets [68].

5.6.2 Two-Dimensional Thiophene

Inspired by the thought that conducting polymer (or transition metal oxide) nanosheets may lead to a significant improvement in the pseudo-capacitance, one promising approach for fabricating superior all-solid-state supercapacitors (ASSSs), that is manufacturing graphene-based heterostructure film electrodes assembled with 2D redox-active nanosheets in a controllable stacked order, has been proposed by Wu et al. [69]. They first fabricated a high-rate and flexible MSCs (TP/EG-MSCs) using stacked-layer heterostructure films (denoted by TP/EG) producing by alternating deposition of electrochemically EG nanosheets (≤3 layers) and redox-active conducting TP nanosheets (thickness of 3.5 nm) in sequence (Fig. 5.13a). Then the as-fabricated heterostructure films were directly transferred on Au-coated PET substrates as the working electrode, in which the gel of PVA/H_2SO_4 was sandwiched between them leading to an electrode-separator-electrolyte integrated structure (Fig. 5.13b). As demonstrated in Fig. 5.13c,d, the electrochemical performance of the TP/EG-ASSSs under different bending conditions was analyzed in order to further evaluate the potential for flexible devices. It was observed that the bending had almost no effect on the capacitive performance from the CVs. Notably, the TP/EG-ASSSs showed an ultrahigh volumetric capacitance of ≈375 F/cm^3, much higher than the ASSSs fabricated based on EG films (139 F/cm^3), activated graphene (60~100 F/cm^3) [14], laser-scribed graphene films (9.7~10.6 F/cm^3) [17], and liquid-mediated graphene films (261.3 F/cm^3) [70], and most conducting polymer-based films, which they listed in "Supporting Information."

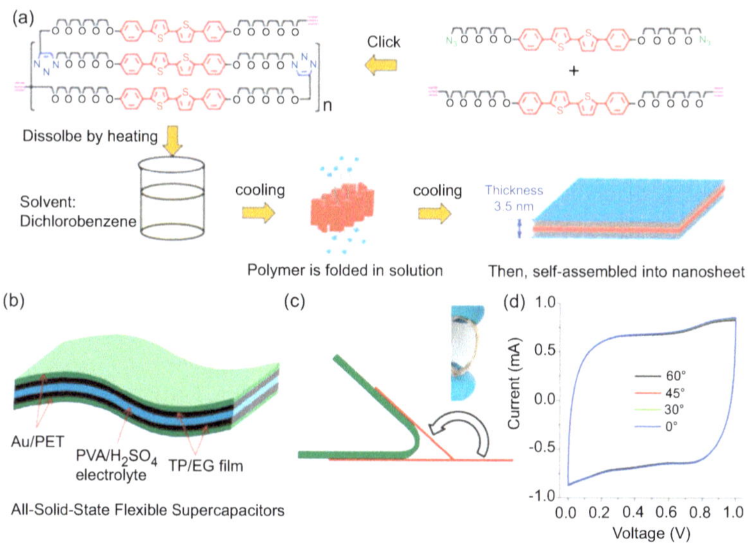

Figure 5.13 (a) Fabrication of TP nanosheets by self-assembly of the polymer consisting of Ph2TPh monomers linked by 1,2,3-triazole units in DCB. (b) The scheme of flexible TP/EG-ASSSs. (c) The bending scheme and (d) the CVs of TP/EG-ASSSs tested under different bending degrees. Reprinted with permission from Ref. [69], Copyright 2016, John Wiley & Sons.

5.7 Summary and Outlook

Lightweight, miniaturization, and flexibility are the main trends of future electronics, which are ideal for dynamic or high-flex applications. MSCs are expected to play a crucial role as high-power energy storage devices for miniaturized and flexible electronics. Two-dimensional materials have shown great potential in micro energy devices. As they provide a high surface area accessible to ions, the availability of multiple reaction sites and interlayer spaces, and the possibility to modify their surface chemistry, 2D materials exhibit a variety of energy storage methods containing electrical double-layer capacitance, redox and ion intercalation pseudo-capacitance. Up to now, with different plastic substrates, a great number of outstanding achievements of flexible MSCs have emerged based on various 2D materials, including graphene, a series of 2D inorganic materials, and 2D organic materials. Studies on graphene-based and inorganic/organic 2D nanomaterials showed that the optimum energy storage is achieved with porous and curved nanosheets that allow fast ion transport through 2D channels and across nanosheets.

Although the drawbacks of some 2D materials, especially 2D nanomaterials—such as poor electrical conductivity of many of them and a tendency to re-stack, which obstructs ion transport—limit further enhancement in energy storage, they can be tackled through hybridization [71–73]. Graphene as well as carbon nanotubes is a promising component for providing extra electronic channel, while multiple composites with graphene show an excellent electronic performance. The design of 3D hierarchical structures combining 2D nanomaterials with other nanomaterials has shown great enhancement in energy storage [74]. The 2D nature of high packing density, with no doubt, leads to high volumetric and areal capacitances, however with the possibility that reduces the active surface of nanosheets accessible to electrolyte. To block re-stacking, pre-intercalation of electrolytes in between 2D nanosheets or pillaring with large molecules prior to electrode manufacturing was employed.

There are still several challenges ahead of 2D materials applied in MSCs: intrinsic electrical performance, stability, and manufacturability. While the interdigital geometry of in-plane electrodes and the porous structure of 2D materials enhance storage and faster ion transport compared to traditional sandwich device architecture, intrinsic electrical conductivity mainly determines the electrical performance affecting the full utilization of the available electrochemically active sites of poorly conducting 2D nanomaterials.

Partial reduction of oxides, chemical surface modification, intralayer doping, and lattice strain can be adopted to modify the electronic structure of 2D nanomaterials [75].

Chemical and electrochemical stability of the active substance in electrodes remains a challenge. Many of the emerging new 2D nanomaterials, such as phosphorene [76] and silicene [77], are unstable in air. Insulating materials, such as boronitrene [78], can only be used in separators. Some other materials are difficult to synthesize [77] and are unavailable in amounts required for supercapacitor applications. Even MXenes oxidize easily under anodic potential in aqueous electrolytes, requiring coupling with more stable positive electrodes in hybrid devices. In addition to the stability of the material itself, the thermal performance of most 2D organic materials assembled as electrodes in integrated circuits should be taken into account.

Methods for fabricating 2D nanomaterials as electrodes for MSCs are commonly CVD, LBL, sputtering, photolithography, electrochemical and electrophoretic deposition, spray deposition, and printing. In many cases, deposition is followed by a patterning step using a standard mask, which complicates the process and lowers production efficiency. Currently, the cost involved in the synthesis of 2D nanomaterials using sophisticated methods such as direct laser writing is much higher than other production processes. In pursuit of mass production, economic efficiency cannot be ignored. The scalability of manufacturing 2D-nanomaterials-based electrodes has been demonstrated for graphene [27]. Printing of 2D nanomaterials is promising, while electrode fabrication and patterning are done simultaneously [79, 80]. And contemporary research seeks to improve composition of inks, printing speed, and resolution. Findings described here point to clear research venues that should lead to further progress in the future.

References

1. Rogers, J. A., Someya, T., and Huang, Y. (2010). Materials and mechanics for stretchable electronics, *Science*, **327**, pp. 1603–1607.
2. Xu, J., Wang, S., Wang, G.-J. N., Zhu, C., Luo, S., Jin, L., Gu, X., Chen, S., Feig, V. R., To, J. W. F., et al. (2017). Highly stretchable polymer semiconductor films through the nanoconfinement effect, *Science*, **355**, pp. 59–64.

3. Sagu, J. S., York, N., Southee, D., and Wijayantha, K. G. U. (2015). Printed electrodes for flexible, light-weight solid-state supercapacitors: A feasibility study, *Circuit World*, **41**, pp. 80–86.
4. Nishide, H. and Oyaizu, K. (2008). Materials science: Toward flexible batteries, *Science*, **319**, pp. 737–738.
5. Kyeremateng, N. A., Brousse, T., and Pech, D. (2017). Microsupercapacitors as miniaturized energy-storage components for on-chip electronics, *Nat. Nanotechnol.*, **12**, pp. 7–15.
6. Ghidiu, M., Lukatskaya, M. R., Zhao, M. Q., Gogotsi, Y., and Barsoum, M. W. (2014). Conductive two-dimensional titanium carbide 'clay' with high volumetric capacitance, *Nature*, **516**, pp. 78–81.
7. Chmiola, J., Largeot, C., Taberna, P.-L., Simon, P., and Gogotsi, Y. (2010). Monolithic carbide-derived carbon films for micro-supercapacitors, *Science*, **328**, pp. 480–483.
8. Shao, Y., El-Kady, M. F., Wang, L. J., Zhang, Q., Li, Y., Wang, H., Mousavi, M. F., and Kaner, R. B. (2015). Graphene-based materials for flexible supercapacitors, *Chem. Soc. Rev.*, **44**, pp. 3639–3665.
9. Levi, M. D., Lukatskaya, M. R., Sigalov, S., Beidaghi, M., Shpigel, N., Daikhin, L., Aurbach, D., Barsoum, M. W., and Gogotsi, Y. (2015). Solving the capacitive paradox of 2D MXene using electrochemical quartz-crystal admittance and in situ electronic conductance measurements, *Adv. Energy Mater.*, **5**, pp. 1400815.
10. Qi, D., Liu, Y., Liu, Z., Zhang, L., and Chen, X. (2016). Design of architectures and materials in in-plane micro-supercapacitors: Current status and future challenges, *Adv. Mater.*, **29**, pp. 1602802.
11. Bonaccorso, F., Colombo, L., Yu, G., Stoller, M., Tozzini, V., Ferrari, A. C., Ruoff, R. S., and Pellegrini, V. (2015). 2D materials. Graphene, related two-dimensional crystals, and hybrid systems for energy conversion and storage, *Science*, **347**, pp. 1246501.
12. Xia, J., Chen, F., Li, J., and Tao, N. (2009). Measurement of the quantum capacitance of graphene, *Nat. Nanotechnol.*, **4**, pp. 505–509.
13. Liu, C., Yu, Z., Neff, D., Zhamu, A., and Jang, B. Z. (2010). Graphene-based supercapacitor with an ultrahigh energy density, *Nano Lett.*, **10**, pp. 4863–4868.
14. Zhu, Y., Murali, S., Stoller, M. D., Ganesh, K. J., Cai, W., Ferreira, P. J., Pirkle, A., Wallace, R. M., Cychosz, K. A., Thommes, M., et al. (2011). Carbon-based supercapacitors produced by activation of graphene, *Science*, **332**, pp. 1537–1541.

15. Miller, J. R., Outlaw, R. A., and Holloway, B. C. (2010). Graphene double-layer capacitor with ac line-filtering performance, *Science*, **329**, pp. 1637–1639.

16. Wu, Z. S., Winter, A., Chen, L., Sun, Y., Turchanin, A., Feng, X., and Mullen, K. (2012). Three-dimensional nitrogen and boron co-doped graphene for high-performance all-solid-state supercapacitors, *Adv. Mater.*, **24**, pp. 5130–5135.

17. El-Kady, M. F., Strong, V., Dubin, S., and Kaner, R. B. (2012). Laser scribing of high-performance and flexible graphene-based electrochemical capacitors, *Science*, **335**, pp. 1326–1330.

18. Wu, Z.-S., Zhou, G., Yin, L.-C., Ren, W., Li, F., and Cheng, H.-M. (2012). Graphene/metal oxide composite electrode materials for energy storage, *Nano Energy*, **1**, pp. 107–131.

19. Xiong, G., Meng, C., Reifenberger, R. G., Irazoqui, P. P., and Fisher, T. S. (2014). A review of graphene-based electrochemical microsupercapacitors, *Electroanalysis*, **26**, pp. 30–51.

20. Dubal, D. P., Chodankar, N. R., Kim, D.-H., and Gomez-Romero, P. (2018). Towards flexible solid-state supercapacitors for smart and wearable electronics, *Chem. Soc. Rev.*, DOI: 10.1039/C7CS00505A.

21. Dreyer, D. R., Park, S., Bielawski, C. W., and Ruoff, R. S. (2010). The chemistry of graphene oxide, *Chem. Soc. Rev.*, **39**, pp. 228–240.

22. Bottari, G., Herranz, M. A., Wibmer, L., Volland, M., Rodriguez-Perez, L., Guldi, D. M., Hirsch, A., Martin, N., D'Souza, F., and Torres, T. (2017). Chemical functionalization and characterization of graphene-based materials, *Chem. Soc. Rev.*, **46**, pp. 4464–4500.

23. Wu, Z. S., Parvez, K., Feng, X., and Mullen, K. (2013). Graphene-based in-plane micro-supercapacitors with high power and energy densities, *Nat. Commun.*, **4**, pp. 2487.

24. Yang, S., Lohe, M. R., Mullen, K., and Feng, X. (2016). New-generation graphene from electrochemical approaches: Production and applications, *Adv. Mater.*, **28**, pp. 6213–6221.

25. Li, J., Sollami Delekta, S., Zhang, P., Yang, S., Lohe, M. R., Zhuang, X., Feng, X., and Ostling, M. (2017). Scalable fabrication and integration of graphene microsupercapacitors through full inkjet printing, *ACS Nano*, **11**, pp. 8249–8256.

26. Liu, Z., Wu, Z. S., Yang, S., Dong, R., Feng, X., and Mullen, K. (2016). Ultraflexible in-plane micro-supercapacitors by direct printing of solution-processable electrochemically exfoliated graphene, *Adv. Mater.*, **28**, pp. 2217–2222.

27. El-Kady, M. F. and Kaner, R. B. (2013). Scalable fabrication of high-power graphene micro-supercapacitors for flexible and on-chip energy storage, *Nat. Commun.*, **4**, pp. 1475.

28. Peng, Z., Ye, R., Mann, J. A., Zakhidov, D., Li, Y., Smalley, P. R., Lin, J., and Tour, J. M. (2015). Flexible boron-doped laser-induced graphene microsupercapacitors, *ACS Nano*, **9**, pp. 5868–5875.

29. Wu, Z., Li, L., Yan, J.-M., and Zhang, X.-B. (2017). Materials design and system construction for conventional and new-concept supercapacitors, *Adv. Sci.*, **4**, pp. 1600382.

30. Zhang, P., Zhu, F., Wang, F., Wang, J., Dong, R., Zhuang, X., Schmidt, O. G., and Feng, X. (2017). Stimulus-responsive micro-supercapacitors with ultrahigh energy density and reversible electrochromic window, *Adv. Mater.*, **29**, pp. 1604491.

31. Wu, Z. S., Parvez, K., Li, S., Yang, S., Liu, Z., Liu, S., Feng, X., and Mullen, K. (2015). Alternating stacked graphene-conducting polymer compact films with ultrahigh areal and volumetric capacitances for high-energy micro-supercapacitors, *Adv. Mater.*, **27**, pp. 4054–4061.

32. Heon, M., Lofland, S., Applegate, J., Nolte, R., Cortes, E., Hettinger, J. D., Taberna, P.-L., Simon, P., Huang, P., Brunet, M., et al. (2011). Continuous carbide-derived carbon films with high volumetric capacitance, *Energy Environ. Sci.*, **4**, pp. 135–138.

33. Hsia, B., Kim, M. S., Vincent, M., Carraro, C., and Maboudian, R. (2013). Photoresist-derived porous carbon for on-chip micro-supercapacitors, *Carbon*, **57**, pp. 395–400.

34. Shen, C., Wang, X., Zhang, W., and Kang, F. (2013). Direct prototyping of patterned nanoporous carbon: A route from materials to on-chip devices, *Sci. Rep.*, **3**, pp. 2294.

35. Pech, D., Brunet, M., Taberna, P.-L., Simon, P., Fabre, N., Mesnilgrente, F., Conédéra, V., and Durou, H. (2010). Elaboration of a microstructured inkjet-printed carbon electrochemical capacitor, *J. Power Sources*, **195**, pp. 1266–1269.

36. Pech, D., Brunet, M., Durou, H., Huang, P., Mochalin, V., Gogotsi, Y., Taberna, P. L., and Simon, P. (2010). Ultrahigh-power micrometre-sized supercapacitors based on onion-like carbon, *Nat. Nanotechnol.*, **5**, pp. 651–654.

37. Gao, W., Singh, N., Song, L., Liu, Z., Reddy, A. L., Ci, L., Vajtai, R., Zhang, Q., Wei, B., and Ajayan, P. M. (2011). Direct laser writing of micro-supercapacitors on hydrated graphite oxide films, *Nat. Nanotechnol.*, **6**, pp. 496–500.

38. Li, L., Zhang, J., Peng, Z., Li, Y., Gao, C., Ji, Y., Ye, R., Kim, N. D., Zhong, Q., Yang, Y., et al. (2016). High-performance pseudocapacitive microsupercapacitors from laser-induced graphene, *Adv. Mater.*, **28**, pp. 838–845.

39. Naguib, M., Mochalin, V. N., Barsoum, M. W., and Gogotsi, Y. (2014). 25th anniversary article: MXenes: a new family of two-dimensional materials, *Adv. Mater.*, **26**, pp. 992–1005.

40. Naguib, M., Kurtoglu, M., Presser, V., Lu, J., Niu, J., Heon, M., Hultman, L., Gogotsi, Y., and Barsoum, M. W. (2011). Two-dimensional nanocrystals produced by exfoliation of Ti_3AlC_2, *Adv. Mater.*, **23**, pp. 4248–4253.

41. Kurra, N., Ahmed, B., Gogotsi, Y., and Alshareef, H. N. (2016). MXene-on-paper coplanar microsupercapacitors, *Adv. Energy Mater.*, **6**, pp. 1601372.

42. Li, H., Hou, Y., Wang, F., Lohe, M. R., Zhuang, X., Niu, L., and Feng, X. (2017). Flexible all-solid-state supercapacitors with high volumetric capacitances boosted by solution processable MXene and electrochemically exfoliated graphene, *Adv. Energy Mater.*, **7**, pp. 1601847.

43. Jiang, H., Ma, J., and Li, C. Z. (2012). Mesoporous carbon incorporated metal oxide nanomaterials as supercapacitor electrodes, *Adv. Mater.*, **24**, pp. 4197–4202.

44. He, Y., Chen, W., Li, X., Zhang, Z., Fu, J., Zhao, C., and Xie, E. (2013). Freestanding three-dimensional graphene MnO_2 composite networks as ultralight and flexible supercapacitor electrodes, *ACS Nano*, **7**, pp. 174–182.

45. Dong, X.-C., Xu, H., Wang, X.-W., Huang, Y.-X., Chan-Park, M. B., Zhang, H., Wang, L.-H., Huang, W., and Chen, P. (2012). 3D graphene–cobalt oxide electrode for high-performance supercapacitor and enzymeless glucose detection, *ACS Nano*, **6**, pp. 3206–3213.

46. Hu, C.-C., Chang, K.-H., Lin, M.-C., and Wu, Y.-T. (2006). Design and tailoring of the nanotubular arrayed architecture of hydrous RuO_2 for next generation supercapacitors, *Nano Lett.*, **6**, pp. 2690–2695.

47. Cao, X., Shi, Y., Shi, W., Lu, G., Huang, X., Yan, Q., Zhang, Q., and Zhang, H. (2011). Preparation of novel 3D graphene networks for supercapacitor applications, *Small*, **7**, pp. 3163–3168.

48. Zhai, T., Wang, F., Yu, M., Xie, S., Liang, C., Li, C., Xiao, F., Tang, R., Wu, Q., Lu, X., et al. (2013). 3D MnO_2-graphene composites with large areal capacitance for high-performance asymmetric supercapacitors, *Nanoscale*, **5**, pp. 6790–6796.

49. Cheng, Y., Lu, S., Zhang, H., Varanasi, C. V., and Liu, J. (2012). Synergistic effects from graphene and carbon nanotubes enable flexible and robust electrodes for high-performance supercapacitors, *Nano Lett.*, **12**, pp. 4206–4211.
50. Tao, J., Liu, N., Li, L., Su, J., and Gao, Y. (2014). Hierarchical nanostructures of polypyrrole@MnO_2 composite electrodes for high performance solid-state asymmetric supercapacitors, *Nanoscale*, **6**, pp. 2922–2928.
51. Toupin, M., Brousse, T., and Bélanger, D. (2004). Charge storage mechanism of MnO_2 electrode used in aqueous electrochemical capacitor, *Chem. Mater.*, **16**, pp. 3184–3190.
52. Wang, Y., Takahashi, K., Lee, K. H., and Cao, G. Z. (2006). Nanostructured vanadium oxide electrodes for enhanced lithium-ion intercalation, *Adv. Funct. Mater.*, **16**, pp. 1133–1144.
53. Chou, S.-L., Wang, J.-Z., Sun, J.-Z., Wexler, D., Forsyth, M., Liu, H.-K., MacFarlane., D. R., and Dou, S.-X. (2008). High capacity, safety, and enhanced cyclability of lithium metal battery using a V_2O_5 nanomaterial cathode and room temperature ionic liquid electrolyte, *Chem. Mater.*, **20**, pp. 7044–7051.
54. Cheah, Y. L., von Hagen, R., Aravindan, V., Fiz, R., Mathur, S., and Madhavi, S. (2013). High-rate and elevated temperature performance of electrospun V_2O_5 nanofibers carbon-coated by plasma enhanced chemical vapour deposition, *Nano Energy*, **2**, pp. 57–64.
55. Shao, M., Zhang, R., Li, Z., Wei, M., Evans, D. G., and Duan, X. (2015). Layered double hydroxides toward electrochemical energy storage and conversion: Design, synthesis and applications, *Chem. Commun.*, **51**, pp. 15880–15893.
56. Tang, C., Wang, H. S., Wang, H. F., Zhang, Q., Tian, G. L., Nie, J. Q., and Wei, F. (2015). Spatially confined hybridization of nanometer-sized NiFe hydroxides into nitrogen-doped graphene frameworks leading to superior oxygen evolution reactivity, *Adv. Mater.*, **27**, pp. 4516–4522.
57. Shao, M., Li, Z., Zhang, R., Ning, F., Wei, M., Evans, D. G., and Duan, X. (2015). Hierarchical conducting polymer@clay core-shell arrays for flexible all-solid-state supercapacitor devices, *Small*, **11**, pp. 3530–3538.
58. Wang, Q. and O'Hare, D. (2012). Recent advances in the synthesis and application of layered double hydroxide (LDH) nanosheets, *Chem. Rev.*, **112**, pp. 4124–4155.
59. Peng, L., Peng, X., Liu, B., Wu, C., Xie, Y., and Yu, G. (2013). Ultrathin two-dimensional MnO_2/graphene hybrid nanostructures for high-

performance, flexible planar supercapacitors, *Nano Lett.*, **13**, pp. 2151–2157.

60. Xiao, X., Ding, T., Yuan, L., Shen, Y., Zhong, Q., Zhang, X., Cao, Y., Hu, B., Zhai, T., Gong, L., et al. (2012). WO_{3-x}/MoO_{3-x}Core/Shell nanowires on carbon fabric as an anode for all-solid-state asymmetric supercapacitors, *Adv. Energy Mater.*, **2**, pp. 1328–1332.

61. Liu, R. and Lee, S. B. (2008). MnO_2 poly4-ethylenedioxythiophene coaxial nanowires by one-step coelectrodeposition for electrochemical energy storage, *J. Am. Chem. Soc.*, **130**, pp. 2942–2943.

62. Zhao, J., Xu, S., Tschulik, K., Compton, R. G., Wei, M., O'Hare, D., Evans, D. G., and Duan, X. (2015). Molecular-scale hybridization of clay monolayers and conducting polymer for thin-film supercapacitors, *Adv. Funct. Mater.*, **25**, pp. 2745–2753.

63. Chang, J., Jin, M., Yao, F., Kim, T. H., Le, V. T., Yue, H., Gunes, F., Li, B., Ghosh, A., Xie, S., et al. (2013). Asymmetric supercapacitors based on graphene/MnO_2 nanospheres and graphene/MoO_3 nanosheets with high energy density, *Adv. Funct. Mater.*, **23**, pp. 5074–5083.

64. Kim, D., Yun, J., Lee, G., and Ha, J. S. (2014). Fabrication of high performance flexible micro-supercapacitor arrays with hybrid electrodes of MWNT/V_2O_5 nanowires integrated with a SnO_2 nanowire UV sensor, *Nanoscale*, **6**, pp. 12034–12041.

65. Zhuang, X., Mai, Y., Wu, D., Zhang, F., and Feng, X. (2015). Two-dimensional soft nanomaterials: A fascinating world of materials, *Adv. Mater.*, **27**, pp. 403–427.

66. Yang, M. and Zhou, Z. (2017). Recent breakthroughs in supercapacitors boosted by nitrogen-rich porous carbon materials, *Adv. Sci.*, **4**, pp. 1600408.

67. Yang, C., Schellhammer, K. S., Ortmann, F., Sun, S., Dong, R., Karakus, M., Mics, Z., Loffler, M., Zhang, F., Zhuang, X., et al. (2017). Coordination polymer framework based on-chip micro-supercapacitors with AC line-filtering performance, *Angew. Chem. Int. Ed.*, **56**, pp. 3920–3924.

68. Lin, J., Zhang, C., Yan, Z., Zhu, Y., Peng, Z., Hauge, R. H., Natelson, D., and Tour, J. M. (2013). 3-Dimensional graphene carbon nanotube carpet-based microsupercapacitors with high electrochemical performance, *Nano Lett.*, **13**, pp. 72–78.

69. Wu, Z. S., Zheng, Y., Zheng, S., Wang, S., Sun, C., Parvez, K., Ikeda, T., Bao, X., Mullen, K., and Feng, X. (2017). Stacked-layer heterostructure films of 2D thiophene nanosheets and graphene for high-rate all-solid-state pseudocapacitors with enhanced volumetric capacitance, *Adv. Mater.*, **29**, pp. 1602960.

70. Yang, X., Cheng, C., Wang, Y., Qiu, L., and Li, D. (2013). Liquid-mediated dense integration of graphene materials for compact capacitive energy storage, *Science*, **341**, pp. 534–537.

71. Roldan, R., Chirolli, L., Prada, E., Silva-Guillen, J. A., San-Jose, P., and Guinea, F. (2017). Theory of 2D crystals: Graphene and beyond, *Chem. Soc. Rev.*, **46**, pp. 4387–4399.

72. Yu, J., Wang, Q., O'Hare, D., and Sun, L. (2017). Preparation of two-dimensional layered double hydroxide nanosheets and their applications, *Chem. Soc. Rev.*, **46**, pp. 5950–5974.

73. Zhang, X. Y., Hou, L. L., Ciesielski, A., and Samori, P. (2016). 2D materials beyond graphene for high-performance energy storage applications, *Adv. Energy Mater.*, **6**, pp. 1600671.

74. Mao, J., Iocozzia, J., Huang, J., Meng, K., Lai, Y., and Lin, Z. (2018). Graphene aerogels for efficient energy storage and conversion, *Energ Environ Sci.*, DOI: 10.1039/C7EE03031B.

75. Fu, Q. and Bao, X. (2017). Surface chemistry and catalysis confined under two-dimensional materials, *Chem. Soc. Rev.*, **46**, pp. 1842–1874.

76. Castellanos-Gomez, A. (2015). Black phosphorus: Narrow gap, wide applications, *J. Phys. Chem. Lett.*, **6**, pp. 4280–4291.

77. Peplow, M. (2015). Silicene makes its transistor debut, *Nature*, **581**, pp. 17.

78. Auwa¨rter, W., Kreutz, T. J., Greber, T., and Osterwalder, J. (1999). XPD and STM investigation of hexagonal boron nitride on Ni (111), *Surf. Sci.*, **429**, pp. 229–236.

79. Kelly, A. G., Hallam, T., Backes, C., Harvey, A., Esmaeily, A. S., Godwin, I., Coelho, J., Nicolosi, V., Lauth, J., Kulkarni, A., et al. (2017). All-printed thin-film transistors from networks of liquid-exfoliated nanosheets, *Science*, **356**, pp. 69–73.

80. McManus, D., Vranic, S., Withers, F., Sanchez-Romaguera, V., Macucci, M., Yang, H., Sorrentino, R., Parvez, K., Son, S. K., Iannaccone, G., et al. (2017). Water-based and biocompatible 2D crystal inks for all-inkjet-printed heterostructures, *Nat. Nanotechnol.*, **12**, pp. 343–350.

Chapter 6

Flexible On-Chip Interdigital Micro-Supercapacitors: Efficient Power Units for Wearable Electronics

Guozhen Shen,[a] Kai Jiang,[b] and Di Chen[c]
[a]*State Key Laboratory for Superlattices and Microstructures, Institute of Semiconductors, Chinese Academy of Sciences, Beijing 100083, China*
[b]*Institute & Hospital of Hepatobiliary Surgery, Key Laboratory of Digital Hepatobiliary Surgery of Chinese PLA, Chinese PLA Medical School, Chinese PLA General Hospital, Beijing 100853, China*
[c]*College of Physics and Mathematics and Beijing Key Laboratory for Magneto-Photoelectrical Composite and Interface Science, University of Science and Technology Beijing, Beijing 100083, China*
gzshen@semi.ac.cn

Fast development of flexible electronics has greatly stimulated research in flexible, miniaturized energy storage devices to get fully flexible integrated systems. With the remarkable features of excellent electrochemical performance, miniaturization, high security, and

Flexible and Stretchable Electronics: Materials, Designs, and Devices
Edited by Run-Wei Li and Gang Liu

ISBN 978-981-4800-46-4 (Hardcover), 978-0-429-05890-5 (eBook)
www.jennystanford.com

easy integration to build an all-in-one integrated system, flexible on-chip interdigital micro-supercapacitors (MSCs) are thought to be excellent power-unit candidates for meeting the request of micro-portable electronic equipment. This chapter focuses on the recent developments and integration of flexible on-chip interdigital MSCs. Typical device fabrication techniques, including photolithography, laser scribing, printing, etc. are highlighted. Besides, stretchable MSCs and flexible MSCs integrated sensing systems are also included in this chapter. Demonstrations show that flexible on-chip interdigital MSCs are promising candidates as power units for future flexible electronics.

6.1 Introduction

Flexible electronics contain emerging technologies that build electronic or optoelectronic devices on flexible substrates. Compared with conventional microelectronic devices fabricated on rigid substrates, such as silicon, flexible electronic devices not only show comparable performance but also have some unique features beyond rigid devices, including excellent flexibility, ultrathin thickness, etc. Many types of flexible electronic/optoelectronic devices have been designed and fabricated in the past several years, such as flexible displays, flexible sensors and detectors, electronic skins, integrated circuit smart cards, and so on [1–10]. With the rapid development of flexible electronic devices, there has been an increasing and urgent demand for flexible power sources that can provide not only the desired energy and power densities, but also excellent flexibility to match these electronic devices [11–18].

Currently, flexible electronic devices mainly rely on batteries as the power sources, including thin-film batteries or micro-batteries, which have been produced commercially. Unfortunately, conventional batteries are usually too bulky and thick to fit for future request of fully flexible systems. Although researchers have tried many techniques for making batteries flexible, several drawbacks still exist. For example, it usually requires complicated procedures to fabricate flexible batteries. It is hard to reduce the size of flexible batteries down to the scale of microelectronic devices. It is hard to fabricate all-in-one systems as well [19–22].

As an alternative to batteries, supercapacitors, also called electrochemical capacitors, have drawn considerable attention because they can provide an energy density orders of magnitude higher than that of conventional dielectric capacitors, a higher power density, and a longer cycle life than batteries [23, 24]. Based on the electrode materials used, supercapacitors can be divided into two types: electrical double-layer capacitors (EDLCs) and pseudocapacitors. EDLCs store the charge by the surface charge separation at the electrode–electrolyte interface, while pseudocapacitors use fast and reversible surface faradic redox reactions for charge storage [25–29].

Conventional supercapacitors, assembled into button cells or spiral-wound configurations, are inflexible and heavy, which make them unfit for flexible electronic devices. Thus, in the past several years, flexible MSCs with the features of miniaturized size, excellent flexibility, and excellent electrochemical performance developed very fast and were considered excellent power source for flexible electronics. In its early stage, MSCs were usually fabricated with similar device structure as conventional supercapacitors, where two thin-film electrodes are stacked on top of each other with solid or gel-like electrolyte in between, as shown in Fig. 6.1a [30]. Although this type of MSCs exhibit improved electrochemical performance, some obvious drawbacks still need to be overcome. The first one is that it is hard to accurately control the small distance between two electrodes, resulting in the increase in ion transport length and loss of power. The second one is that short circuit and dislocation of electrodes usually happened, which affect the practical applications of the devices. Considering these drawbacks, researchers later developed MSCs with on-chip interdigital electrodes, as depicted in Fig. 6.1b. In this design, the ion transport path is very short, so the device exhibited ultrahigh power capability.

The work on on-chip MSCs was pioneered by Gogotsi et al. in 2010 [31]. They reported the fabrication of on-chip MSCs with interdigital electrodes using carbide-derived carbon (CDC) as electrodes. Figures 6.2a–d show the schematic illustration of fabrication of on-chip MSCs integrated onto a silicon chip. In this process, the TiC film was first sputtered on silicon substrate. Chlorination was then performed by using the chlorine-containing plasma etching process at elevated temperature. After performing standard photolithography techniques and deposition of gold current collector, on-chip MSCs based on CDC film were fabricated. With the interdigital configuration, the as-fabricated on-chip MSCs exhibited very high volumetric capacity, exceeding that of stacked MSCs and macroscale supercapacitors.

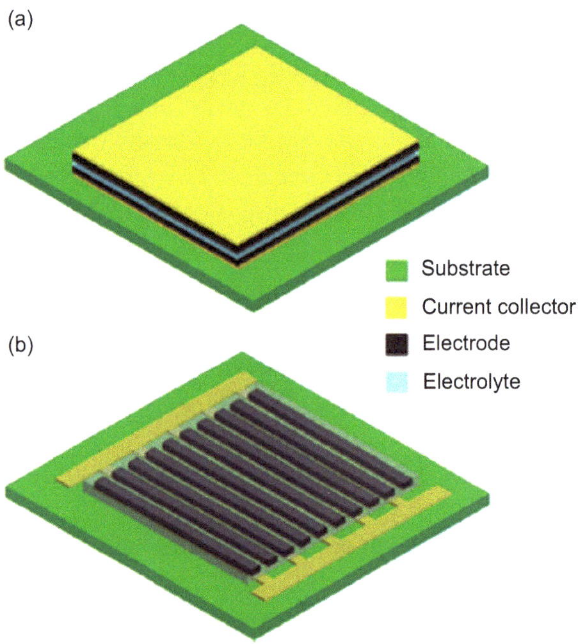

Figure 6.1 Schematic illustration of the electrode configuration of on-chip micro-supercapacitors with (a) conventional 2D architecture and (b) on-chip interdigital architecture. Reproduced with permission of the Royal Society of Chemistry 2014.

Inspired by this work, remarkable efforts have been made in designing on-chip MSCs based on different active electrodes with different fabrication methods. In this chapter, we will primarily focus on the fabrication methods for flexible on-chip MSCs. Fully flexible electronic systems powered by flexible on-chip MSCs will also be discussed in detail.

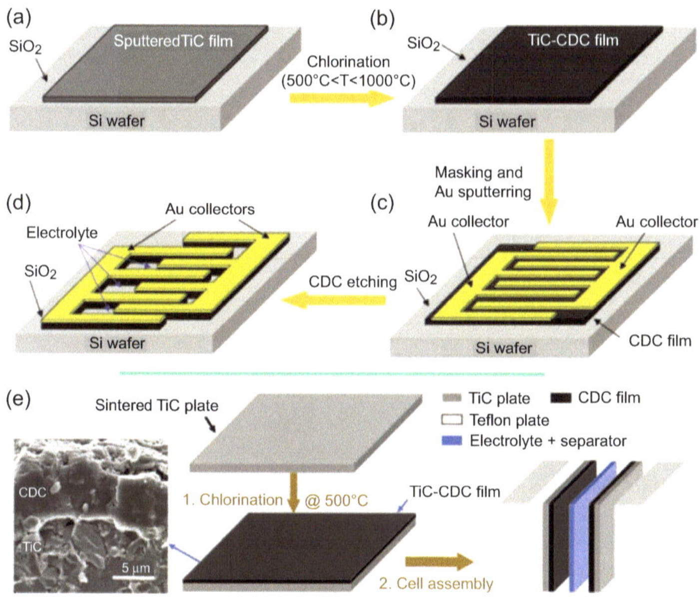

Figure 6.2 Schematic of the fabrication of a micro-supercapacitor integrated onto a silicon chip based on the bulk carbide-derived carbon film process. Reproduced with permission from Ref. [31], Copyright 2010, The American Association for the Advancement of Science.

6.2 Fabrication Methods

Since the first report by Gogotsi et al. in 2010, there have been several fabrication methods such as conventional photolithography method, laser-scribing method, and printing method to build flexible on-chip MSCs, classified into two categories: technologies based on the transfer of an existing electrode material in powder form onto the miniaturized cell, and technologies based on direct synthesis of electrode materials during the manufacture of microdevices (Fig. 6.3) [32, 33].

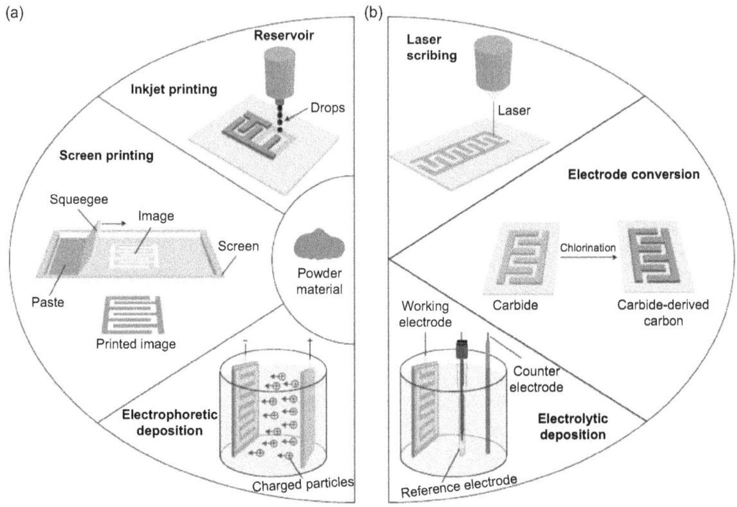

Figure 6.3 Typical microfabrication processes for the realization of planar on-chip micro-supercapacitors. Reproduced with permission from Ref. [32], Copyright 2016, Springer Nature.

6.2.1 Conventional Photolithography Method

Photolithography is the most commonly used method to produce microelectronic devices. Patterns with precisely controlled dimensions down to several nanometers can be obtained using the photolithography method. For the fabrication of flexible on-chip MSCs, the most efficient approach is combining photolithography with etching technology, similar to typical fabrication of micro electronic devices [34–43]. Briefly, an active electrode material film is first prepared on a substrate, and then a patterned metallic current collector is synthesized by performing photolithography, followed with evaporation or sputtering. After etching the excessive materials, on-chip MSCs with desired interdigital electrodes are obtained.

For example, Fig. 6.4 shows the typical fabrication of on-chip MSCs using the photolithography method. Briefly, a graphite oxide (GO) film was spin coated on flexible PET substrate [34].

Photolithography was then performed, followed with oxygen plasma treatment to remove the unwanted GO. After removing the extra resist, being reduced with hydrazine, a patterned reduced GO (rGO) film was obtained. The gold electrode was then deposited followed by the coating of gel electrolytes. Figure 6.4a shows the optical image of the obtained on-chip MSC. The thickness of the rGO electrode could be easily controlled, and Fig. 6.4c gives the planar optical image of the border between the rGO film and the PET substrate. For the MSC with an rGO film of thickness 1.2 μm, the areal capacitance and volume capacitance were 781.7 $\mu F/cm^2$ and 6.98 F/cm^3, respectively. The as-fabricated MSC delivered higher energy density than that of traditional supercapacitors.

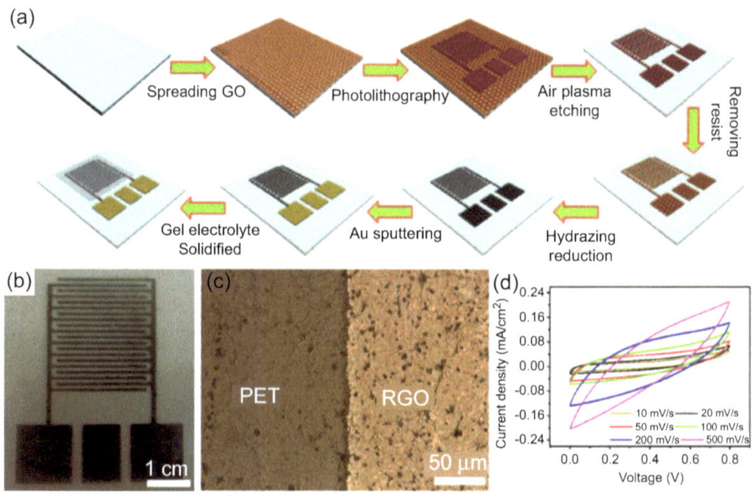

Figure 6.4 Schematic illustration of the fabrication process of flexible on-chip micro-supercapacitors with rGO electrodes. Reproduced with permission from Ref. [34], Copyright 2015, Elsevier.

Besides the above photolithography fabrication process, a photolithography process combined with other techniques was also proposed to produce flexible on-chip MSCs [35]. Figure 6.5 shows the schematic illustration of fabrication of on-chip MSCs by combining photolithography with a transfer method, involving conventional photolithography followed by rGO layer transfer, electrochemical deposition of pseudocapacitve materials [MnO_2, $Co(OH)_2$, PANI], and finally resist lift-off. The fabricated MSCs delivered higher energy densities in the range of 1.3–5.3 mWh/cm^3 and power densities in the range of 178–533 mW/cm^3, superior to the lithium film batteries and conventional supercapacitors.

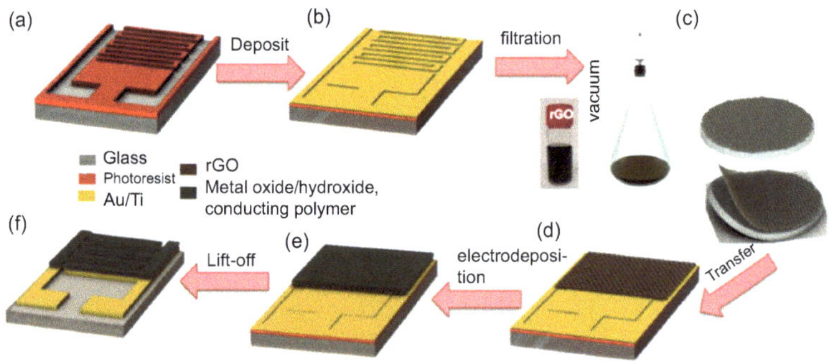

Figure 6.5 Schematic illustration of the fabrication of on-chip micro-supercapacitors with rGO-based heterostructures. Reproduced with permission from Ref. [35], Copyright 2015, Elsevier.

Lin et al. fabricated 3D graphene carbon nanotube carpet-based on-chip MSCs [36]. In their process, a Cr/Ni electrode was first patterned via the photolithography process, where Ni acted as both the current collector and catalysts for further graphene growth. After the growth of graphene via the CVD method, Fe/Al_2O_3 catalysts were then patterned on the graphene film to grow CNT carpets via the CVD method. The whole process is depicted in Fig. 6.6. Studies found that the fabricated on-chip MSCs exhibited significantly improved electrochemical performance, which showed excellent frequency response with an impedance phase angle as large as 81.5°. In addition, the device acquired a real capacitance of 2.16 mF/cm^2 in aqueous electrolytes and 3.93 mF/cm^2 in ionic liquids.

Photolithography process has been extensively used to fabricate flexible on-chip MSCs. Many kinds of electrodes, including CNTs, graphene, metal oxides, and polymers, have been explored. All the results proved that methods containing photolithography are efficient ways to get flexible on-chip MSCs with improved electrochemical performances.

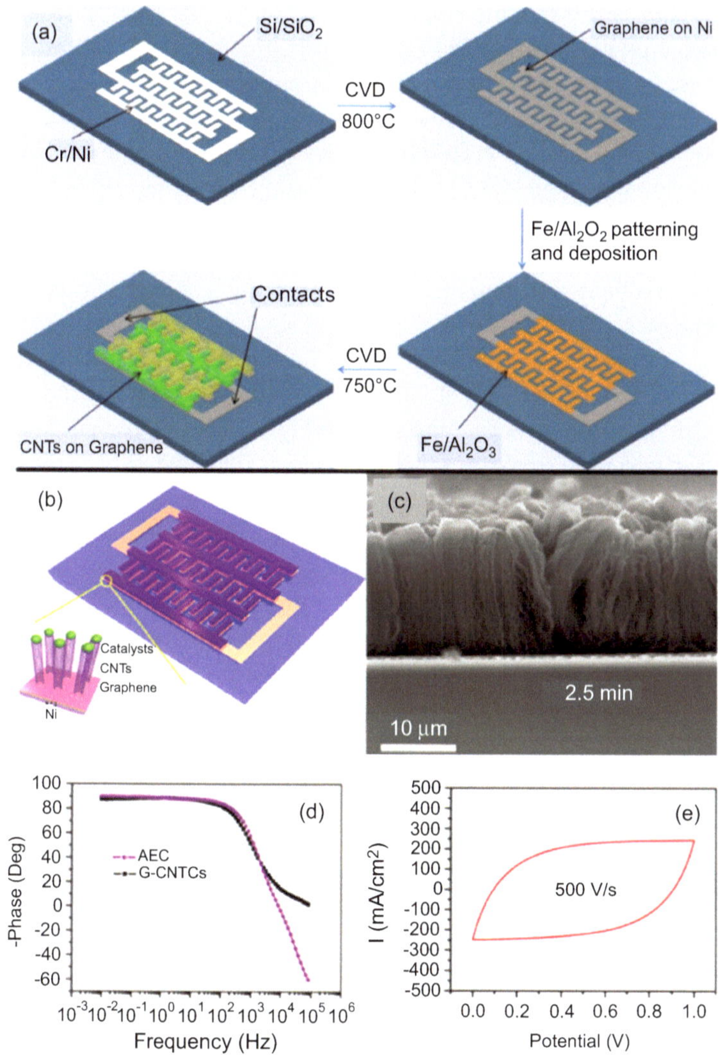

Figure 6.6 Schematic illustration of the fabrication of on-chip micro-supercapacitors with in situ CVD-grown CNTs. Reprinted with permission from Ref. [36], Copyright 2013, American Chemical Society.

6.2.2 Laser-Scribing Method

As a simple, scalable, and cost-effective microfabrication technique, the laser-scribing method has been employed to fabricate flexible on-chip MSCs starting from the reduction of GO to graphene due to the photothermal effect. During this process, the length of laser pulse, power, and scanning speed should be carefully controlled [44–49]. A typical example work was done by El-kady, where a household LightScribe DVD burner was used for scalable fabrication of flexible on-chip MSCs [44]. In their work, more than 100 MSCs can be produced on a single disc in less than 30 min. The as-fabricated flexible MSCs can deliver a high area capacitance of about 2.32 mF/cm^2 and excellent cycle stability while preserving 96% of its initial value even after 10,000 cycles.

With mushroom-derived carbon electrodes, Ogale et al. reported the fabrication of highly stable flexible on-chip MSCs using the laser-scribing method, as demonstrated in Fig. 6.7 [45]. A low-energy CO_2 laser was used to pattern the electrode as depicted in Fig. 6.7a. The thickness of the electrode was calculated to be about 7 μm from the SEM image in Fig. 6.7b. The fabricated MSCs showed excellent flexibility with outstanding electrochemical performance. An 89% capacitance retention up to 1000 bending (60°) cycles and 88% performance retention under 120° bent condition were obtained for the devices. Besides, the device also shows excellent cyclic stability (95% retention) over 15k cycles. The highest volumetric capacitance obtained for the devices is 12.92 F/cm^3 at 0.05 mA/cm^2.

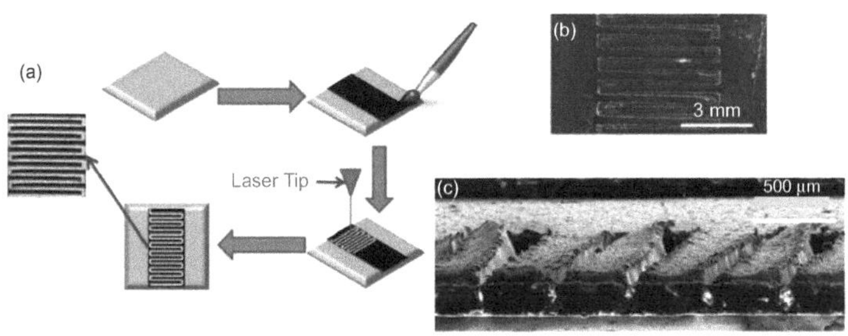

Figure 6.7 Fabrication of on-chip micro-supercapacitors using laser tip scribing method. Reprinted with permission from Ref. [45], Copyright 2016, John Wiley & Sons.

The direct laser-writing method was also used to fabricate flexible MSCs on hydrated GO films [46]. The ability to laser reduce GO into conducting rGO allows the facile and nontoxic writing of rGO–GO–rGO patterns in various configurations to build flexible on-chip MSCs, as shown in Fig. 6.8. The obtained MSCs exhibited an areal capacitance of about 0.51 mF/cm^2, nearly twice that of the conventional sandwich supercapacitors. After 10,000 cycles, about 35% drop in capacitance was observed for the device, showing relatively good cycling stability.

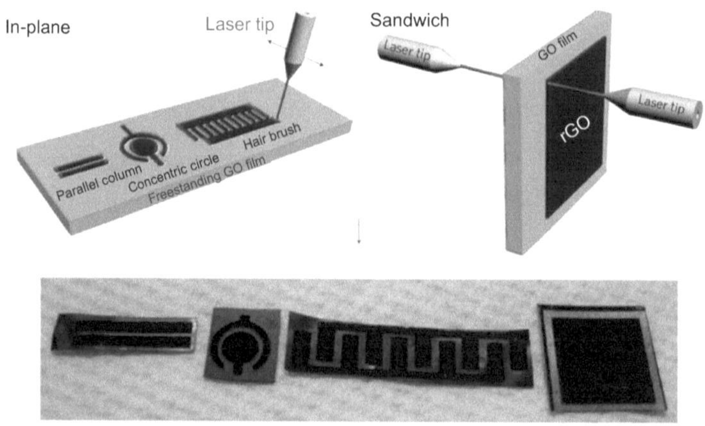

Figure 6.8 Direct laser writing of on-chip micro-supercapacitors. Reproduced with permission from Ref. [46], Copyright 2011, Springer Nature.

Different electrode materials are tested to get flexible on-chip MSCs in the laser-scribing process. For example, Ogal et al. synthesized MOF-based metal-decorated and heteroatom-doped porous graphene and then used CO_2 laser to fabricate flexible on-chip MSCs, as shown in Fig. 6.9 [47]. The MOF slurry was first coated on flexible Kapton substrate, and then laser was used to get laser-induced MOF-derived graphene. Gel electrolyte was finally coated on the patterned electrodes to get the final flexible devices. With the MOF-derived graphene, the fabricated MSCs showed extremely high cycling stability. Near 100% retention after 200,000 cycles as well as stability under 150° bending was obtained for the devices. The highest areal capacitance value obtained for the MSCs was 1.36 mF/cm^2 at a current density 0.15 mA/cm^2.

The laser-scribing process is proved to be an effective method for scalable fabrication of flexible on-chip MSCs with controlled densities, configurations, and sizes. A good example by Wu et al. reported the scalable fabrication of MSCs using UV light irradiation (Fig. 6.10) [48]. In the process, a thin film of GO/TiO_2 nanoparticles was first transferred onto a flexible PET substrate. A 500 W high-pressure Hg lamp was then used for synchronous photoreduction and patterning of GO films with the help of a photomask by UV irradiation for 4 h. A thin layer of gold was then deposited on the top surface of the photoreduced graphene oxide (PRG) patterns as current collectors, and a suitable external electrolyte was dropped to finish the fabrication of MSCs.

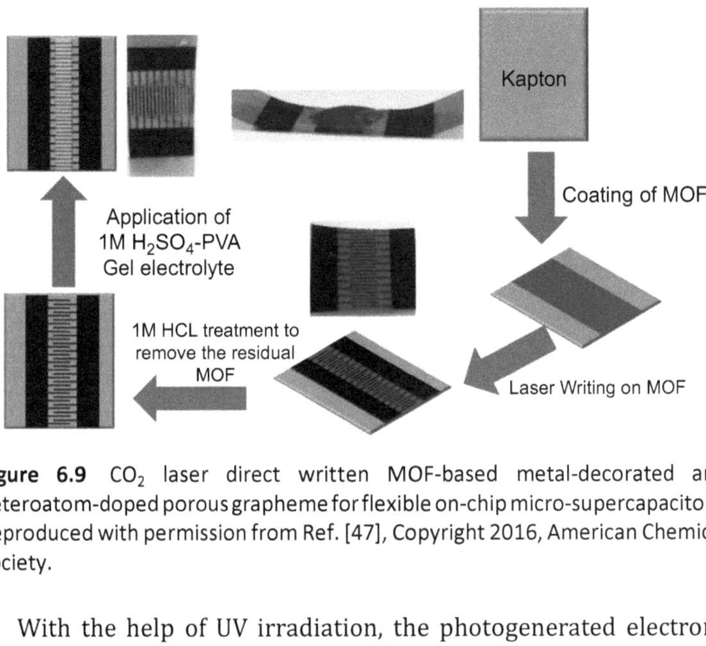

Figure 6.9 CO_2 laser direct written MOF-based metal-decorated and heteroatom-doped porous grapheme for flexible on-chip micro-supercapacitors. Reproduced with permission from Ref. [47], Copyright 2016, American Chemical Society.

With the help of UV irradiation, the photogenerated electrons transfer from TiO_2 nanoparticles to GO sheets, resulting in efficient reduction of the GO film and restoration of the sp^2-C network, as shown in Fig. 6.10b. Interestingly, using the UV irradiation method, the authors could control the shape of the MSCs in different desirable geometries and sizes, including interdigital fingers, concentric circular, parallel strips, folding line fingers, etc., as shown in Fig. 6.11. The fabricated devices showed multiple features of excellent flexibility, scalable integration, and superior electrochemical performance.

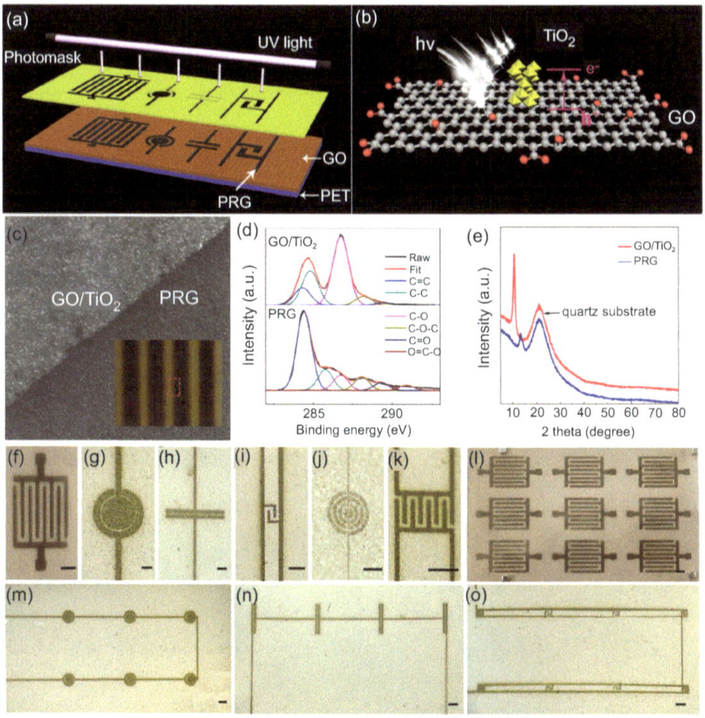

Figure 6.10 One-step UV-light-assisted fabrication of on-chip micro-supercapacitors with different device configurations. Reproduced with permission from Ref. [48], Copyright 2017, American Chemical Society.

6.2.3 Printing Method

Printed electronics have attracted huge attention in the recent years because using different printing methods, researchers can directly produce flexible electronic devices on a large scale. Flexible on-chip MSCs can also be produced using printing methods [50–52].

The screen-printing method is a simple and low-cost printing method, which has been widely used to print electronic devices. Recently, Wang et al. reported the printing of flexible on-chip MSCs using the screen-printing method, as shown in Fig. 6.11 [50]. During this process, interdigital silver electrodes were first printed on flexible substrate using the screen-printing method, acting as the current collectors. In the second step, slurry MnO_2/

onion-like carbon composite electrodes were in situ printed on the patterned silver current collectors. After dropping a gel electrolyte, flexible on-chip MSCs were obtained, as seen in Fig. 6.11a. Using the screen-printing method, researchers could print MSCs on different substrates, including both rigid and flexible substrate (Fig. 6.11b).

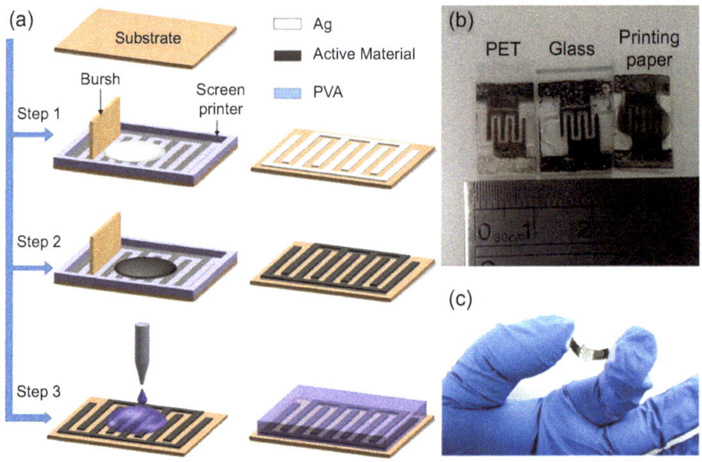

Figure 6.11 Fabrication of flexible on-chip micro-supercapacitors with the screen-printing technology. Reproduced with permission from Ref. [50], Copyright 2014, IOP Publishing.

The obtained flexible MSCs showed a capacity of 7.04 mF/cm^2 at a current density of 20 mA/cm^2, and an excellent cycling stability with 80% retention of its initial capacitance after 1000 cycles. Since the whole process was done in ambient conditions without high cost metal deposition, the screen-printing method is thought to be a general and low-cost way for scalable production of flexible on-chip MSCs with high performances. By choosing suitable electrode materials, devices with further improved electrochemical performance should, in principle, be prepared with this method.

As is well known, inkjet printing can reproduce digital images by pushing droplets of liquid precursor materials onto various substrates, including paper and plastic. Owing to the intrinsic features such as fast process, low cost, and scalability, inkjet printing also shows great potential in printed electronics besides traditional printing. Inkjet printing is another typical technology used to fabricate flexible on-chip MSCs. Recently, through full inkjet printing, graphene-based flexible on-chip MSCs were fabricated in a scalable way, as shown in Fig. 6.12 [51]. Electrochemically exfoliated graphene (EEG) inks were first prepared through a solvent exchange technique, which makes it possible to print thick graphene films, acting as both the active electrodes and the current collectors. Poly (4-styrenesulfonic acid) (PSSH) was also printed as the solid-state electrolyte. Interdigitial MSCs can be printed on different substrates, such as glass slide, Kapton substrate, and silicon wafer. This approach greatly increases the areal capacitance of printed all-solid-state graphene MSCs as single devices and enables effective integration of large-scale MSC arrays to attain excellent collective electrochemical performance.

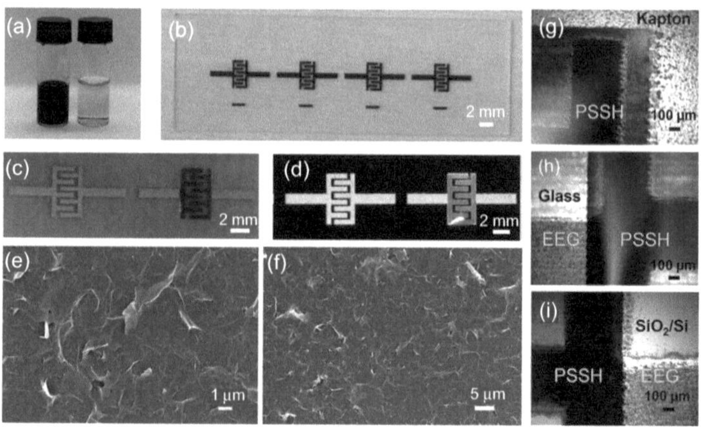

Figure 6.12 Large-scale fabrication of flexible on-chip micro-supercapacitors using the inkjet printing method. Reproduced with permission from Ref. [51], Copyright 2017, American Chemical Society.

6.3 Stretchable On-Chip MSCs

During the past years, stretchable electronics, as a kind of "soft" electronic devices that can be stretched, deformed, and wrapped onto nonplanar curved surfaces, have attracted tremendous attention due to their potential applications in wearable electronics, implantable biomedical devices, and artificial electronic skin. Besides flexibility, on-chip MSCs must also have the feature of stretchability in some cases, to fit for stretchable electronic systems.

An efficient way to get stretchable on-chip MSCs is to design special electrode structures, such as wavy electrodes, to endure shape changes, such as stretching, twisting, crimping, and winding [53]. Li et al. recently fabricated stretchable MSCs with wave-shaped hybrid multiwalled carbon nanotubes/polyaniline electrodes. To get the wave-shaped pattern, a mode with a convex wavy electrode based on photosensitive acrylate resin was first produced using a 3D printer. PDMS mixture was then casted on the mold and then peeled off to serve as a stretchable substrate for MSCs. After sputtering Au electrode on the PDMS substrate, injecting MWNT/PANI slurry and gel electrolyte, the final stretchable on-chip MSCs were obtained, as shown in Fig. 6.13. The as-fabricated designed wavy electrode structure helped to achieve the stretchability of MSCs. The fabricated MSCs exhibited nearly invariable electrochemical performance under stretching conditions.

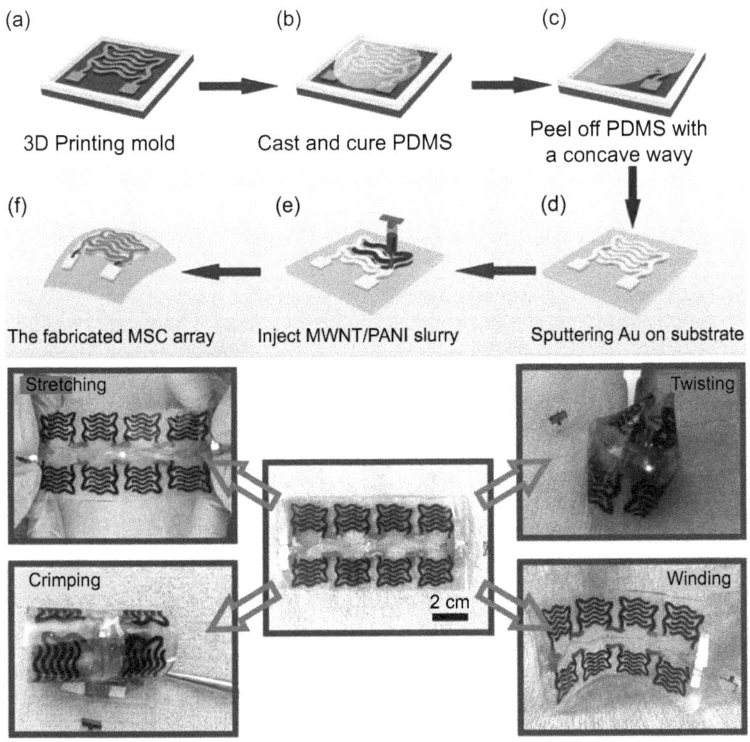

Figure 6.13 Electrochemcial performance and stretchability of stretchable on-chip micro-supercapacitors. Reprinted with permission from Ref. [53], Copyright 2017, John Wiley & Sons.

Komvopoulos et al. present another interesting example of stretchable MSCs, where they produced MSCs with honeycomb structures (Fig. 6.14) [54]. The specifically designed honeycomb-like structure of the PDMS substrate makes it possible to accommodate large deformation upon stretching, bending, and twisting without producing excessive deformation in MSCs. The stretchability of the MSC array system is controlled by the deformation of the honeycomb PDMS structure. The device capacitance remained stable even up to 150% stretching, 180° bending, and 60° twisting.

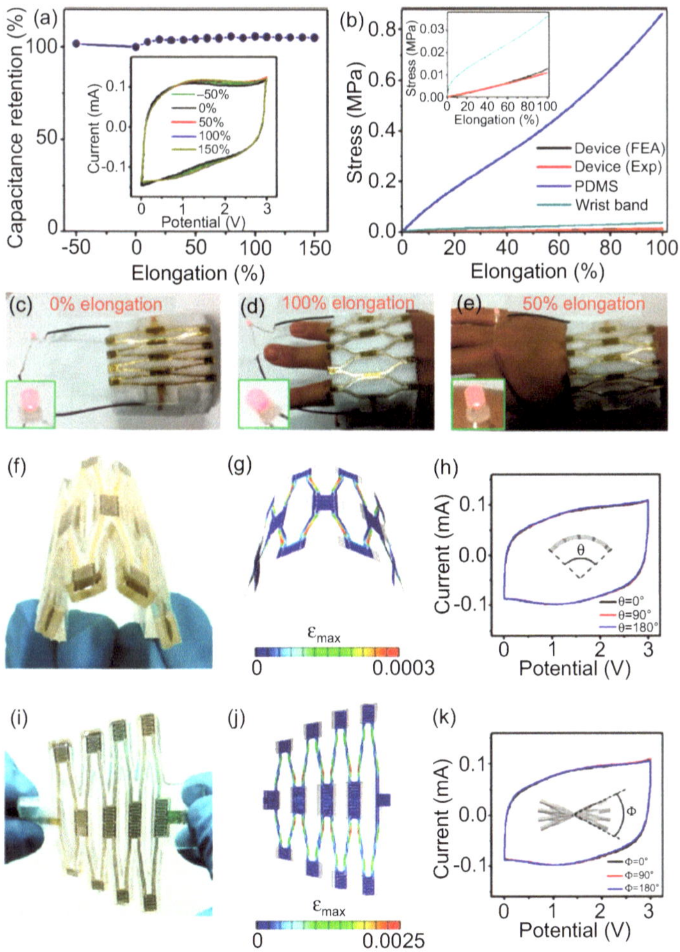

Figure 6.14 Highly stretchable on-chip micro-supercapacitor arrays with honeycomb structures. Reproduced with permission from Ref. [54], Copyright 2016, American Chemical Society.

For the practical application of stretchable on-chip MSCs, especially in the case of wearable electronic applications, sometimes it is desirable to encapsulate the entire MSCs to protect the human body from being contaminated and protect the devices from environmental impacts. Recently, Ha et al. reported the fabrication of an encapsulated stretchable array of stacked planar MSCs as a wearable energy storage device for waterproof applications, as demonstrated in Fig. 6.15 [55]. The design of a stretchable Ecoflex substrate with embedded long Galinstan microchannel interconnections exhibited high mechanical stability over various deformations of bending, twisting, and stretching. Encapsulation of the entire device using Ecoflex film made it useful as a wearable energy storage device, even in water.

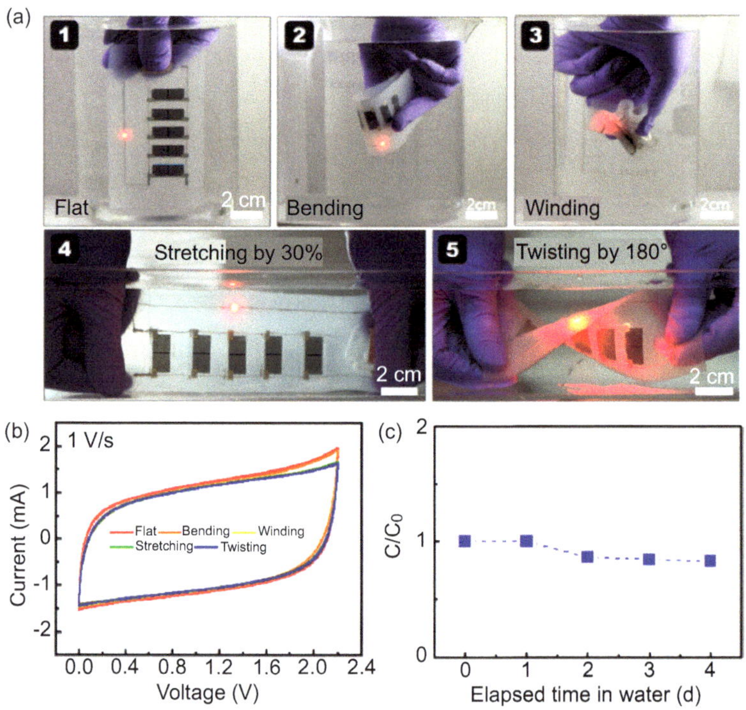

Figure 6.15 Encapsulated stretchable arrays of on-chip micro-supercapacitors. Reproduced with permission from Ref. [55], Copyright 2016, American Chemical Society.

6.4 Integrated Systems

With the rapid popularization of electronic products, wearable electronic devices, and portable devices, multifunctional integrated systems that can integrate devices with different functions in one device and miniaturize their volume are constantly evolving in recent years. Integrating flexible on-chip MSCs with other functional devices is one efficient way to get miniaturized integrated systems. This field developed very fast in the past several years, and many types of integrated systems with flexible on-chip MSCs have been obtained.

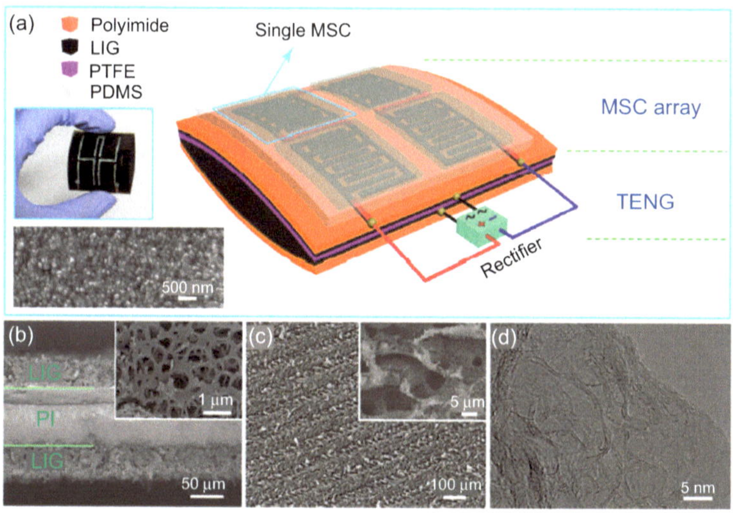

Figure 6.16 Flexible self-charging power unit with integrated micro-supercapacitors and triboelectric nanogenerators. Reproduced with permission from Ref. [56], Copyright 2015, Tsinghua University Press and Springer-Verlag Berlin Heidelberg.

Integrating flexible on-chip MSCs with triboelectric nanogenerators (TENG) provides ways to get flexible self-charging power unit to meet the increasing demand for flexible and efficient energy harvesting and storage units. Wang et al. recently produced such a power unit using a simple and cost-effective laser-engraving technique [56]. Figure 6.16 shows the integrated power unit. On-chip MSC arrays were fabricated using laser-induced graphene. A high degree of integration was realized in their work by double-faced laser engraving of a PI substrate; the two sides of which were used to fabricate TENG and MSC array, respectively, as shown in Fig. 6.16a. After tuning the structure of TENG, the output voltage of TENG was matched to MSC arrays. When working in energy-harvesting mode, the integrated power unit converted mechanical energy to electrical energy and stored it in the MSC arrays. Then in the energy storage mode, the unit can be used to power other functional devices. For example, two LEDs can be simultaneously powered, as demonstrated in the work.

With the successful fabrication of high performance flexible on-chip MSCs, it is then possible to get integrated sensing system. Recently, Ha et al. designed a photosensing system by integrating an MSC array with an SnO_2 nanowire UV sensor [57]. The MSC array was patterned by photolithography with MWNT/V_2O_5 nanowire composites as the active electrodes. By integrating SnO_2 nanowire UV sensor with the MSC array on the same substrate, the researchers were able to detect UV light by measuring the discharge rate of the MSC. Our group also fabricated an integrated sensing system driven by flexible on-chip MSCs. For example, using graphene-based MSCs, we fabricated integrated CdS nanowires visible photodetectors [34]. With rGO/Fe_2O_3 hollow nanospheres composite as electrodes, the performance of the flexible on-chip MSCs was further improved, which can be integrated with nanowire photodetector to get self-power-like on-chip integrated systems [43].

Flexible on-chip MSCs were also used as power units to integrate with other sensors, for example, gas sensors and pressure sensors [58, 39].

Later, Gao et al. introduced a wireless charging coil into the photodetecting system, as shown in Fig. 6.17 [59]. The whole system was built on the same substrate with integrated on-chip MSCs, photodetectors, and the wireless charging coil. In the system, the MSCs are of asymmetric configurations with MnO_2-PPy and V_2O_5-PANI as the electrodes. The photodetectors were built on $CH_3NH_3PbI_3$ nanowires. MSCs were first fully charged through the wireless charging coil, and then the photodetector was driven to detect light. The system exhibited comparable performance with the external power-driven system.

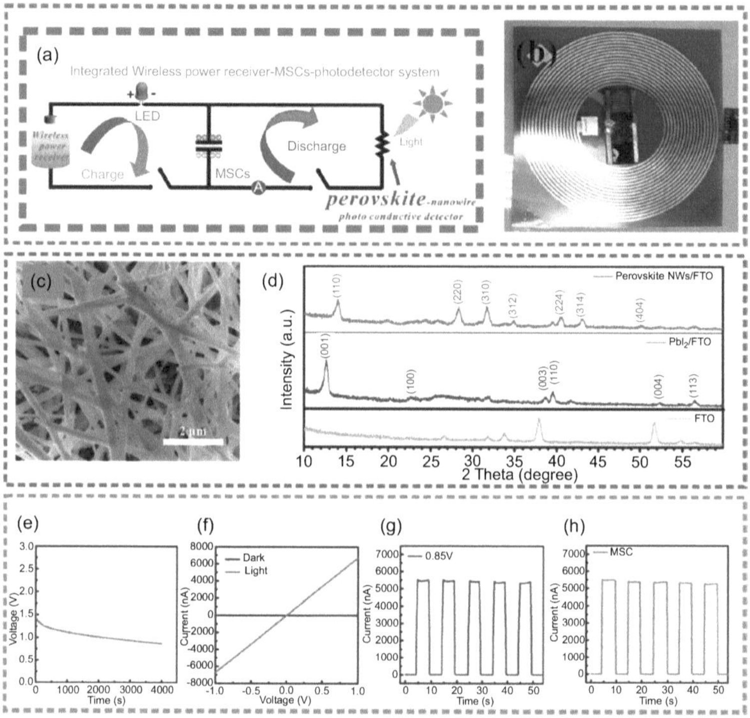

Figure 6.17 A flexible integrated system with micro-supercapacitors-powered photodetectors and wireless charging coil. Reproduced with permission from Ref. [59], Copyright 2016, American Chemical Society.

As researchers could fine-tune the device configurations and the performance of flexible on-chip MSCs, they could be used as power units to be integrated with complicated sensing systems [60–62]. Recently, Ha et al. developed a body-attachable and stretchable multisensor integrated with wirelessly rechargeable energy storage devices [60]. Figure 6.18a showed the photograph of the stretchable system, which consists of an RF power receiver, a nine-MSC array, a strain sensor, and UV/NO_2 gas sensor. The corresponding circuit diagram of the integrated system was depicted in Fig. 6.18b.

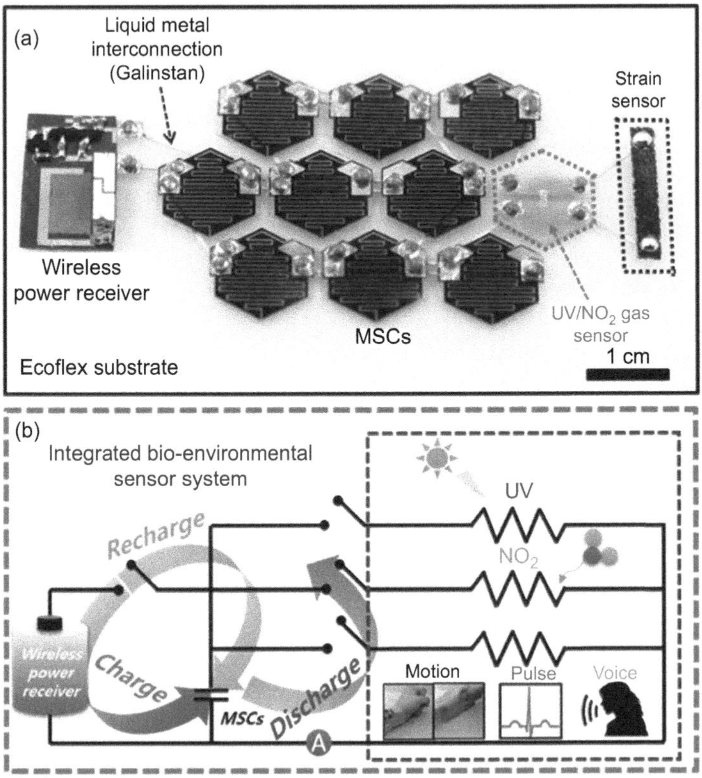

Figure 6.18 Body-attachable and stretchable multisensors with integrated on-chip micro-supercapacitors. Reproduced with permission from Ref. [60], Copyright 2015, John Wiley & Sons.

All these components are integrated on a single stretchable substrate with an embedded liquid metal interconnection [Galinstan, a eutectic alloy liquid metal consisting of gallium (68.5%), indium (21.5%), and tin (10%)]. The power receiver consists of a receiving antenna and six voltage multipliers. Nine hexagonal MSCs with interdigital electrode are connected in parallel to form a close-packed array to increase the fill factor. The active electrode material used to fabricate the MSCs is MWNTs. And the MWNTs/SnO_2 nanowire hybrid film was used to fabricate the NO_2 gas sensors and UV sensors. The whole integrated multisensory system enabled the operation of multisensors by integrated MSCs that could be repeatedly recharged with the integrated wireless RF power receiver. The fabricated stretchable FGF sensor attached to the skin could successfully detect biosignals such as a neck pulse, saliva swallowing, voice, and body movement. The MWNT/SnO_2 nanowire hybrid film sensor exhibited a stable sensing signal upon exposure to both NO_2 gas and UV light even under a uniaxial strain up to 50%. Such an interesting work showed the great potential of highly compacted integrated system for application to next-generation body-attached healthcare and environmental sensor devices.

6.5 Conclusion

In conclusion, this chapter reviews the recent progress made on flexible on-chip MSCs. Techniques used to fabricate the flexible on-chip MSCs were highlighted in detail, focusing on photolithography-related process, laser-scribing method, and printing method. In fact, besides these mentioned methods, there are also several other types of techniques that were developed to get flexible MSCs. Sometimes, combining several types of fabrication techniques is necessary to get flexible MSCs with desired configurations and electrochemical performance.

The successful fabrication of flexible on-chip MSCs makes it possible to fabricate integrated systems driven by these MSCs. Several types of integrated systems, including gas-sensing systems, photodetecting systems, and multifunctional systems, were introduced. The integrated systems have the features of highly compacted integrations, small sizes, thin thickness, and comparable performance with the conventional system driven by external power units. All these demonstrations confirmed that flexible on-chip MSCs are ideal power units for future wearable and flexible electronics.

For future research in this field, although many kinds of active materials have been used to fabricate flexible on-chip MSCs, the performance of MSCs still needs further improvement. Controlled synthesis of active materials on a large scale is still very important and needs further investigation. For fabricating MSCs, a new low-cost, scalable, and efficient technique is vital for their real applications. Besides, further reduction in the size of integrated systems and integration of functional devices on the same chip to get complicated systems for multifunctional applications are also important and need further study.

Acknowledgments

The work was supported by the National Natural Science Foundation of China (NSFC, Grant No. 61625404, 51672308), and the Key Research Program of Frontier Sciences, CAS (QYZDY-SSW-JWC004). The authors are indebted for the kind permission from the corresponding publishers and authors to reproduce their materials, especially figures, used in this chapter.

References

1. Lee, J. W., Xu, S., Lee, S., Jang, K. I., Yang, Y., Banks, A., Yu, K. J., Kim, J., Xu, S., Ma, S., Jang, S. W., Won, P., Li, Y., Kim, B. H., Choe, J. Y., Huh, S., Kwon, Y. H., Huang, Y., Paik, U., and Rogers, J. A. (2016). Soft, thin skin-mounted power management systems and their use in wireless thermography, *Proc. Natl. Sci.*, **113**, pp. 6131–6136.
2. Gao, W., Emaninejad, S., Nyein, H. Y. Y., Challa, S., Chen, K., Peck, A., Fahad, H. M., Ota, H., Shiraki, H., Kiriya, D., Lien, D. H., Brooks, G. A., Davis, R. W., and Javey, A. (2016). Fully integrated wearable sensor arrays for multiplexed in situ perspiration analysis, *Nature*, **529**, pp. 509–514.
3. Park, S. and Jayaraman, S. (2013). Smart textiles: Wearable electronic systems, *MRS Bull.*, **28**, pp. 585–591.
4. Shim, B. S., Chen, W., Doty, C., Xu, C., and Kotov, N. A. (2008). Smart electronic yarns and wearable fabrics for human biomonitoring made by carbon nanotube coating with polyelectrolytes, *Nano Lett.*, **8**, pp. 4151–4157.
5. Wang, L., Chen, D., Jiang, K., and Shen, G. Z. (2017). New insights and perspectives into biological materials for flexible electronics, *Chem. Soc. Rev.*, **46**, pp. 6764.
6. Liu, Z., Xu, J., Chen, D., and Shen, G. Z. (2015). Flexible electronics based on inorganic nanowires, *Chem. Soc. Rev.*, **44**, pp. 161–192.
7. Chen, S., Lou, Z., Chen, D., and Shen, G. Z. (2018). An artificial flexible visual memory system based on ultraviolet-motivated memristor, *Adv. Mater.*, **30**, pp. 1705400.

8. Li, L., Lou, Z., and Shen, G. Z. (2018). Flexible broadband image sensors with SnS quantum dots/Zn_2SnO_4 nanowires hybrid nanostructures, *Adv. Funct. Mater.*, **18**, pp. 1705389.

9. Lou, Z., Chen, S., Wang, L., Shi, R., Li, L., Jiang, K., Chen, D., and Shen, G. Z. (2017). Ultrasensitive and ultraflexible e-skins with dual functionalities for wearable electronics, *Nano Energy*, **38**, pp. 28–35.

10. Wang, X., Liu, B., Liu, R., Wang, Q., Hou, X., Chen, D., Wang, R., and Shen, G. Z. (2014). Fiber-based flexible all-solid-state asymmetric supercapacitors for integrated photodetecting system, *Angew. Chem. Int. Ed.*, **53**, pp. 1849–1853.

11. Liu, B, Zhang, J., Wang, X., Chen, G., Chen, D., Zhou, C., and Shen, G. Z. (2012). Hierarchical 3D $ZnCo_2O_4$ nanowire arrays/carbon cloth anodes for a novel class of high-performance flexible lithium-ion batteries, *Nano Lett.*, **12**, pp. 3005–3011.

12. Wang, X., Liu, B., Wang, Q., Song, W., Hou, X., Cheng, Y. B., and Shen, G. Z. (2013). Three-dimensional hierarchical $GeSe_2$ nanostructures for high performance flexible all-solid-state supercapacitors, *Adv. Mater.*, **25**, pp. 1479–1486.

13. Wang, X., Lu, X., Liu, B., Chen, D., Tong, Y., and Shen, G. Z. (2014). Flexible energy storage devices: Design consideration and recent progress, *Adv. Mater.*, **26**, pp. 4763–4782.

14. Wang, X., Jiang, K., and Shen, G. Z. (2015). Flexible fiber energy storage and integrated devices: Recent progress and perspectives, *Mater. Today*, **18**, pp. 265–272.

15. Liu, B., Zhang, J. G., and Shen, G. Z. (2016). Pursuing two-dimensional nanomaterials for flexible lithium-ion batteries, *Nano Today*, **11**, pp. 82–97.

16. Wen, L., Li, F., and Cheng, H. M. (2016). Carbon nanotubes and graphene for flexible electrochemical energy storage: From materials to devices, *Adv. Mater.*, **28**, pp. 4306–4337.

17. Wang, X. L. and Shi, G. (2015). Flexible graphene devices related to energy conversion and storage, *Energy Environ. Sci.*, **8**, pp. 790–823.

18. Liu, Q. C., Xu, J. J., Xu, D., and Zhang, X. B. (2015). Flexible lithium-oxygen battery based on a recoverable cathode, *Nat. Commun.*, **6**, pp. 7892.

19. Wang, X., Liu, B., Xiang, Q., Wang, Q., Hou, X., Chen, D., and Shen, G. Z. (2014). Ultralong-life and high-rate web-like $Li4Ti_5O_{12}$ anodes for high-performance flexible lithium-ion battery, *Nano Res.*, **7**, pp. 1073–1082.

20. He, Y., Matthews, B., Wang, J., Song, L., Wang, X., and Wu, G. (2018). Innovation and challenges in materials design for flexible rechargeable batteries: From 1D to 3D, *J. Mater. Chem.* A, **6**, pp. 735–753.

21. Cha, H., Kim, J., Lee, Y., Cho, J., and Park, M. (2017). Issues and challenges facing flexible lithium-ion batteries for practical application, *Small*, DOI:10.1002/smll.201702989.

22. Du, C. F., Liang, Q., Luo, Y., Zheng, Y., and Yan, Q. (2017). Recent advances in printable secondary batteries, *J. Mater. Chem. A*, **5**, pp. 22442–22458.

23. Simon, P. and Gogotsi, Y. (2008). Materials for electrochemical capacitors, *Nat. Mater.*, **7**, pp. 845–854.

24. Zhang, L. and Zhao, X. S. (2009). Carbon-based materials for supercapacitor electrodes, *Chem. Soc. Rev.*, **38**, pp. 2520–2531.

25. Peng, X., Peng, L. L., Wu, C. Z., and Xie, Y. (2014). Two dimensional nanomaterials for flexible supercapacitors, *Chem. Soc. Rev.*, **44**, pp. 3303–3323.

26. Yan, J., Wang, Q., Wei, T., and Fan, Z. J. (2014). Recent advances in design and fabrication of electrochemical supercapacitors with high energy densities, *Adv. Energy Mater.*, **4**, pp. 1300816.

27. Beguin, F., Presser, V., Balducci, A., and Frackowiak, E. (2014). Carbons and electrolytes for advanced supercapacitors, *Adv. Mater.*, **26**, pp. 2219–2251.

28. Cao, Z. Y. and Wei, B. Q. (2013). A perspective: Carbon nanotube macro-films for energy storage, *Energy Environ. Sci.*, **6**, pp. 3183–3201.

29. Zhai, Y. P., Dou, Y. Q., Zhao, D. Y., Fulvio, P. F., Mayes, R. T., and Dai, S. (2011). Carbon materials for chemical capacitive energy storage, *Adv. Mater.*, **23**, pp. 4828–4850.

30. Beidaghi, M. and Gogotsi, Y. (2014). Capacitive energy storage in micro-scale devices: Recent advances in design and fabrication of microsupercapacitors, *Energy Environ. Sci.*, **7**, pp. 867–884.

31. Chmiola, J., Largeot, C., Taberna, P. L., Simon, P., and Gogotsi, Y. (2010). Monolithic carbide-derived carbon films for micro-supercapacitors, *Science*, **328**, pp. 480–483.

32. Kyeremateng, N. A., Brousse, T., and Pech, D. (2016). Microsupercapacitors as miniaturized energy-storage components for on-chip electronics, *Nat. Nanotechnol.*, **12**, pp. 7–15.

33. Wu, Z. S., Parvez, K., Feng, X. L., and Mullen, K. (2013). Graphene-based in-plane micro-supercapacitors with high power and energy densities, *Nat. Commun.*, **4**, pp. 2487.

34. Xu, J. and Shen, G. Z. (2015). A flexible integrated photodetector system driven by on-chip microsupercapacitors, *Nano Energy*, **13**, pp. 131–139.

35. Kurra, N., Jiang, Q., and Alshareef, H. N. (2015). A general strategy for the fabrication of high performance microsupercapacitors, *Nano Energy*, **16**, pp. 1–9.

36. Lin, J., Zhang, C., Yan, Z., Zhu, Y., Peng, Z., Hauge, R. H., Natelson, D., and Tour, J. M. (2013). 3-dimensional graphene carbon nanotube carpet-based microsupercapacitors with high electrochemical performance, *Nano Lett.*, **13**, pp. 72–78.

37. Hu, H. B., Pei, Z. B., and Ye, C. H. (2015). Recent advances in designing and fabrication of planar micro-supercapacitors for on-chip energy storage, *Energy Storage Mater.*, **1**, pp. 82–102.

38. Dinh, T. M., Armstrong, K., Guay, D., and Pech, D. (2014). High-resolution on-chip supercapacitors with ultra-high scan rate ability, *J. Mater. Chem. A.*, **2**, pp. 7170–7174.

39. Li, L., Lou, Z., Han, W., and Shen, G. Z. (2016). Flexible in-plane microsupercapacitors with electrospun $NiFe_2O_4$ nanofibers for portable sensing applications, *Nanoscale*, **8**, pp. 14986–14991.

40. Sun, L, Wang, X., Zhang, K., Zhou, J., and Zhang, Q. (2016). Metal-free SWNT/carbon/MnO_2 hybrid electrode for high performance coplanar micro-supercapacitors, *Nano Energy*, **22**, pp. 11–18.

41. Liu, Z., Tian, X., Xu, X., He, L., Yan, M., Han, C., Li, Y., Yang, W., and Ma, L. (2017). Capacitance and voltage matching between MnO_2 nanoflake cathode and Fe_2O_3 nanoparticle anode for high-performance asymmetric micro-supercapacitors, *Nano Res.*, **10**, pp. 2471–2481.

42. Yun, J., Kim, D., Lee, G., and Ha, J. S. (2014). All-solid-state flexible micro-supercapacitor arrays with patterned graphene/MWNT electrodes, *Carbon*, **79**, pp. 156–164.

43. Gu, S., Lou, Z., Li, L., Chen, Z., Ma, X., and Shen, G. Z. (2016). Fabrication of flexible reduced graphene oxide/Fe_2O_3 hollow nanospheres based on-chip micro-supercapacitors for integrated photodetecting applications, *Nano Res.*, **9**, pp. 424–434.

44. El-kady, M. F. and Kaner, R. B. (2013). Scalable fabrication of high-power graphene micro-supercapacitors for flexible and on-chip energy storage, *Nat. Commun.*, **4**, pp. 1475.

45. Yadav, P., Basu, A., Suryawanshi, A., Game, O., and Ogale, S. (2016). Highly stable laser-scribed flexible planar microsupercapacitor using

mushroom derived carbon electrodes, *Adv. Mater. Interfaces*, **1**, pp. 1600057.

46. Gao, W., Singh, N., Song, L., Liu, Z., Reddy, A. L. M., Ci, L. J., Vajtai, R., Zhang, Q., Wei, B. Q., and Ajayan, P. M. (2011). Direct laser writing of micro-supercapacitors on hydrated graphite oxide films, *Nat. Nanotechnol.*, **6**, pp. 496–500.
47. Basu, A., Roy, K., Sharma, N., Nandi, S., Vaidhyanathan, R., Rane, S., Rode, C., and Ogale, S. (2016). CO_2 laser direct written MOF-based metal-decorated and heteroatom-doped porous grapheme for flexible all-solid-state microsupercapacitor with extremely high cycling stability, *ACS Appl. Mater. Interfaces*, **8**, pp. 31841–31848.
48. Wang, S., Wu, Z. S., Zheng, S., Zhou, F., Sun, C., Cheng, H. M., and Bao, X. (2017). Scalable fabrication of photochemically reduced graphene-based monolithic micro-supercapacitors with superior energy and power densities, *ACS Nano*, **11**, pp. 4283–4291.
49. Li, L., Zhang, J., Peng, Z., Li, Y., Gao, C., Ji, Y., Ye, R., Kim, N. D., Zhong, Q., Yang, Y., Fei, H., Ruan, G., and Tour, J. M. (2016). High-performance pseudocapacitive microsupercapacitors form laser-induced graphene, *Adv. Mater.*, **28**, pp. 838–845.
50. Wang, Y., Shi, Y., Zhao, C. X., Wong, J. I., Sun, X. W., and Yang, H. Y. (2014). Printed all-solid flexible microsupercapacitors: Towards the general route for high energy storage devices, *Nanotechnology*, **25**, pp. 094010.
51. Li, J., Delekta, S. S., Zhang, P., Yang, S., Lohe, M. R., Zhuang, X., Feng, X. L., and Ostling, M. (2017). Scalable fabrication and integration of graphene microsupercapacitors through full inkjet printing, *ACS Nano*, **11**, pp. 8249–8256.
52. Hu, H., Zhang, K., Li, S., Ji, S., and Ye, C. (2014). Flexible, in-plane, and all-solid-state micro-supercapacitors based on printed interdigital Au/polyaniline network hybrid electrodes on a chip, *J. Mater. Chem. A.*, **2**, pp. 20916–20922.
53. Li, L., Lou, Z., Han, W., Chen, D., Jiang, K., and Shen, G. Z. (2017). Highly stretchable micro-supercapacitor arrays with hybrid MWCNT/PANI electrodes, *Adv. Mater. Technol.*, **2**, pp. 1600282.
54. Pu, J., Wang, X., Xu, R., and Komvopoulos, K. (2016). Highly stretchable microsupercapacitor arrays with honeycomb structures for integrated wearable electronic system, *ACS Nano*, **10**, pp. 9306–9315.
55. Kim, H., Yoon, J., Lee, G., Paik, S. H., Choi, G., Kim, D., Kim, B. M., Zi, G., and Ha, J. S. (2016). Encapsulated, high-performance, stretchable array of stacked planar micro-supercapacitors as waterproof wearable energy storage devices, *ACS Appl. Mater. Interfaces*, **8**, pp. 16016–16025.

56. Luo, J., Fan, F. R., Jiang, T., Wang, Z., Tang, W., Zhang, C., Liu, M., Cao, G., and Wang, Z. L. (2015). Integration of micro-supercapacitors with triboelectric nanogenerators for a flexible self-charging power unit, *Nano Res.*, **8**, pp. 3934–3943.

57. Kim, D., Yun, J., Lee, G., and Ha, J. S. (2014). Fabrication of high performance flexible microsupercapacitor arrays with hybrid electrodes of MWNT/V_2O_5 nanowires integrated with a SnO_2 nanowire UV sensors, *Nanoscale*, **6**, pp. 12034–12041.

58. Yun, J., Lim, Y., Jang, G. N., Kim, D., Lee, S. J., Park, H., Hong, S. Y., Lee, G., Zi, G., and Ha, J. S. (2016). Stretchable patterned graphene gas sensor driven by integrated micro-supercapacitor array, *Nano Energy*, **19**, pp. 401–414.

59. Yue, Y., Yang, Z, Liu, N., Liu, W., Zhang, H., Ma, Y., Yang, C., Su, J., Li, L., Long, F., Zou, Z., and Gao, Y. (2016). A flexible integrated system containing a microsupercapacitor, a photodetector and a wireless charging coil, *ACS Nano*, **10**, pp. 11249–11257.

60. Kim, D., Kim, D., Lee, H., Jeong, Y. R., Lee, S. J., Yang, G., Kim, H, Lee, G., Jeong, S., Zi, G., Kim, J., and Ha, J. S. (2016). Body-attachable and stretchable multisensors integrated with wirelessly rechargeable energy storage devices, *Adv. Mater.*, **28**, pp. 748–756.

61. Li, L., Fu, C., Lou, Z., Chen, S., Han, W., Jiang, K., Chen, D., and Shen, G. Z. (2017). Flexible planar concentric circular micro-supercapacitor arrays for wearable gas sensing applications, *Nano Energy*, **41**, pp. 261–268.

62. Ai, Y., Lou, Z., Wang, Z. M., Jiang, K., and Shen, G. Z. (2017). All rGO-on-PVDF-nanofibers based self-powered electronic skins, *Nano Energy*, **35**, pp. 121–127.

Chapter 7

Flexible and Stretchable Sensors

Tie Li, Yudong Cao, Chunyan Qu, and Ting Zhang
i-Lab, Suzhou Institute of Nano-Tech and Nano-Bionics (SINANO), Chinese Academy of Sciences (CAS), 398 Ruoshui Road, Suzhou 215123, China
tli2014@sinano.ac.cn

Currently, flexible electronics provide the most exciting frontier for next-generation devices in both circuits and wearable/portable devices due to their unique properties, such as strong stretchability and flexibility, which allows the board to conform to any aspiring shapes for space saving. Specially, smart and multifunctional, flexible and stretchable sensors broaden the place where electronic devices would be used, making the device usable in wearable health-monitoring system, cost-effective disease diagnosis in telemedicine, and medicine robots and prosthetics. To meet those special characters and requirements, in the past 10 years, various strategies of structural design and functional materials have been employed for the realization of flexibility and stretchability in sensors. To showcase an overview of the development of flexible electronics that have advanced in the past years and have made great changes in our

Flexible and Stretchable Electronics: Materials, Designs, and Devices
Edited by Run-Wei Li and Gang Liu

ISBN 978-981-4800-46-4 (Hardcover), 978-0-429-05890-5 (eBook)
www.jennystanford.com

daily life, we discuss classes of structure, materials, and applications covering several critical aspects of flexible and stretchable sensors.

7.1 Introduction

Flexible and/or stretchable electronics, such as stretchable transistors, smart flexible sensors, stretchable energy harvesters and energy storage devices, have intensively obtained a great deal of attraction due to their high performance and elastic mechanical responses, which possess vast potential demand for flexible, lightweight, and mobile electronics that can be used in various areas containing robotic sensory skins, healthcare monitoring, bio-integrated devices, human–machine interfaces, wearable communication devices, and so on [1–8]. Especially, the rapid advances in the design of various flexible and/or stretchable sensors, which can be integrated in curved surfaces such as touchscreen, flexible displays, and even human body, have tremendously broadened the scope of flexible electronics to new classes of soft electronic systems [9, 10]. For example, these flexible sensors could help the electronic devices understand how real-world objects "feel" during interactions and also obtain biosignals such as finger touching and body motions [11, 12]. Further, compliant wearable flexible sensor systems have the potential to interface better with human skin, thus improving the sensitivity of detection of health indicators [13, 14]. In addition, flexible and/or stretchable sensors in large-area arrays have also been developed to mimic the properties of natural existed structure or feature with multiple functionalities, which may enable promising applications in emerging Internet of Things such that people, processes, data, and devices are integrated and connected to improve the quality of life [15–17].

The most crucial challenges in the development of flexible and stretchable sensors are the simultaneous achievement of both favorable mechanically durable materials and robust flexible and stretchable substrates, deformable electrodes and circuits, novel processing methods, and system integration [18]. In the past decade, rapid developments in new nanomaterials, new concepts in structural design, fabrication techniques, and original applications have contributed to significant progress in the achievement of flexible and stretchable sensors [19]. Specifically, constituent materials and devices in stretchable integrated systems must be designed so that their mechanical and electrical functionality is preserved under high strain values [20]. For example, the emerging nanostructured material, graphene, which possesses unique mechanical, electrical, and optical properties, is an attractive candidate for combination with flexible and stretchable polymeric substrates such as poly(ethylene terephthalate) (PET), polyethylenimine (PEI), or poly(dimethylsiloxane) (PDMS) for producing various flexible and/or stretchable sensors such as flexible pressure sensor, stretchable fiber strain sensor, flexible optoelectronic and energy-harvesting devices, and so on [21–28]. Moreover, the combination of the aforementioned graphene-based flexible and stretchable sensors with low-power silicon-based electronics can also lead to the development of new wearable electronics, which are driven by improved three-dimensional (3D) interpenetration ability of soft electronic systems and flexible or stretchable electronic systems [29, 30].

To provide an overview of flexible and stretchable electronics that have advanced and made great changes in our daily life in the past few years, we present this chapter covering the following critical aspects of flexible and stretchable sensors.

7.2 Classes of Architectural Strategies for Flexible and Stretchable Sensors

To realize high flexibility and stretchability, the authors investigated the fabrication of stretchable electronic devices and device arrays via various approaches, among which the most popular strategies include the following: (1) "Wavy structural configuration" methods that use rigid semiconductor nanowires, nanomembranes, and nanoribbons designed as "wavy" shapes obtained by geometric engineering, which can accommodate large applied strains without fracturing the materials relying on nonlinear buckling phenomena of a thin stiff layer on a soft substrate. For instance, two-dimensional (2D) horseshoe metal or 3D metal interconnects can effectively sustain maximum (>100%) elongation [31]. (2) "Stretchable interconnects" methods that use combinations or stacking strategy to interconnect stretchable components with non-stretchable components or other stretchable components to obtain integral structural stretchability by well-known processes. Various processes are employed to achieve the stretchable interconnects on elastic substrates, which include transfer and screen printing, vacuum evaporation, photolithographic patterning, and so on. However, this strategy is limited to low-aspect-ratio features such as 2D topography that should be supported by substrates [32]. (3) "Wrinkle structure" method that uses stretching/releasing of pre-strained oxide/polymer bilayers and swelling films confined on a rigid substrate to obtain surface wrinkle patterns to realize mechanical stretchability. When the system is endured in the non-equilibrium and equilibrium status, some soft materials, including hydrogels, elastomers, and elastoplastic polymers, display elastic instabilities in nature, which can disappear/reappear by an external applied force [33]. In practical work, the aforementioned methods are often reciprocally used to avoid instability: For instance, "wavy configuration" is usually employed with "stretchable interconnects" to ensure integrated mechanical stretchability. Here, we summarize the architectural strategies for flexibility and stretchability from the view of structural configuration features of fiber, plane, and blocks structures depending on the morphological features of elastomeric substrates, respectively [34].

7.2.1 One-Dimensional Fibrous Configuration

Flexible and stretchable one-dimensional (1D) fibrous configuration may be ideal for realizing flexibility and stretchability because elastic fibers are lighter in weight and smaller in size with high length–diameter ratio, which can be discretionarily intertwined with themselves or woven with clothes to prepare electronic fabrics that conform well to human bodies such as wrists and fingers. Additionally, the low dimensionality and a relatively high freedom of fibrous configuration make it the building blocks for higher dimensional structures such as 2D woven fabric. This can further increase the stretchability of materials. Hence, the fiber structure may be robust candidates for the next-generation flexible electronics such as sensors, actuators, and artificial muscles [35].

There are two typical fibrous configurations to achieve flexibility and stretchability: One is the sheath–core structure, when employing the pre-strain strategy, arising from the heterogeneous stress distribution existed in the interior of composite materials. In a typical study reported by Baughman et al., they fabricated a highly stretchable (up to 1320%) sheath–core conducting fiber sensor with distinct short- and long-period sheath wrinkle structure created by wrapping carbon nanotube sheets oriented in the fiber direction on stretched rubber fiber cores, which enable a resistance change of less than 5% for a 1000% stretch (Fig. 7.1a) [36]. To downsize the low electrical conductivity effect caused by the high core-to-sheath volume ratio, they further prepared a superplastic conducting sheath–core fiber sensor as thin as 40 μm, showing high sensitivity to applied strains that can exceed 300% [37]. On the other hand, many other materials and polymer substrates can also be used to construct the sheath–core structure by the pre-strain method. Moon et. al. prepared a stretchable and foldable conductive cord based on Ag NWs depositing on the pre-strained polyurethane cord, which resulted in the conductive fiber exhibiting highly stable electroconductivity even under strain values of 167% with favorable stretching cyclic performance [38]. To satisfy the application of sensing high strain (>5%) with excellent flexibility and stretchability in human-friendly interactive electronics, Hu et al. reported a core–sheath fiber sensor composed of graphene@PVA hollow nanotubes by massively fabricating from ultralong chemical-vapor-deposition-grown graphene bundles, which displays superior conductivity and mechanical properties [39]. Although such carbon nanotubes, Ag NWs, and graphene-based functional materials have been made to fabricate flexible and stretchable fibrous sensors, it is difficult to simultaneously satisfy high stretchability, conductivity, and sensitivity. Liu et. al presented a kind of core–shell fiber sensor with multiscale wrinkled microstructure by simply writing Ag NWs ink on pre-strained commercial polyurethane fiber, showing high stretchability (400%), durability (over 1200 s ultrasonic treatment), and conductivity (10^{-4} to 10^{-5} Ωcm), which obtained desirable fast response (35 ms), detection limit (10 mg), and sensitivity (0.12 kPa^{-1} and 0.012 rad^{-1}) to pressure and bending deformations [40].

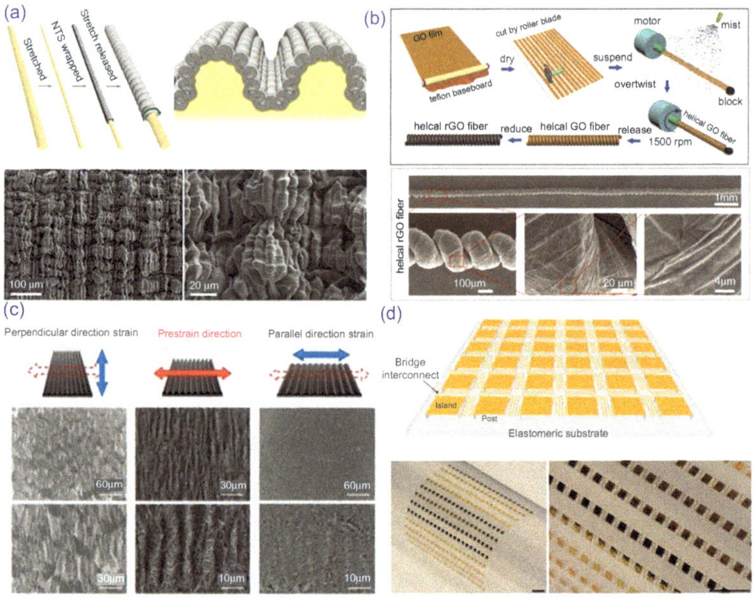

Figure 7.1 (a) Steps in the fabrication and structural illustration of an NTSm@ fiber and its low- and high-resolution SEM images showing period buckles at 100% applied strain. Reprinted with permission from Ref. [36], Copyright 2015, The American Association for the Advancement of Science. (b) Illustration of the fabrication process and SEM images of helical graphene oxide (GO) and reduced graphene oxide (rGO) fibers. Reprinted with permission from Ref. [41], Copyright 2016, The Royal Society of Chemistry. (c) SEM image showing the surface of the Ag NW network strain sensor when strain is applied in the perpendicular and parallel direction. Reprinted with permission from Ref. [49], Copyright 2015, American Chemical Society. (d) Schematic and optical image of the island-bridge structure with serpentine geometry. Reprinted with permission from Ref. [53], Copyright 2013, The Royal Society of Chemistry.

The second fibrous configuration is the twisted or helical structures that employ nature's gyroscopic principle. For the intractable mechanical properties, many materials cannot form sheath-core structure under the above-mentioned pre-strain strategy, hence, novel stretchable methods that can produce bent or twisted structures need to be developed. Previously, only straight graphene oxide yarns have been limitedly applied as stretchable nano-textiles; the fibers made from graphene or carbon nanotubes are strong and conductive enough to encode helical structures that may render high stretchability. Hua et al. reported self-coiled helical graphene oxide fibers with uniformly tunable arranged loops by a freestanding strip-scrolling process under controlled environmental humidity (Fig. 7.1b), enabling fabrication of a fiber-shaped stretchable temperature sensor, which works at large strains (up to 50%) and high temperature (up to 300°C) with a reliable deformation-independent response [41]. Liquid metal is an inartificial functional material that has emerged in recent years, and the low toxicity and viscosity allows it to flow in response to deformation. Composites with decoupled electrical and mechanical properties can be obtained by embedding liquid metal in elastomer, which provides a promising method to create flexible and stretchable sensors. Dickey et al. injected liquid metal into the core of hollow and extremely stretchable elastomeric intertwined fibers with a helix structure to fabricate stretchable capacitive sensors of torsion, strain, and touch [42].

Except the twistable types, the two-ply or parallel constitutions have also been employed to prepare stretchable working devices, especially for the fibers integrated with electrochemical sensors and fibrous energy electrodes. Baughman et al. introduced a weavable fiber with the multifunctional properties of redox supercapacitors and strain sensors, which contained a sandwiched dielectric layer between buckled carbon nanotube electrodes, generating strain sensing performance of approximately 115.7% and 26% capacitance changes during 200% stretching/1700 rad/m giant twist and electrochemical energy storage of high linear and areal capacitances (2.38 mF/cm and 11.88 mF/cm^2) with mechanical stability, respectively [43]. On the other hand, while the emerging stretchable electronic sensors have been demonstrated as promising wearable functional devices, challenges in achieving highly stretchable and self-powered fiber-like sensors still exist. Zhou et al. fabricated an active fiber-based self-powered strain sensor system depending on coiling a fiber-based generator around a stretchable silicone fiber, which shows the ability to detect strain up to 25% and was used to detect finger motion states [44]. Gong et al. developed a stretchable sheath–core structural triboelectric fiber with advanced built-in wavy structure design to serve as a self-powered multifunctional kinematic fiber-like sensor, exhibiting an ultrahigh working strain (100%) and displaying high sensitivity in response to not only stretching but also bending and compressing [45]. These studies on fibrous devices provide new opportunities for flexible and stretchable sensors with full potential in human wearable electronics.

7.2.2 Two-Dimensional Planar Configuration

Instabilities such as sea wave surface, cloud waves, sand dunes, and wrinkles on human skin commonly exist in the natural environments, which are present in materials systems in compatible states of equilibrium and non-equilibrium [46]. The elastic instabilities, which can disappear/reappear under an outside force and usually considered failure mechanisms in soft materials such as elastoplastic polymers and hydrogels, have been employed to fabricate complex patterns to assemble novel flexible and stretchable electronic devices. When processed as planar 2D wrinkled configuration, the physical properties of elastoplastic instability materials will deviate substantially from their bulk counterparts, which can produce the wanted flexibility and stretchability to self-organize on the flexible film surface [47].

In the past, many experimental and theoretical researches have been used to comprehend and manipulate various elastic instabilities such as wrinkling and crumpling to prepare flexible and stretchable sensors. Duan et al. presented a stretchable piezoelectric sensor by direct-writing straight nanofibers onto a pre-strained poly(dimethylsiloxane) (PDMS) substrate using mechanoelectrospinning method, resulting in a tuned wrinkled morphology with a highly synchronized serpentine structure array, which exhibits stable piezoelectric performance up the failure strain of PDMS (about 110% in this study) [48]. Additionally, for most of the previously reported strain sensors, it is difficult to realize multiaxial strain sensing for the strong coupled electrical responses between the major axis and the corresponding perpendicular direction caused by Poisson's effect, especially under large or accurate strain condition. To exploit a novel technique to measure multidimensional strain, a stretchable and sensitive multidimensional strain sensor possessing a pre-strained wrinkle structure is adopted. Kim et al. presented a highly stretchable and sensitive strain sensor capable of detecting in real time "skin-like" multiaxial strain loadings using pre-strained Ag NW percolation networks (Fig. 7.1c) [49]. A similar tactic was employed by Shin et al. to assemble a wavy-structured stretchable conductor using a simple pre-straining approach based on polydimethylsiloxane (PDMS)-infiltrated vertically aligned carbon nanotube (VACNT) films, showing a stable resistance change ratio and of high reversibility [50]. Further, uniform large-area patterning with three-dimensional wrinkles can be obtained by controlling the conformal contact between material and substrate during strain relief when a soft skin layer exists. For instance, a multiaxial strain stretchable configuration formed along two axes was fabricated by Lee et al. on pre-strained thermoplastic sheets with surface hierarchical patterning of multiscale graphene wrinkles with predetermined orientations enabled by varying thicknesses [51].

On the other hand, a popular and important approach to realize flexible and stretchable electronic devices involves an island–bridge layout, in which inorganic semiconductors reside at the islands and the electrical interconnects form the bridges. A serpentine shape composed of arcs and straight lines is introduced to fully bond such compositions to the pre-strained elastomer substrate, which can yield significantly enhanced stretchability of more than two times to offer improved stretchability for a given space between adjacent islands. In this field, Rogers, Huang and their coworkers have made significant achievements and opened a doorway to the new exciting field of technological innovation. Their teams demonstrated the possible mechanisms and models of various material and geometric parameters to predict the wavelength of the serpentine wrinkling shape and explain the thickness dependence of buckling behaviors [52]. For instance, they presented a systematic and thorough designed strategy of the stretchable serpentine microstructures via buckling physics models, finite element computations, and quantitative experiments, demonstrating its routes for application in an ultra stretchable electrode biaxial stretchability as large as about 170% (Fig. 7.1d) [53]. Further, they designed and built strain-limiting substrate materials matched with serpentine structural film mesh to fabricate composite stretchable and flexible connected substrates through the bilinear stress–strain relation, offering favorable sensitivity at not only small strain due to a low elastic modulus but also large strain attributed to a high tangent modulus (e.g., >1000 times the elastic modulus) [54]. Based on the serpentine microstructures, Wang et al. underpinned the wearable technology as they first prepared an all-printed and highly stretchable CNT-based electrochemical sensor and biofuel cell array, which can endure strains as high as 500% with no effect on structural integrity and sensor performance under repeated strain (0–500%), torsional twisting (50 repetitions of 180°), and indenting stress (50 cycles of 5 mm depth)[55].

7.2.3 Three-Dimensional Blocks Configuration

Although materials and devices with 1D and 2D structure are favorable for assembling flexible and stretchable sensors as mentioned earlier, these composites also have some limits, such as unstable electrical conductivity and poor reliability arising from the discontinuity of conductive network caused by interface problems such as percolation, aggregation, and sliding. To overcome these defects, 3D block configuration with self-assembled foam network structures is a practical option for lightweight flexible sensors for their ultralow densities, high conductivity, recovered compression or stretch, and novel chemical stability [56]. For instance, a 3D graphene–poly(dimethylsiloxane) hollow structure material was realized by Zhao et al. to develop a flexible sensor, which exhibits high sensitivity of 15.9 kPa^{-1} to compression force in a 60 kPa linear region with fast response time of 1.2 ms and measurement error below 6% under bending radius variations from −25 to +25 mm by optimizing the decoupled preparation process, exploring the potential applications in wearable medical devices and robotic vision [57]. In addition, except conventional downward compression, 3D foam with tensile deformation property is an attractive method for preparing flexible stretchable sensors, but few references have been reported for technical difficulty. Li et al. demonstrated highly reliable all-carbon conductive foams with joint-welded structure that have stable resistance and excellent elasticity under both tensile and compressive deformations [58]. Similarly, Fabian Schütt et al. displayed hierarchical self-entangled carbon nanotube networks with properties advantageous to those existing so far, which have application in flexible tensile sensors [59].

On the other hand, structural strength is one of the most important restricting issues for widespread applications of 3D-structure-based flexible sensors, which require foams to slickly maintain the capacity of mechanical deformation under the cyclic strain of stretching, compression, and torsion. Constructing 3D foam with excellent mechanical properties is still a big challenge. For example, to enhance the flexibility of graphene sponge, Qin et. al. developed a facile way of obtaining mechanically super-flexible 3D graphene aerogel through introducing water-soluble polyimide (PI), in which the synergistic action between PI and graphene endows the elastomer with desirable flexibility, excellent super-elasticity, and reversible compressibility, showing potential application in multifunctional strain sensors under various deformation conditions (Fig. 7.2a) [60]. Similarly, Cheng et. al demonstrated a novel and efficient method to prepare ultralightweight copper nanowires (Cu NWs)-poly(vinyl alcohol) (PVA) aerogel monolith with improved mechanical robustness and elasticity, with resistivity highly responsive to strains and manifested two distinct domains that both followed a power law function consistent with pressure-controlled percolation theory. Interesting, these Cu NWs-PVA aerogels can be cut into any arbitrary 1D, 2D, and 3D shapes when infiltrated into PDMS to form conducting rubber ambers [61]. However, foams composited with elastomers usually possess relatively good stretchability and conductivity and are very sensitive to environmental variations such as temperature and moisture changes. Hence it is highly desirable to produce self-compensated sensors in direct and precise detection. Zhang et al. proposed ideal porous polypyrrole/silver coaxial nanowire elastic aero-sponges for self-temperature-compensated pressure sensors with nearly zero temperature coefficient resistance (TCR) by Incipient Network Conformal Growth (INCG) technology, showing high sensitivity (0.33 kPa^{-1}) and minimum detectable pressure (4.93 Pa) with rapid response time (1 ms) and low TCR (0.86×10^{-3}/°C), which was employed to prepare a stress-triggered Joule heater with nearly zero temperature coefficient (Fig. 7.2b)

[62].

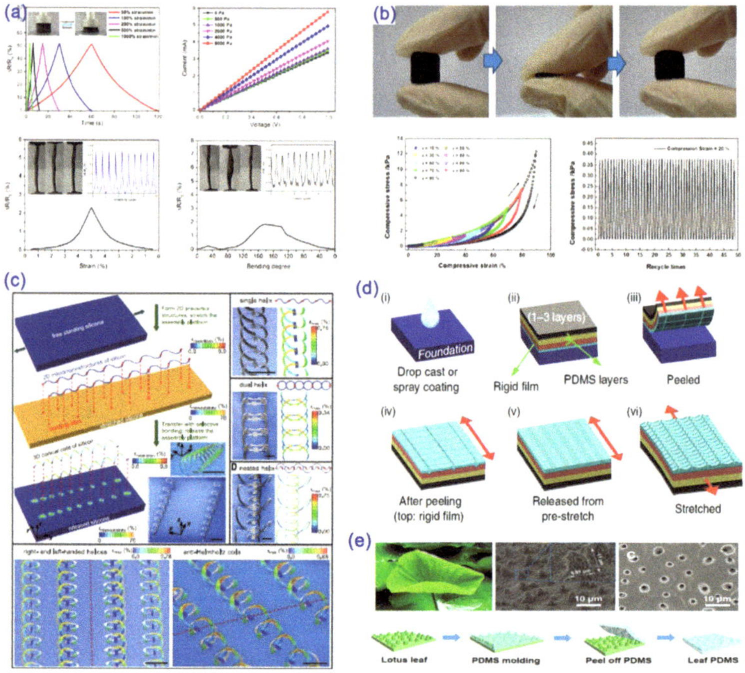

Figure 7.2 (a) Variation in the resistance and current–voltage of the rGO/PI composite under different strain rates, applied stress, stretching strain, and applied torsion. Reprinted with permission from Ref. [60], Copyright 2015, American Chemical Society. (b) Digital pictures and compression tests of elasticity of PPy-Ag NW aero-sponges. Reprinted with permission from Ref. [62], Copyright 2015, American Chemical Society. (c) Process for deterministic assembly of 3D mesostructures of monocrystalline silicon from 2D precursors. Reprinted with permission from Ref. [63], Copyright 2015, The American Association for the Advancement of Science. (d) Preparation and digital photos of mechanochromisms and strain-responsive properties of transparency change mechanochromism. Reprinted from Ref. [68] under Creative Commons license, Copyright 2016, Springer Nature. (e) Schematic of the fabrication process for micropatterned PDMS film from the bionic microstructure of a lotus leaf. Reprinted with permission from Ref. [3], Copyright 2016, John Wiley & Sons.

A special strategy employed by J. A. Rogers has revealed a novel and interesting perspective to assemble highly stretchable sensors that employs the construction and regulation of 3D functional mesostructures. Although lack of practical methods for desired 3D layouts has hindered development to date, attractive opportunities existed in exploiting 3D configurations of human-made devices for application in sensing, biomedicine, energy field, electronics, and so on. Rogers et al. have demonstrated brand-new tactics to previously inaccessible classes of 3D architectures, in which the extended 3D layouts were transformed through compressive buckling from 2D geometric micro-/nanostructures. They have demonstrated more than 40 representative geometries with various levels of configurations by theoretical and experimental research, including helices, conical spirals, spherical baskets, starbursts, flowers, scaffolds, and so on (Fig. 7.2c) [63]. Even an animal shape can be formed by this mechanically guided 3D assembly method, showing the potential for providing the required access to wide-ranging structural geometries to assemble stretchable sensors except the existed state-of-the-art 2D microsystems [64].

7.2.4 Nature-Inspired Structure for Flexibility and Stretchability

Many natural organisms can rapidly respond to changes in the outside environment by controlling physiological functions such as surface structures with outstanding reversibility. Hence, to satisfy the development of a multifunctional sensor with the requirements of ultrahigh flexibility, stretchability, and durability, bio-inspired structures are a feasible choice to prepare sensitive and flexible sensors. Among them, bio-inspired sensitive and reversible strain-dependent cracks and folds are also practical methods to endow the high-sensitive sensors with flexible and stretchable performances [65, 66]. For example, spiders can use their crack-shaped slit organs near their leg joints to sense infinitesimal mechanical stress variations. According to this phenomenon, Kang et al. demonstrated a flexible sensor based on nanoscale crack junctions inspired from the aforementioned spider's slit geometry, which can obtain ultrahigh sensitivity to strain (over 2000 gauge factor in strain range of 0–2%) and vibration (detecting amplitudes of about 10 nm) and serve multiple purposes [67]. Moreover, numerous marine organisms achieve rapid changes in color and transparency through muscle-controlled surface structures. For instance, Zeng et al. developed an analogous deformation-controlled surface-engineering method to assemble multifunctional sensors that can realize four mechanochromic roles via strain-dependent cracks and folds inspired from these display tactics, which display the reversibly comprehensive ability of transparent and opaque switch, patterns revealing/concealing, fluorescence emitting and turning under stretched strain, respectively (Fig. 7.2d) [68]. Inspired from the bionic strategy, the micro-/nanostructures of natural plants can be employed as an effective soft mold to prepare micropatterned electrodes. Based on it, a novel flexible capacitive tactile sensor was prepared by Zhang et al. based on the micropatterned flexible substrate duplicated from a lotus leaf, which possesses uniform surface microstructures and is easy to obtain and duplicate (Fig. 7.2e). The sensor shows optimum performances of sensitivity (0.815 kPa^{-1}) with a short response time (≈38 ms) and low detection limit (17.5 Pa)[3].

7.3 Classes of Functional Materials for Flexible and Stretchable Sensors

For the difficulties in evolving new flexible and stretchable electronic materials, the present strategy in attempting to acquire flexibility and stretchability is to engineer new structural constructs from currently sophisticated materials [23]. However, in contrast to the traditional sensing materials hindered by their rigidity, sensors based on flexible and stretchable materials can capture high-quality signals more efficiently through their conformal characters, which provide advantages for their application in wearable electronics [37]. The production of high-performance flexible and stretchable sensors is needed to exploit various novel approaches such as structure design, micropattern substrate, electrodes, and especially the selection or preparation of active sensing materials [2–10]. Among emerging candidates for flexible and stretchable sensors, active sensing materials with various dimensions and/or nature-inspired structures including plastic materials, metal nanostructures, 2D nanostructure-based semiconductors (graphene, black phosphorus, etc.), and transition-metal dichalcogenides (TMDs) offer distinctive sensing performance characteristics. Here we will review progress in active materials for flexible and stretchable sensors from the views of structures [44–46].

7.3.1 One-Dimensional Nanowire Materials

One-dimensional nanotubes, nanofibers, and nanowires have been widely employed as potential building components for nanoscale electronics, sensors, and energy devices benefited from their size-related intriguing physical properties and relatively high carrier mobility. Compared with other configurations, through sliding and buckling of the individual elements, 1D-materials-based sensing networks can accommodate strain and retain electrical properties during rigid stretching cycles. Moreover, assembly of functional materials on stretchable substrates is also a feasible approach to fabricate stretchable devices, in which the materials are fully exposed to strain, offering a unique ability to monitor the external strain-dependent variations.

Among 1D materials, the typical case is CNTs for high field-effect mobility and large tensile modulus (about 1 TPa), which show good performance stability while confronting mechanical forces, leading furthermore to multifunctional applications in electronic devices that are unattainable with conventional Si-based devices. For instance, Yamada et al. demonstrated a new class of flexible and stretchable nanomaterial sensors prepared from aligned single-walled-CNTs-based thin film that can perceive the changes in strain through bridging the fractured gaps and islands (Fig. 7.3a), which possess excellent performances of measuring strains up to 280% under fast response of 14 ms with high durability (10,000 cycles at 150% strain) and low creep (3.0% at 100% strain) [69]. The aforementioned characters permit integration of CNT-based sensors with clothing such as stockings and gloves to assembly wearable systems that can be used to precisely detect various human motions, including typing, breathing, and speech. However, these wearable monitoring systems require sensors to be permanent enough to resist unpredictable impacts and wear under uncontrolled conditions; many of the flexible and stretchable sensors prepared to date display fragile factors such as crack in

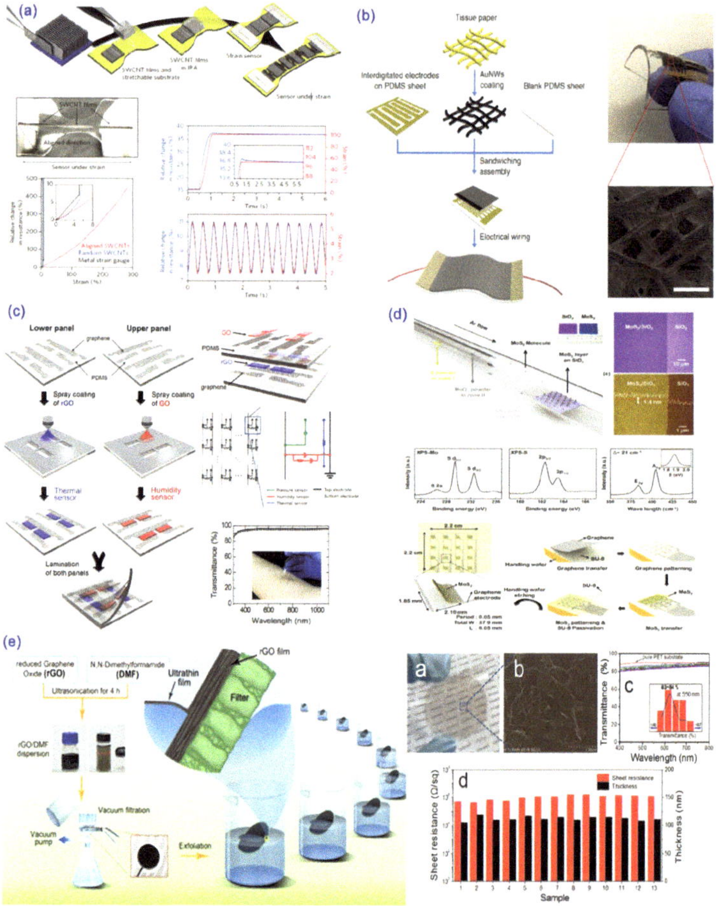

Figure 7.3 (a) Key fabrication, photograph under strain, and relative change in resistance versus strain of SWCNT-film strain sensor. Reprinted with permission from Ref. [69], Copyright 2011, Springer Nature. (b) Schematic illustration and practical photograph of the fabricated flexible sensor and the SEM image of Au NWs coated tissue fibers (scale bar, 100 mm). Reprinted with permission from Ref. [75], Copyright 2014, Springer Nature. (c) Fabrication schematic and optical transmittance of the stretchable and multimodal all-graphene E-skin sensor matrix with four pixels (2 × 2) and its integrated circuit diagram. Reprinted with permission from Ref. [78], Copyright 2016, John Wiley & Sons. (d) Schematic illustration, microscopic, Raman spectrum, and AFM images of the synthesis of an MoS2 atomic layer by three-zone CVD, and the fabrication steps for the derived strain gauge sensor. Reprinted with permission from Ref. [79], Copyright 2016, John Wiley & Sons. (e) Schematic of the reproducible exfoliation process for the fabrication of freestanding r-GO ultrathin films and its typical photograph. Reprinted with permission from Ref. [9], Copyright 2014, John Wiley & Sons.

the thin film. Bao et al. demonstrated a stretchable transistor sensor that can be resistant to impacts, punctures, and tears, based on the durable material composed of polymer-sorted CNTs and thermoplastic polyurethane (TPU), exhibiting stretchability up to 100% strain with programmable characteristics and stable stretching cycles performance [70]. As a derived morphology of CNTs, CNT fibers integrate the comprehensive properties of CNTs, and carbon fibers display high mechanical strength, extraordinary structural flexibility, and novel corrosion and oxidation resistivities, which make them a very feasible candidate for new-generation wearable smart textiles and devices. So far several wearable electronics based on CNT fibers have been reported, such as fiber sensors and fiber energy storage devices. For example, Sun et al. fabricated a fiber-based nonvolatile memristor through two cross-stacking CNT fibers covered by graphene, showing an ON/OFF switch from low to high resistance by a sudden current increase when the positive sweep voltages over 3.5 V [71]. On the other hand, the development of printing technologies has provided large size, low-coat preparation probabilities for meter-scale flexible electronics and sensors that can conformally attach to various curved surfaces, which can be applied as smart wallpapers, health monitoring, wearable interactive surfaces, and so on. For instance, a pressure sensor array for tactile pressure mapping was fabricated by gravure printing of a 20 × 20 CNT–TFT array [72, 73].

Significant attention has been devoted to CNTs, but several integration process technologies, such as separation of semi-conducting CNTs from metallic ones, have restricted further development of CNTs-based devices. In contrast, inorganic NWs have attracted wide attention in the field of flexible and stretchable electronics because of their intriguing physical properties such as excellent mechanical flexibility, large specific surface, size-related optoelectronic/electrical properties, and controllable higher carrier mobility. These highly desirable properties enable the inorganic NWs to exhibit high sensitivity toward changes in outer environments, such as pressure differential, optical radiation, temperature variation, pH value, proteins, and DNA. Experiments have demonstrated the applications of NWs as various sensors [74]. For example, ultrathin gold (Au) NWs are mechanically flexible and robust, enabling their use as superlattice nanomembranes in building blocks with potential application in future flexible devices. Cheng et al. fabricated an ultra-sensitive flexible and robust pressure sensor by sandwiching ultrathin Au-NWs-impregnated tissue paper between two blank PDMS sheets (Fig. 7.3b), which can be operated under low energy consumption (<30 μW, supplied by a battery voltage of 1.5 V) to detect pressing forces, showing the low limit of detection to 13 Pa with fast response time (<17 ms) and high sensitivity (>1.14 kPa^{-1}). Additionally, this sensor displayed high stability of >50,000 cycles and can resolve the differences between various forces such as bending, pressing, torsional and acoustic vibrations, enabling real-time monitoring of blood pulses as well as small vibrational forces from music [75]. Wang fabricated a stretchable strain sensor (up to about 50% strain) with fast response, high sensitivity and durability based on ZnO NWs-polystyrene nanofiber hybrids on PDMS substrate, which can be employed to measure small and rapid human motions driven by solar cells [76].

7.3.2 Two-Dimensional Planar Materials

Two-dimensional materials (2DMs) are currently intensively investigated under the strong motivation derived from the significant star power of graphene, which produced a Nobel Prize. According to different chemical classes, the most popular 2DMs contain typical layered structure, such as graphene, black phosphorous (BP) monolayers, 2D clay materials, layered transition-metal dichalcogenides (TMDCs)/dioxides (TMDOs), hexagonal boron nitride (h-BN), graphitic C3N4 (g-C3N4), all showing extraordinary physicochemical characters, especially related to the high electronic and thermal conductivity in their planar morphology. These performances offer many applications of 2DMs in sensors, electronics, energy storage, and catalysis [22–40].

As described earlier, graphene, a typical structured prototype of 2DMs, possesses excellent mechanical and physical properties of 25% fracture strains, 1 TPa Young's modulus, optical transmittance, and high carrier mobility, which is highly suitable for preparing flexible and stretchable electronic devices, including touch panels, high-speed transistor, transparent electrodes, and flexible energy devices. Especially, the piezoresistive and semi-metallic characteristics of graphene arising from strain-triggered shift in band structure make it a high gauge factor (GF) of ca. 1.9–6.0, which can be applied to flexible and stretchable strain/tactile sensors [77]. For instance, Ho et al. fabricated a transparent and stretchable multifunctional sensing matrix by integrating humidity, pressure, and thermal sensors through a simple layer-by-layer lamination process, in which CVD-grown graphene was employed as not only the active sensing materials but also the interconnected electrodes, as shown in Fig. 7.3c. In the device, the top and bottom PDMS substrates were laminated in a crisscross structure with the same geometry, forming the humidity and the temperature sensor array, respectively. And the top ones that were sandwiched by two CVD-graphene electrodes acted as the capacitive pressure and strain sensors. Each sensor of the integrated matrix can exhibit simplex sensing performance separately to monitor a variety of daily life sensations, which are collected as 2D color mappings with excellent sensitivity, such as a hot wind blowing, breathing, finger touching, and so on [78].

Further, a broad interdisciplinary study is being pursued on the various applications of other 2DMs beyond graphene, due to their unique physicochemical and electronic properties. With the downsizing of electronic devices, among various 2D semiconductor materials, molybdenum disulfide (MoS_2) has emerged as a promising candidate for post-silicon electronic devices complementing graphene with a gapless semi-metallic nature. For instance, for preparing practical multicell devices, Park et al. demonstrated a tactile sensing array with an area of 2.2 cm × 2.2 cm based on integrating ultrathin MoS_2 nanosheet with graphene electrode (Fig. 7.3d), which achieved excellent mechanical flexibility over a strain of 1.98% and optical transmittance over 80%, presenting high sensitivity and stable cycle even after being pasted on unusual surfaces such as human fingertip [79]. The fabrication of large-sized MoS_2 layers with spatial homogeneity is a prerequisite for its practical application in flexible sensor arrays. Lim et al. introduced a facile strategy to synthesize wafer-scale MoS_2 layers with homogeneous stoichiometry, which employed a two-step thermal decomposition process under a low temperature of 450°C, enabling direct growth of high-quality MoS_2 on a polymer matrix. Based on this, a homogeneous MoS_2-based sensor array was fabricated on a 4-inch wafer plastic substrate for visible-light photodetection, showing well-photoelectrical properties and mechanical stability [80].

Besides, other 2DMs and film morphologies are also used to fabricate flexible and stretchable sensors. For example, a new 2DM first successfully synthesized in 2011, derived from transition-metal carbides and carbonitrides, MXenes have been widely applied in field-effect transistors, electromagnetic shielding, energy storage, conductive electrodes, and so on. Based on the essential features of greatly changed interlamellar intervals under an external force, Ma et al. first illustrated a highly flexible piezoresistive sensor by employing these novel MXenes as the active component, which shows high sensitivity (Gauge Factor ~180.1) with fast response of <30 ms and excellent reversible compressibility and was used to sense weak pressure like human being's subtle expression activities [81]. Similarly, with layered 2DMs, thin films can be utilized to develop a new class of stretchable electronics for their mechanical flexibility and tailorability. Typically, Hirsch et al. prepared a new-style stretchable electronic conductor based on a biphasic solid–liquid thin metal films network composed of clusters of the solid intermetallic alloy $AuGa_2$ and supercool liquid gallium. And they demonstrated that this epidermal sensor can monitor the changes in the positions of the metacarpophalangeal and proximal interphalangeal joints of human fingers [82]. Moreover, it is still a challenge to realize a flexible and stretchable sensor for adapting to customizations under different low-cost occasions. Zhong et al. introduced a novel paper-based active tactile sensor array (PATSA) by an easy fabrication method, which is permitted to tailor the sizes/shapes as needed. The PATSA can be used to record position and pressure information, exhibiting high sensitivity of 0.35 V/N and durability under 5 Hz stimuli [83]. Further, Zhang et al. developed a facile fabrication method for highly transparent freestanding ultrathin films based on graphene (Fig. 7.3e) and CNTs. The freestanding films can be reproducibly exfoliated from the dense film on the filter by surface tension of deionized water, which were easily transferred onto PET substrates to make highly sensitive flexible sensors, such as chemiresistive gas sensors and humidity sensors [9, 11].

7.3.3 Semiconductors

To address the development of flexible and stretchable sensors with excellent sensitivity and durability, the aforementioned strategies and materials have been contributed, including various dimensional structure and nanomaterials such as graphene, carbon nanotubes, and hybrid composites. However, several drawbacks existing in sensors, such as high hysteresis and nonlinearity, still restrict high performance when responding to applied strain. In contrast, semiconductor materials, widely used in various applications such as field-effect transistors (FETs), catalysts, and energy harvesting, give the opportunity to fabricate sensors with high sensitivity and stable operation. Though the specific sensing theory is still ambiguous, partially admissive mechanisms of semiconductors are described as trapping of electrons through energy band adjustment, which is responsible for changes in conductivity.

Nowadays, in addition to the previously described typical semiconductors such as CNTs and 2DMs, other semiconductors are also employed to prepare flexible sensors. For example, metal oxide semiconductors are especially used to realize chemical, optical, and mechanical sensors. These kinds of sensing materials can be divided into two categories: transition-metal oxides and non-transition-metal oxides. The representatives of the former type are V_2O_5, TiO_2, and WO_3, which are applied as catalysts, photoanodes, and chemical sensors. Members of the second type, such as Ga_2O_3, In_2O_3, ZnO, and SnO_2, show good conductivity and high sensitivity, which can form the channel layer for transistors and are useful for multifunction detectors [84]. In addition, as compared to traditional Si technologies, metal compound semiconductors also have the advantage of using facile printable methods to rapidly from film deposition, which enable the preparation of low-cost sensors with uniform reliability and sensitivity [85]. For instance, Zhang, Liu, and co-authors reported an electrochemical shock method to prepare uniform c-BN NCs with a lateral size of only 3.4 ± 0.6 nm and a thickness of only 0.74 ± 0.3 nm at ambient temperature and pressure. Most interestingly, the NCs exhibit excellent piezoelectric performance with a large recordable piezoelectric coefficient of 25.7 pC/N, which is almost six times larger than that from bulk c-BN and even competitive to conventional piezoelectric materials [15].

It is worth mentioning that self-powered semiconductors can convert mechanical energy into electrical signals, which are usually divided into piezoelectricity and triboelectricity, exhibiting favorable potentials in fabricating self-powered sensors due to the strain-caused voltage/current. Lang et al. introduced poly(vinylidene fluoride) piezoelectric nanofiber webs to prepare flexible pressure sensor devices (Fig. 7.4a), exhibiting an acoustic-to-electric transfer capacity that can transmit vibration directly to the nanofiber layer. These features make them precisely distinguish the frequency of sound waves from middle to low region with an optimal sensitivity as high as 266 mV/Pa [86]. However, these piezoelectricity-induced pressure sensors have challenges in measuring static signals. Chen et al. demonstrated a static measurement sensor with nanowires/graphene heterostructures due to the synergistic effect of carrier scattering in graphene and the strain-induced polarization in piezoelectric nanowires, showing a sensitivity of 9.4×10^{-3} kPa^{-1} with a response time of 5–7 ms [87]. Similarly, triboelectric nanogenerators (TENGs) can be used for the next generation of artificial intelligent products. Wang et al. developed an ultrahigh stretchable (1160% strain) and transparent (96.2% visible light), skin-like TENG, which could realize tactile sensing depending on self-harvesting biomechanical energy, in which the instantaneous peak of power density (35 mW/m^2) converted from human motions can drive wearable devices such as artificial electronic skin and electronic watch for touch/pressure perception [88]. Further, they reported a flexible and thin tactile sensor based on dual-mode TENGs (Fig. 7.4b), indicating a linear pressure detection ranging from 40 to 140 N with a high-pressure sensing sensitivity up to 28 mV/N [89]. On the other hand, ultrathin piezoelectric materials may pave the way for positioning an object with extreme precision. Using ultrathin CdS thin films, Zhang, Liu, and co-authors demonstrated vertical piezoelectricity in atomic scale (three to five space lattices) through in situ scanning Kelvin force microscopy and single and dual resonance tracking piezoelectric force microscopy [22], showing the vertical piezoelectric coefficient (d33) up to 33 pm/V for CdS ultrathin films. These findings shed light on the design of next-generation sensors and microelectromechanical devices.

On the other hand, organic semiconductors are suitable for application in flexible and stretchable sensors, because their inherent mechanical properties of flexibility and compatibility permit them to form compact and stable interface structure with plastic substrates, as the resulting charge transport can render the materials stiff and brittle. Flexible electronics based on these have been widely investigated and employed in sensors, light-emitting diodes, identification tags, etc. [90]. For instance, Oh et al. prepared a modified viscoelastic PEDOT: PSS by compositing with excess nonvolatile surfactant plasticizer called Triton X-100 ($C_{14}H_{22}O(C_2H_4O)_n$)(n = 9–10), which displayed repeated stretchability similar to rubber play dough without degradation of conductivity. When covered by metallic circuits, PEDOT: PSS can maintain metallic conductivity under extreme folding cycles. Curing the cracks in the metal may open a new avenue for deformable flexible electronics designed by customers [91]. However, an existing defect prevents the popularization of organic electronics, namely, the restriction of mobility and stability [92].

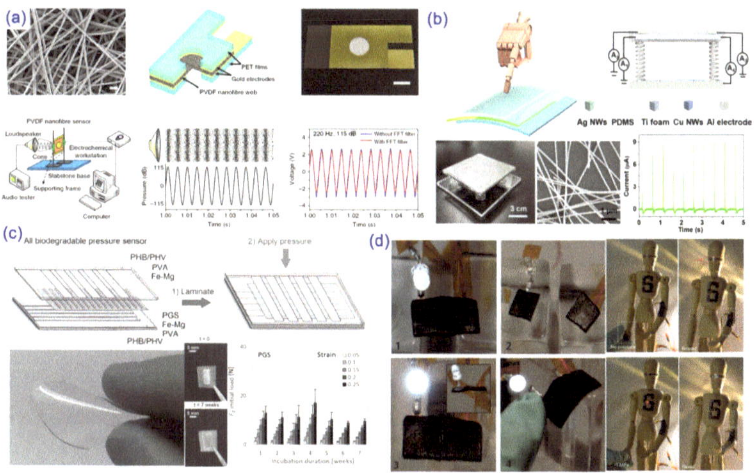

Figure 7.4 (a) Pressure sensor derived from PVDF nanofibers and its voltage outputs of the sound wave sensing performance without FFT treatment. Reprinted from Ref. [86] under Creative Commons license, Copyright 2016, Springer Nature. (b) The application diagram, working principle, and short-circuit current of an Ag-NWs-derived triboelectric sensor. Reprinted with permission from Ref. [89], Copyright 2017, American Chemical Society. (c) The fabrication schematic and practical structure of the fully biodegradable pressure sensor and its initial load F_0 as a function of the incubation duration. Reprinted with permission from Ref. [99], Copyright 2015, John Wiley & Sons. (d) Demonstration of the healing process for a conductive composite material with an LED in series with a self-healing electrical conductor, the materials composed of a supramolecular organic polymer with embedded nickel nanostructured microparticles. Reprinted with permission from Ref. [102], Copyright 2012, Springer Nature.

7.3.4 Other Special Functional Materials

For the required features of high flexibility and stretchability, lightweight, facile and inexpensive fabrication, the current developments in flexible and stretchable sensor technologies are dominated by inorganic/organic compound-based nanomaterials, such as carbon materials, metal oxide semiconductors, and polymers, and applications of these inorganic/organic materials are frequently restricted to some particular industrial situations due to cytotoxicity, high cost, harsh synthesis, and non-degradability. Hence, there is still increasing attention on flexible electronics prepared by specially functional sensing materials that are biocompatible, environmentally friendly, and self-healing [93].

Inspired by the increased comprehension of operating principles of functional organs at play in biological matters with the ability to convert their natural characteristics, great efforts have been made toward mimicking such functions and architectures in artificial materials to fabricate flexible and stretchable biomimetic sensors [94]. For example, Wang et al. employed a new biologically inspired approach to fabricate a flexible superhydrophobic sensor array (12 × 8 pixels with area of 18 × 12 cm^2) based on the incorporation of natural elastic hollow microcapsules within a biomimetic hierarchical structure of sunflower pollen, showing the lowest detecting limit of 1.6 Pa with well stability of over 25,000 cycles and a special capacity of discriminating spatiotemporal dynamic and static tactile stimuli [95]. Moreover, biocompatibility is a necessary consideration for such flexible sensors as they are closely related to biological interfaces. An ideal strategy to achieve biocompatible devices is to employ naturally occurring materials, such as chitosan, silk, starch, and DNA. Among them, silk fibroin has been widely used for preparing the substrate of flexible electronics. In 2017, Zhang et al. adopted a cost-effective and large-area-capable method to fabricate flexible and biocompatible pressure sensors by integrating transparent carbonized silk nanofiber membranes and unstructured PDMS films, which obtained superior performances of ultralow detection limit of 0.8 Pa with ultrahigh sensitivity of 34.47 kPa^{-1}, rapid response time of <16.7 ms, and good durability of >41,000 cycles for a broad pressure range. These advantageous properties allowed monitoring of human physiological signals, perceived micro-touches, and spatial distribution of pressure [96].

In addition, except the aforementioned novel functional materials that are biocompatible enough for incorporation into tissues, the developments ranging from nanomedicine to bioelectronics also need biodegradable "green" electronic materials, ultimately establishing created paths to produce human- and environmentally friendly electronics for integration with living tissue, which reduce waste and improve the efficacy of implantable systems associated with wearable and point-of-care diagnostics [97, 98]. Bao et al. developed a single-use biodegradable pressure sensor (Fig. 7.4c), which can dissolve when immersed in phosphate-buffered saline solution at 37°C for 7 weeks. These sensors exhibit an ultrahigh sensitivity of 0.76 kPa^{-1} and have a capacity of distinguishing a weight of 5 mg, which can be applied to facilitate fast and easy positioning for cardiovascular health monitoring [99]. To avoid invasive removal surgery, biodegradable and flexible pressure sensors are often directly implanted with a soft biological system to measure physiological internal forces, such as eye pressure and brain pressure. Curry et al. employed an approved implant polymer called poly-L-lactide (PLLA) to create a biodegradable piezoelectric sensor for monitoring biological internal pressures, showing a wide sensing range of 0–18 kPa and maintaining a persistent expression for a period of 4 days under aqueous condition. Implanted inside the abdominal cavity of a mouse, the PLLA-based sensor monitored the pressure of diaphragmatic contraction [100].

Moreover, for future applications in emerging fields such as biomimetic prostheses, soft robotics, and artificial skin, the development of a flexible and electrically conducting material possessing the ability of repeated self-healing will be widely encountered in nature for sensing mechanical forces [101]. Bao et al. prepared a pressure- and flexion-sensitive material composited of nickel nanostructured microparticles embedded into supramolecular organic polymer, as shown in Fig. 7.4d, showing mechanical and electrical self-healing properties at ambient conditions, which is suitable for applications in artificial/electronic skin. On adjusting the ratios of nickel particles, the electrical conductivity can reach values as high as 40 S/cm, and the resistance can vary inversely with the applied flexion and tactile forces. After self-healing for 15 s, the initial conductivity and mechanical properties can be repeatably restored with ~90% efficiency, and subsequently completely restored for about 10 min. [102]. However, combining all these properties of skin's pressure sensitivity and mechanical self-healing in a human-made material remains a challenging task. Hou et al. described a strong and stretchable film prepared by integration of graphene and polymers, which mimics both the pressure sensitivity and mechanical self-healing characters of natural skin with self-activated feature, demonstrating highly stable sensitivity to a very light touch (0.02 kPa) even under stretching or bending conditions [103].

Due to the presence of irreversible cross-links and water, permanently cross-linked hydrogels also have self-healing function. For example, Phadke et al. prepared a cross-linked polymer hydrogel formulated from acryloyl-6-aminocaproic acid (A6ACA) precursors, which exhibited self-healing in an aqueous environment [104]. Additionally, hydrogels can also be engineered into flexible and stretchable electronics and devices for their hydrogels–polymer networks infiltrated with water. Li et al. demonstrated a set of new biocompatible hydrogel-based electronics and devices with high stretchability and mechanical robustness by integrating stretchable conductors, rigid electronic components, and reservoirs into tough hydrogel matrices, showing multiple novel functions such as mock drugs and stretchable light-emitting diode (LED) array [105]. Ionic conductors, composited by gels swollen with ionic liquids, are a special modality of hydrogel, which show a solid form with stretchable and transparent features. Suo et al. introduced a series of devices derived from ionic conductors with high stretchability and full transparency, which can be operated at >10 kV voltages and over 10 kHz frequencies and were used to assemble a transparent large strain actuator and a transparent loudspeaker that generates voice higher that the whole audible range [106].

7.4 Flexible and Stretchable Sensors for Human Information Detection

Electronic skin: Human skin is considered a benchmark for the performance of artificial tactile systems. Benefiting from its capacity to perceive tactile pressure, flexible and stretchable artificial electronic skin (e-skin) is capable of imitating, even improving the features of human skin through integrating various sensing materials with flexible and stretchable elastomeric and/or plastic substrates, imitating the information capture and performance of the mechanical and other receptors of skin [55–60]. In a previous work by our group, Zhang et al. developed a series of flexible sensing devices by combining uniform micropatterned flexible polydimethylsiloxane (PDMS) substrate with nanostructured materials, such as CNTs, rGO, and graphene composite (Fig. 7.5a). The flexible sensing device constructed with two layers of patterned CNTs/PDMS films demonstrated ultrahigh sensitivity for the detection of minute forces with fast response time (<10 ms), great stability and repeatability (>67,500 cycles), and with a detection limit of the device as low as 0.6 Pa. The flexible device demonstrated the capability of monitoring human physiological signals such as wrist pulse and muscle movement when a person speaks, which may broaden their potential applications for disease diagnosis and voice recognition [107]. In another study, Gerratt et al. reported an integrated strategy for capacitive tactile pressure sensors and resistive sensors to assemble a wearable elastomer-based e-skin, which was applied to monitor pressure distribution and articulation of the entire length of finger. The as-fabricated pressure sensors possessed a linear sensitivity of 0.001 to 0.01 kPa^{-1} corresponding to a range of 5–405 kPa, which covers most of the human physiological sensing scope. Through the decoupling analysis of strain and pressure cross-sensitivity, the e-skin was stuck on a textile glove to enable precise grasp analysis for controlling the grasp of objects, which can be applied in a human-in-the-loop system toward hand prosthesis with biomimetic tactile capabilities [108].

Tactile fingerprint: Although several interesting features of electronic skin have been discovered, such as self-healing, self-cleaning, and self-powered features, it is still an intriguing challenge to make electronic skin having real tactile sense like human fingerprint. Human perception has a wide pressure range (100 Pa to 100 kPa), high-frequency response (~400 Hz), and can recognize the surface textures of different materials. It does not have high enough sensitivity or cannot accurately quantify the detailed patterns of delicate objects. Efforts have been devoted to optimize flexible electronic tactile skin by simulating the human fingertip, because the tactile sensing ability of fingertips is usually considered more sensitive and distinguishable than other parts of the body with amplified tactile information [109]. Chun et al. reported a tactile sensor to detect the low-pressure response based on a relatively complex device structure, and then the sensor was employed to test a broad Si cuboid array instead of the surface morphology [110]. Further, Park et al. demonstrated a fingertip-skin-inspired micro-structured ferroelectric sensor based on hierarchical and independent elementary layer, which can discriminate the multiple spatial tactile stimuli containing vibration, temperature, and static and dynamic pressure. This flexible sensor exhibited multifunctional sensing capacities to detect versatile information, including surface textures, acoustic sounds, pulse, and temperature of artery vessels [111].

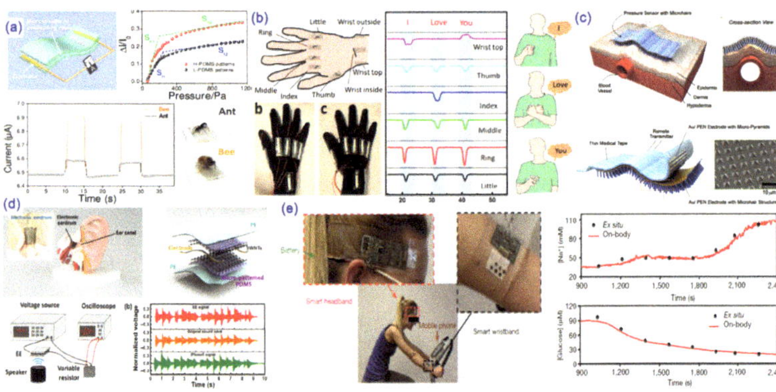

Figure 7.5 (a) Real-time *I–t* curve and the sensitivities of the e-skin constructed with H-PDMS for detection of a bee (40 mg) and an ant (10 mg). Reprinted with permission from Ref. [107], Copyright 2013, John Wiley & Sons. (b) Schematic for the ionic sensor arrays derived smart glove for different sign language with six sensors (wrist top, thumb, index, middle, ring, little) and eight sensors (wrist top, wrist outside, wrist inside, thumb, index, middle, ring, little), respectively. Reprinted with permission from Ref. [113], Copyright 2017, American Chemical Society. (c) Schematic illustration of a microhair-structured sensor to detect pulse on a human neck. The pyramid-shaped PDMS dielectric layer was placed between the two Au electrodes on PEN plastic substrates. Reprinted with permission from Ref. [118], Copyright 2014, John Wiley & Sons. (d) Image and schematic configuration of the flexible electronic eardrum placed in an ear model and its filtered detection signal of sound compared with iPhone5 signal. Reprinted with permission from Ref. [122], Copyright 2017, Springer Nature. (e) Photographs of a subject wearing a "smart headband" and a "smart wristband" during stationary cycling, and the ex situ calibration data of the sodium and glucose sensors from the collected sweat samples with the on-body readings during the stationary cycling exercise. Reprinted with permission from Ref. [127], Copyright 2016, Springer Nature.

Body motions: Physical signals, such as bodily motion, joint motion, and muscle movement produced from regular human activities, generate a serial range of strain or stress change but are significant health indicators, which offer a probability for healthy monitoring, disease diagnosis, motion control, and human–computer interactive activity. These require the applied sensor to possess synthetical properties of a large pressure range with a high sensitivity. For instance, a pressure less than 20 kPa will be produced when a person walks on ground; therefore, it is necessary to introduce a kind of pressure sensor that possesses a large range (up to 20 kPa), meanwhile having a high sensitivity to prolong its application in various wearable electronics. For this purpose, Tao et al. fabricated a graphene-paper-based pressure sensor that possesses an optimized working pressure range of 0–20 kPa with a sensitivity of up to 17.2 kPa^{-1}, which was further applied in voice recognition, respiration and pulse detection, and other intense motion detections [112]. Real-time detection is still a challenge in these monitoring activities. To realize this goal, Zhao et al. developed a smart glove integrated with several ionic sensors for human motion monitoring, which shows a passive and space discriminating capacity in the complex geometry of hand configurations (Fig. 7.5b) [113]. Further, the results from Zhang's groups showed that a multifunctional superhydrophobic smart coating with a thicknesses of only 1 μm, which possesses high sensitivity (gauge factor of 5.4–80) and resolution (1° of bending) and wide sensing ranges (bending: 0–140°, torsion: 0–350 rad/m) with a rapid response time (<8 ms) and a well stretching–relaxing durability (over 5000 cycles), can be directly used to spray on clothing for a full-range and real-time detection of human motions under extreme wet and corrosive conditions, such as humidity, alkali and acid environment [114].

Physiological signals: Monitoring human physiological signals is essential in routine biomedical research and clinical diagnoses. Real-time tracking of these signals is an effective way for disease diagnosis and health assessment, and it is crucial for a person suffering from cardiovascular disease. Conventional methods, such as intrusive detection and Si-based rigid devices, are costly and bring discomfort to patients in clinical diagnoses and treatments. Also their poor biocompatibility could cause infection and secondary damage. To improve comfortability, convenience, and security of health monitoring in clinical diagnoses and daily life, a new type of portable and wearable flexible and stretchable sensors is yet to be proposed [115].

For instance, in traditional Chinese medicine, wrist pulse is a significant indicator of physical condition and supplies many important information for non-invasive medical diagnosis. In our previous work [107], the flexible e-skin devices sticking around the radial artery of wrist can distinguish tiny pressure distinction between different persons with various body statuses. In another work, Schwartz et al. fabricated a flexible low-power (<1 mW) pressure sensor based on the structure of organic thin-film transistors, showing a high sensitivity of 8.4 kPa^{-1} with a fast response time of <10 ms and high stability. As the change in capacitance can be strongly amplified through dielectric compression, this sensor was used for non-invasive monitoring of weak physiological signals of continuous radial artery pulse wave. Three obvious, differentiable peaks of pulse waves can be obtained with a high resolution of small features and a clear revealation of the diastolic tail, which most sensors applied in arterial tonometry cannot test. Integrated with long-term mobile biomonitoring, the sensors may be used for mobile and remote medical diagnostics in cardiovascular medicine [116].

Interior body pressures are external symbolic manifestations of the general internal health of a person. Real-time monitoring of body pressures can obtain important and valuable information for preliminarily screening of people suffering from serious diseases. For instance, the clinical diagnosis of cardiac impulse caused by jugular venous pulses (JVPs) has close relation to the risk of heart failure, such as arrhythmias, right heart hemodynamics, and pericardial disease. The development of highly sensitive flexible sensors has given the capability of detecting JVP waveforms with ease compared with the traditional sphygmomanometers, which have defects such as time delay, stationary setup, and expensiveness [117]. Pang et al. demonstrated a facile wearable bandage-like capacitive pressure sensor that adopted bio-inspired designs (Fig. 7.5c). After being integrated with a wireless transmitter, the system was used to measure weak waveforms of arterial tonometry or pulses in the neck epidermis, as caused from JVPs related to cardiovascular diseases from the human body. Results indicated that the waveforms detected from a healthy person showed quite different patterns with those retrieved from diseased ones, displaying that this wearable system may be helpful for fast diagnosis of cardiac and cardiovascular diseases [118].

Another significant interior pressure used as a health indicator is the intraocular pressure produced from human eyes. Glaucoma is the largest risk factor for human blindness, which can be assessed by intraocular pressure. However, it is difficult to diagnose glaucoma early because of its slow and symptomless progression. Wearable contact lens sensors have got much attention due to the ability of direct inspection of intraocular pressure, but previous lens sensor generally obstructs the field of vision and cannot detect the marked signals on persistent work status. Kim et al. integrated graphene- and Ag-NWs-based sensors with a soft actual contact lens to demonstrate a wearable smart sensor system with the features of transparency, reliability, and robustness to overcome some of the aforementioned limitations, which can be employed for real-time testing of glucose within tears based on graphene sensors and in vitro wireless detection of intraocular pressure. The live rabbit and bovine eyeballs approved its reliable multifunctional operation with simultaneous and independent electrical response activity [119].

Bionic organ functions: Improved understanding of various basic functions of living organs in biology, such as optesthesia, smell, and auditory sense, provides possibilities for the creation of bionic electronics that employ the analogous engineering mechanisms and operational characteristics beyond those existing technologies. Flexible and stretchable biomimetic sensors, developed with merging electronics that adopt the bionics principle, provide opportunities for application in bionic monitoring systems like robots. For instance, eyes are the most important organ for living. Especially, some insects such as fly are well known for their compound eyes with high sensitivity to motion, wide fields of view, and infinite depth of field. The analogous bionic imaging technique has generated great interest due to the potential applications in tools for endoscopy and surveillance devices. Rogers et al. presented a complete set of elastomeric optical elements, thin silicon photodetectors, and elastical planar schemes for digital cameras, which is named as electronic eye and mimics hemispherical apposition compound eyes of arthropods. This bionic eye with compressible and stretchable design can realize full hemispherical shapes (about 160°) and a total field of view of 180° [120]. To mimic the olfactory function of discriminating smell, an electronic nose, a technical device usually connected with sensing of odors or the ability to "scent," has been developed for disease diagnosis, environmental monitoring, and food and beverage inspection. Henning et al. displayed a NWs-based field-effect transistor as a robust gas sensor with high sensitivity, working at ambient temperature. Through a careful design of NW sensing materials, this device can capably sense 100 parts per million of ethanol gas without any surface modification [121]. For flexible electronics to "hear" sound, through optimization of a flexible mechano-sensor, a kind of ultrathin flexible mechano-sensors, called flexible electronic eardrum, were fabricated by Zhang et al. to detect sound signal based on single-walled CNTs, composited polymer films, and flexible microstructure substrates (Fig. 7.5d), which can detect and record sound signals in a wide frequency domain of 20–13,000 Hz with high signal-to-noise ratio (~55 dB), fast response time (76.9 μs), and stable durability (150,000 cycles). The characteristics of the flexible electronic eardrum provide the opportunity for implantable acoustic bioelectronics in clinical prosthetics applications [122].

Human atmospheres: Along with the development and application of flexible electronic devices, the rapid advances in the design of various flexible sensors have received increasing attention in individual medical care. Particularly, a comfortable living environment and body health information are the most intensively measured variables in our daily life. For example, body temperature is one of the most commonly sensed factors in humans and animals, which is an important character in detecting body activities and determining the health status of humans. Kim et al. introduced a flexible Pt RTD-based temperature sensor, which was integrated inside a suture strip, which can be sutured, wrapped, and even folded around a fingertip with good deformability and stability under mechanical strain. This Pt-temperature sensor showed a linear response to temperature and was further used to monitor temperature by partially embedding it in the place of an incision on an animal model [123]. On the other hand, humidity is also one of the most important and useful factors in daily environmental monitoring, such as biological facilities, agricultural activities, automobile industries, and luxury indoor surroundings. Though flexible humidity sensors were rarely reported until now, Zhang et al. fabricated a new type of multifunctional flexible humidity sensor based on porous ionic membrane derived from polymer gel-electrolyte, showing the RH sensing range from 10.89% to 81.75% with rapid response and recovery times of 0.4 s and 2.6 s. Significantly, the sensor was insensitive to temperature (0–95°C) and pressure (0–6.8 kPa) change, indicating that it is a successful candidate for highly selective flexible humidity sensors. Based on this humidity sensor, a non-contact switch system and an attachable smart label were assembled, showing that the switch responded favorably to RH change and the label displayed a great linear relationship between the sensitivity of sensor and facial water contents. In a word, such high sensing performance makes flexible temperature and humidity sensors a promising candidate for various applications in wearable electronics and artificial intelligent robots [124].

Perspiration analysis: The increased attention on flexible and wearable electronics has influenced studies on development of non-invasive platforms for body fluid monitoring. Among them, wearable biosensors for measurement of human sweat could enable real-time continuous monitoring of an individual's physiological information containing physiologically and metabolically rich information about disease diagnosis, drug abuse, and athletic performance optimization, which provide favorable insight into a user's health state through the detection of biomarkers at molecular levels [125]. Wang and Zhang et al. employed an integrated electrode of a larger-surface-area (~7.23 cm^2) gold nanodendrite (Au ND) array and highly stable (over 3 months' storage) poly(vinyl acetate)/inorganic salt (PVA/KCl) membrane to fabricate a new type of all-solid-state ion-selective electrode for assembling a wearable sweatband sensor platform, which was developed for real-time harvesting and analysis of [Na^+] in sweat during indoor exercise [126]. However, only a single analyte of the physiological state can be monitored at one time by previous sweat sensors without calibration mechanisms and on-site signal-processing circuitry for accurate analysis. Because of the complexity of sweat secretion, integration of a full system is critically required to ensure accurate measurements of target biomarkers on the high level of simultaneous and multiplexed screening. For this purpose, Gao et al. introduced an integrated and multiplexed flexible sensor array for real-time and selective in situ analysis of sweat metabolites, skin temperature, and electrolytes without external analysis (Fig. 7.5e). Through merging flexible silicon-integrated circuits and plastic-based flexible sensors, this technology has helped to fill in the gaps between various complex signal-processing processes in wearable biosensors. And this wearable system was used to assess the physiological status of human subjects through a real-time detection of the detailed sweat character when they engaged in prolonged indoor and outdoor physical activities. This platform enables a wide application of personalized diagnostic and physiological monitoring [127].

7.5 Conclusion

This chapter described classes of architectural strategies (containing 1D, 2D, 3D, and nature-inspired) and functional materials (NWs, planar, 3D, semiconductors, and others) for the realization of flexible and stretchable sensors. Their applications in human information detection, including e-skin, human motions, physiological signals, perspiration analysis, etc. are demonstrated based on the previously reported works. It is worth noting that development of this emerging field would greatly benefit from coordination between materials science, electrical engineering, biomedical engineering, chemistry, physics, and energy research communities. It is hoped that this chapter provides stimulating highlights of this interdisciplinary research frontier.

References

1. Boutry, C. M., Nguyen, A., Lawal, Q. O., Chortos, A., Rondeau-Gagne, S., and Bao, Z. (2015). A sensitive and biodegradable pressure sensor array for cardiovascular monitoring, *Adv. Mater.*, **27**(43), pp. 6954–6961.

2. Yao, Y., Dong, H., and Hu, W. (2016). Charge transport in organic and polymeric semiconductors for flexible and stretchable devices, *Adv. Mater.*, **28**(22), pp. 4513–4523.

3. Li, T., Luo, H., Qin, L., Wang, X. W., Xiong, Z. P., Ding, H. Y., Gu, Y., Liu, Z., and Zhang, T. (2016). Flexible capacitive tactile sensor based on micropatterned dielectric layer, *Small,* **12**(36), pp. 5042–5048.

4. Jang, H., Park, Y. J., Chen, X., Das, T., Kim, M. S., and Ahn, J. H. (2016). Graphene-based flexible and stretchable electronics, *Adv. Mater.,* **28**(22), pp. 4184–4202.

5. Lee, H., Kim, M., Kim, I., and Lee, H. (2016). Flexible and stretchable optoelectronic devices using silver nanowires and graphene, *Adv. Mater.,* **28**(22), pp. 4541–4548.

6. Trung, T. Q. and Lee, N. E. (2017). Recent progress on stretchable electronic devices with intrinsically stretchable components, *Adv. Mater.,* **29**(3), pp. 1603167.

7. Wang, L., Chen, D., Jiang, K., and Shen, G. (2017). New insights and perspectives into biological materials for flexible electronics, *Chem. Soc. Rev.*, **46**(22), pp. 6764–6815.

8. Liu, Y., He, K., Chen, G., Leow, W. R., and Chen, X. (2017). Nature-inspired structural materials for flexible electronic devices, *Chem. Rev.,* **117**(20), pp. 12893–12941.

9. Wang, X. W., Xiong, Z. P., Liu, Z., and Zhang, T. (2015). Exfoliation at the liquid/air interface to assemble reduced graphene oxide ultrathin films for a flexible non-contact sensing device, *Adv. Mater.,* **27**(21), pp. 1370–1375.

10. Jeong, S. H., Zhang, S., Hjort, K., Hilborn, J., and Wu, Z. (2016). PDMS-based elastomer tuned soft, stretchable, and sticky for epidermal electronics, *Adv. Mater.,* **28**(28), pp. 5830–5836.

11. Wang, X. W., Li, G. H., Liu, R., Ding, H. Y., and Zhang, T. (2012). Reproducible layer-by-layer exfoliation for free-standing ultrathin films of single-walled carbon nanotubes, *J. Mater. Chem.,* **22**, pp. 21824–21827.

12. Wang, J., Yan, C., Cai, G., Cui, M., Lee-Sie, Eh. A., and See, L. P. (2016). Extremely stretchable electroluminescent devices with ionic conductors, *Adv. Mater.,* **28**(22), pp. 4490–4496.

13. Hong, S., Lee, H., Lee, J., Kwon, J., Han, S., Suh, Y. D., Cho, H., Shin, J., Yeo, J., and Ko, S. H. (2015). Highly stretchable and transparent metal nanowire heater for wearable electronics applications, *Adv. Mater.,* **27**(32), pp. 4744–4751.

14. Long, Y. Z., Yu, M., Sun, B., Gu, C. Z., and Fan, Z. (2012). Recent advances in large-scale assembly of semiconducting inorganic nanowires and nanofibers for electronics, sensors and photovoltaics, *Chem. Soc. Rev.*, **41**(12), pp. 4560–4580.

15. Chen, Z. G., Li, L. H., Cong, S., Xuan, J. N., Zhang, D. S., Geng, F. X., Zhang, T., and Zhao, Z. G. (2017). Rapid synthesis of sub-5 nm sized cubic boron nitride nanocrystals with high-piezoelectric behavior via electrochemical shock, *Nano Lett.*, **17**, pp. 355–361.

16. Yang, S., Khare, K., and Lin, P-C. (2010). Harnessing surface wrinkle patterns in soft matter, *Adv. Funct. Mater.*, **20**(16), pp. 2550–2564.

17. Ahn, J-H. and Je, J. H. (2012). Stretchable electronics, materials: Architectures and integrations, *J. Phys. D: Appl. Phys.*, **45**(10), pp. 103001.

18. Hu, L. B., Hecht, D. S., and Gruner, G. (2010). Carbon nanotube thin films: Fabrication, properties, and applications, *Chem. Rev.*, **110**(10), pp. 5790–5844.

19. Harvey, J. D., Jena, P. V., Baker, H. A., Zerze, G. H., Williams, R. M., Galassi, T. V., Roxbury, D., Mittal, J., and Heller, D. A. (2017). A carbon nanotube reporter of miRNA hybridization events in vivo, *Nat. Biomed. Eng.*, **1**, pp. 0041.

20. Yamada, T., Hayamizu, Y., Yamamoto, Y., Yomogida, Y., Izadi-Najafabadi, A., Futaba, D. N., and Hata, K. (2011). A stretchable carbon nanotube strain sensor for human-motion detection, *Nat. Nanotechnol.*, **6**(5), pp. 296–301.

21. Lee, S., Reuveny, A., Reeder, J., Lee, S., Jin, H., Liu, Q., Yokota, T., Sekitani, T., Isoyama, T., Abe, Y., Suo, Z., and Someya, T. (2016). A transparent bending-insensitive pressure sensor, *Nat. Nanotechnol.*, **11**(5), pp. 472–478.

22. Wang, X. W., He, X. X., Zhu, H. F., Sun, L. F., Fu, W., Wang, X. L., Hoong, L. C., Wang, H., Zeng, Q. S., Zhao, W., Wei, J., Jin, Z., Shen, Z. X., Liu, J., Zhang, T., and Liu, Z. (2016). Sub-atom deformation driven by vertical piezoelectricity from CdS ultrathin films, *Sci. Adv.*, **2**, pp. e1600209.

23. Choong, C. L., Shim, M. B., Lee, B. S., Jeon, S., Ko, D. S., Kang, T. H., Bae, J., Lee, S. H., Byun, K. E., Im, J., Jeong, Y. J., Park, C. E., Park, J. J., and Chung, U. I. (2014). Highly stretchable resistive pressure sensors using a conductive elastomeric composite on a micropyramid array, *Adv. Mater.*, **26**(21), pp. 3451–3458.

24. Pan, L., Chortos, A., Yu, G., Wang, Y., Isaacson, S., Allen, R., Shi, Y., Dauskardt, R., and Bao, Z. (2014). An ultra-sensitive resistive pressure

sensor based on hollow-sphere microstructure induced elasticity in conducting polymer film, *Nat. Commun.,* **5**, pp. 3002.

25. Park, M., Im, J., Shin, M., Min, Y., Park, J., Cho, H., Park, S., Shim, M. B., Jeon, S., Chung, D. Y., Bae, J., Park, J., Jeong, U., and Kim, K. (2012). Highly stretchable electric circuits from a composite material of silver nanoparticles and elastomeric fibres, *Nat. Nanotechnol.,* **7**(12), pp. 803–809.
26. Yan, C., Wang, J., Kang, W., Cui, M., Wang, X., Foo, C. Y., Chee, K. J., and Lee, P. S. (2014). Highly stretchable piezoresistive graphene-nanocellulose nanopaper for strain sensors, *Adv. Mater.,* **26**(13), pp. 2022–2027.
27. Yao, S. and Zhu, Y. (2015). Nanomaterial-enabled stretchable conductors: Strategies, materials and devices, *Adv. Mater.,* **27**(9), pp. 1480–1511.
28. Ma, Y., Liu, N., Li, L., Hu, X., Zou, Z., Wang, J., Luo, S., and Gao, Y. (2017). A highly flexible and sensitive piezoresistive sensor based on MXene with greatly changed interlayer distances, *Nat. Commun.,* **8**, pp. 1207.
29. Zhang, J., Liu, X., Neri, G., and Pinna, N. (2016). Nanostructured materials for room-temperature gas sensors, *Adv. Mater.,* **28**(5), pp. 795–831.
30. Zhao, J., Li, N., Yu, H., Wei, Z., Liao, M., Chen, P., Wang, S., Shi, D., Sun, Q., and Zhang, G. (2017). Highly sensitive MoS_2 humidity sensors array for noncontact sensation, *Adv. Mater.,* **29**(34), pp. 1702076.
31. Liu, X., Jin, X., and Ma, P. X. (2011). Nanofibrous hollow microspheres self-assembled from star-shaped polymers as injectable cell carriers for knee repair, *Nat. Mater.,* **10**(5), pp. 398–406.
32. Wang, H., Liu, H., Zhao, Q., Cheng, C., Hu, W., and Liu, Y. (2016). Three-component integrated ultrathin organic photosensors for plastic optoelectronics, *Adv. Mater.,* **28**(4), pp. 624–630.
33. Smith, T. T., Stephan, S. B., Moffett, H. F., McKnight, L. E., Ji, W., Reiman, D., Bonagofski, E., Wohlfahrt, M. E., Pillai, S. P. S., and Stephan, M. T. (2017). In situ programming of leukaemia-specific T cells using synthetic DNA nanocarriers, *Nat. Nanotechnol.,* **12**(8), pp. 813–820.
34. Kim, S. M., Jeon, H., Shin, S. H., Park, S. A., Jegal, J., Hwang, S. Y., Oh, D. X., and Park, J. (2018). Superior toughness and fast self-healing at room temperature engineered by transparent elastomers, *Adv. Mater.,* **30**(1), pp. 1705145.
35. Sun, Y., Lopez, J., Lee, H. W., Liu, N., Zheng, G., Wu, C. L., Sun, J., Liu, W., Chung, J. W., Bao, Z., and Cui, Y. (2016). A stretchable graphitic carbon/

Si anode enabled by conformal coating of a self-healing elastic polymer, *Adv. Mater.*, **28**(12), pp. 2455–2461.

36. Liu, Z. F., Fang, S., Moura, F. A., Ding, J. N., Jiang, N., Di, J., Zhang, M., Lepro, X., Galvao, D. S., Haines, C. S., Yuan, N. Y., Yin, S. G., Lee, D. W., Wang, R., Wang, H. Y., Lv, W., Dong, C., Zhang, R. C., Chen, M. J., Yin, Q., Chong, Y. T., Zhang, R., Wang, X., Lima, M. D., Ovalle-Robles, R., Qian, D., Lu, H., and Baughman, R. H. (2015). Hierarchically buckled sheath-core fibers for superelastic electronics: Sensors and muscles, *Science,* **349**(6246), pp. 400–404.
37. Wang, H., Liu, Z., Ding, J., Lepro, X., Fang, S., Jiang, N., Yuan, N., Wang, R., Yin, Q. L W., Liu, Z., Zhang, M., Ovalle-Robles, R., Inoue, K., Yin, S., and Baughman, R. H. (2016). Downsized sheath-core conducting fibers for weavable superelastic wires, biosensors, supercapacitors, and strain sensors, *Adv. Mater.,* **28**(25), pp. 4998–5007.
38. Kim, A., Ahn, J., Hwang, H., Lee, E., and Moon, J. (2017). A pre-strain strategy for developing a highly stretchable and foldable one-dimensional conductive cord based on an Ag nanowire network, *Nanoscale,* **9**(18), pp. 5773–5778.
39. Wang, X., Qiu, Y., Cao, W., and Hu, P. (2015). Highly stretchable and conductive core-sheath chemical vapor deposition graphene fibers and their applications in safe strain sensors, *Chem. Mater.,* **27**(20), pp. 6969–6975.
40. Wei, Y., Chen, S., Yuan, X., Wang, P., and Liu, L. (2016). Multiscale wrinkled microstructures for piezoresistive fibers, *Adv. Funct. Mater.,* **26**(28), pp. 5078–5085.
41. Hua, C., Shang, Y., Li, X., Hu, X., Wang, Y., Wang, X., Zhang, Y., Li, X., Duan, H., and Cao, A. (2016). Helical graphene oxide fibers as a stretchable sensor and an electrocapillary sucker, *Nanoscale,* **8**(20), pp. 10659–10668.
42. Cooper, C. B., Arutselvan, K., Liu, Y., Armstrong, D., Lin, Y., Khan, M. R., Genzer, J., and Dickey, M. D. (2017). Stretchable capacitive sensors of torsion, strain, and touch using double helix liquid metal fibers, *Adv. Funct. Mater.,* **27**(20), pp. 1605630.
43. Choi, C., Lee, J. M., Kim, S. H., Kim, S. J., Di, J., and Baughman, R. H. (2016). Twistable and stretchable sandwich structured fiber for wearable sensors and supercapacitors, *Nano Lett.,* **16**(12), pp. 7677–7684.
44. Zhong, J., Zhong, Q., Hu, Q., Wu, N., Li, W., Wang, B., Hu, B., and Zhou, J. (2015). Stretchable self-powered fiber-based strain sensor, *Adv. Funct. Mater.,* **25**(12), pp. 1798–1803.

45. Gong, W., Hou, C., Guo, Y., Zhou, J., Mu, J., Li, Y., Zhang, Q., and Wang, H. (2017). A wearable, fibroid, self-powered active kinematic sensor based on stretchable sheath-core structural triboelectric fibers, *Nano Energy,* **39**, pp. 673–683.

46. Mannsfeld, S. C., Tee, B. C., Stoltenberg, R. M., Chen, C. V., Barman, S., Muir, B. V., Sokolov, A. N., Reese, C., and Bao, Z. (2010). Highly sensitive flexible pressure sensors with microstructured rubber dielectric layers, *Nat. Mater.,* **9**(10), pp. 859–864.

47. Jang, K. I., Li, K., Chung, H. U., Xu, S., Jung, H. N., Yang, Y., Kwak, J. W., Jung, H. H., Song, J., Yang, C., Wang, A., Liu, Z., Lee, J. Y., Kim, B. H., Kim, J. H., Lee, J., Yu, Y., Kim, B. J., Jang, H., Yu, K. J., Kim, J., Lee, J. W., Jeong, J. W., Song, Y. M., Huang, Y., Zhang, Y., and Rogers, J. A. (2017). Self-assembled three-dimensional network designs for soft electronics, *Nat. Commun.,* **8**, pp. 15894.

48. Duan, Y., Huang, Y., Yin, Z., Bu, N., and Dong, W. (2014). Non-wrinkled, highly stretchable piezoelectric devices by electrohydrodynamic direct-writing, *Nanoscale,* **6**(6), pp. 3289–3295.

49. Kim, K. K., Hong, S., Cho, H. M., Lee, J., Suh, Y. D., Ham, J., and Ko, S. H. (2015). Highly sensitive and stretchable multidimensional strain sensor with prestrained anisotropic metal nanowire percolation networks, *Nano Lett.,* **15**(8), pp. 5240–5247.

50. Shin, U. H., Jeong, D. W., Kim, S. H., Lee, H. W., and Kim, J. M. (2014). Elastomer-infiltrated vertically aligned carbon nanotube film-based wavy-configured stretchable conductors, *ACS Appl. Mater. Inter.,* **6**(15), pp. 12909–12914.

51. Lee, W. K., Kang, J., Chen, K. S., Engel, C. J., Jung, W. B., Rhee, D., Hersam, M. C., and Odom, T. W. (2016). Multiscale, hierarchical patterning of graphene by conformal wrinkling, *Nano Lett.,* **16**(11), pp. 7121–7127.

52. Yin, L., Hwang, K.-C. Rogers, J. A., and Huang, Y. (2014). Experimental and theoretical studies of serpentine microstructures bonded to prestrained elastomers for stretchable electronics, *Adv. Funct. Mater.,* **24**(14), pp. 2028–2037.

53. Zhang, Y., Xu, S., Fu, H., Lee, J., Su, J., Hwang, K. C., Rogers, J. A., and Huang, Y. (2013). Buckling in serpentine microstructures and applications in elastomer-supported ultra-stretchable electronics with high areal coverage, *Soft Matter.,* **9**(33), pp. 8062–8070.

54. Ma, Y., Jang, K. I., Wang, L., Jung, H. N., Kwak, J. W., Xue, Y., Chen, H., Yang, Y., Shi, D., Feng, X. P., Rogers, J. A. P., and Huang, Y. P. (2016). Design of strain-limiting substrate materials for stretchable and flexible electronics, *Adv. Funct. Mater.,* **26**(29), pp. 5345–5351.

55. Bandodkar, A. J., Jeerapan, I., You, J. M., Nunez-Flores, R., and Wang, J. (2016). Highly stretchable fully-printed CNT-based electrochemical sensors and biofuel cells, combining intrinsic and design-induced stretchability, *Nano Lett.*, **16**(1), pp. 721–727.

56. Miyamoto, A., Lee, S., Cooray, N. F., Lee, S., Mori, M., Matsuhisa, N., Jin, H., Yoda, L., Yokota, T., Itoh, A., Sekino, M., Kawasaki, H., Ebihara, T., Amagai, M., and Someya, T. (2017). Inflammation-free, gas-permeable, lightweight, stretchable on-skin electronics with nanomeshes, *Nat. Nanotechnol.*, **12**(9), pp. 907–913.

57. Luo, N., Huang, Y., Liu, J., Chen, S. C., Wong, C. P., and Zhao, N. (2017). Hollow-structured graphene-silicone-composite-based piezoresistive sensors, decoupled property tuning and bending reliability, *Adv. Mater.*, **29**(40), pp. 1702675.

58. Wang, H., Lu, W., Di, J., Li, D., Zhang, X., Li, M., Zhang, Z., Zheng, L., and Li, Q. (2017). Ultra-lightweight and highly adaptive all-carbon elastic conductors with stable electrical resistance, *Adv. Funct. Mater.*, **27**(13), pp. 1606220.

59. Schutt, F., Signetti, S., Kruger, H., Roder, S., Smazna, D., Kaps, S., Gorb, S. N., Mishra, Y. K., Pugno, N. M., and Adelung, R. (2017). Hierarchical self-entangled carbon nanotube tube networks, *Nat. Commun.*, **8**, pp. 1215.

60. Qin, Y. Y., Peng, Q. Y., Ding, Y. J., Lin, Z. S., Wang, C. H., Li, Y., Li, J. J., Yuan, Y., He, X. D., and Li, Y. B. (2015). Lightweight, superelastic, and mechanically flexible graphene/polyimide nanocomposite foam for strain sensor application, *ACS Nano*, **9**(9), pp. 8933–8941.

61. Tang, Y., Gong, S., Chen, Y., Yap, L. W., and Cheng, W. L. (2014). Manufacturable conducting rubber ambers and stretchable conductors from copper nanowire aerogel monoliths, *ACS Nano*, **8**(6), pp. 5707–5714.

62. He, W. N., Li, G. Y., Zhang, S. Q., Wei, Y., Wang, J., Li, Q. W., and Zhang, X. T. (2015). Polypyrrole/silver coaxial nanowire aero-sponges for temperature-independent stress sensing and stress-triggered joule heating, *ACS Nano*, **9**(4), pp. 4244–4251.

63. Xu, S., Yan, Z., Jang, K. I., Huang, W., Fu, H. R., Kim, J., Wei, Z., Flavin, M. McCracken, J., Wang, R., Badea, A., Liu, Y., Xiao, D. Q., Zhou, G. Y., Lee, J., Chung, H. U., Cheng, H. Y., Ren, W., Banks, A., Li, X. L., Paik, U., Nuzzo, R. G., Huang, Y. G., Zhang, Y. H., and Rogers, J. A. (2015). Assembly of micro/nanomaterials into complex, three-dimensional architectures by compressive buckling, *Science*, **347**(6218), pp. 154–159.

64. Yan, Z., Han, M., Yang, Y., Nan, K., Luan, H., Luo, Y., Zhang, Y., Huang, Y., and Rogers, J. A. (2017). Deterministic assembly of 3D mesostructures

in advanced materials via compressive buckling: A short review of recent progress, *Extreme Mechanics Lett.*, **11**, pp. 96–104.

65. Han, S. T., Peng, H., Sun, Q., Venkatesh, S., Chung, K. S., Lau, S. C., Zhou, Y., and Roy, V. A. L. (2017). An overview of the development of flexible sensors, *Adv. Mater.*, **29**(33), pp. 1700375.

66. Hua, Q., Sun, J., Liu, H., Bao, R., Yu, R., Zhai, J., Pan, C., and Wang, Z. L. (2018). Skin-inspired highly stretchable and conformable matrix networks for multifunctional sensing, *Nat. Commun.*, **9**, pp. 244.

67. Kang, D., Pikhitsa, P. V., Choi, Y. W., Lee, C., Shin, S. S., Piao, L., Park, B., Suh, K. Y., Kim, T. I., and Choi, M. (2014). Ultrasensitive mechanical crack-based sensor inspired by the spider sensory system, *Nature*, **516**(7530), pp. 222–226.

68. Zeng, S., Zhang, D., Huang, W., Wang, Z., Freire, S. G., Yu, X., Smith, A. T., Huang, E. Y., Nguon, H., and Sun, L. (2016). Bio-inspired sensitive and reversible mechanochromisms via strain-dependent cracks and folds, *Nat. Commun.*, **7**, pp. 11802.

69. Yamada, T., Hayamizu, Y., Yamamoto, Y., Yomogida, Y., Izadi-Najafabadi, A., Futaba, D. N., and Hata, K. (2011). A stretchable carbon nanotube strain sensor for human-motion detection, *Nat. Nanotechnol.*, **6**(5), pp. 296–301.

70. Chortos, A., Koleilat, G. I., Pfattner, R., Kong, D., Lin, P., Nur, R., Lei, T., Wang, H., Liu, N., Lai, Y. C., Kim, M. G., Chung, J. W., Lee, S., and Bao, Z. (2016). Mechanically durable and highly stretchable transistors employing carbon nanotube semiconductor and electrodes, *Adv. Mater.*, **28**(22), pp. 4441–4448.

71. Di, J., Zhang, X., Yong, Z., Zhang, Y., Li, D., Li, R., and Li, Q. (2016). Carbon-nanotube fibers for wearable devices and smart textiles, *Adv. Mater.*, **28**(47), pp. 10529–10538.

72. Wang, C., Takei, K., Takahashi, T., and Javey, A. (2013). Carbon nanotube electronics-moving forward, *Chem. Soc. Rev.*, **42**(7), pp. 2592–2609.

73. Chen, K., Gao, W., Emaminejad, S., Kiriya, D., Ota, H., Nyein, H. Y., Takei, K., and Javey, A. (2016). Printed carbon nanotube electronics and sensor systems, *Adv. Mater.*, **28**(22), pp. 4397–4414.

74. Liu, Z., Xu, J., Chen, D., and Shen, G. (2015). Flexible electronics based on inorganic nanowires, *Chem. Soc. Rev.*, **44**(1), pp. 161–192.

75. Gong, S., Schwalb, W., Wang, Y., Chen, Y., Tang, Y., Si, J., Shirinzadeh, B., and Cheng, W. (2014). A wearable and highly sensitive pressure sensor with ultrathin gold nanowires, *Nat. Commun.*, **5**, pp. 3132.

76. Xiao, X., Yuan, L., Zhong, J., Ding, T., Liu, Y., Cai, Z., Rong, Y., Han, H., Zhou, J., and Wang, Z. L. (2011). High-strain sensors based on ZnO nanowire/polystyrene hybridized flexible films, *Adv. Mater.,* **23**(45), pp. 5440–5444.

77. Jang, H., Park, Y. J., Chen, X., Das, T., Kim, M. S., and Ahn, J. H. (2016). Graphene-based flexible and stretchable electronics, *Adv. Mater.,* **28**(22), pp. 4184–4202.

78. Ho, D. H., Sun, Q., Kim, S. Y., Han, J. T., Kim, D. H., and Cho, J. H. (2016). Stretchable and multimodal all-graphene electronic skin, *Adv. Mater.,* **28**(13), pp. 2601–2608.

79. Park, M., Park, Y. J., Chen, X., Park, Y. K., Kim, M. S., and Ahn, J. H. (2016). MoS_2-based tactile sensor for electronic skin applications, *Adv. Mater.,* **28**(13), pp. 2556–2562.

80. Lim, Y. R., Song, W., Han, J. K., Lee, Y. B., Kim, S. J., Myung, S., Lee, S. S., An, K. S., Choi, C. J., and Lim, J. (2016). Wafer-scale, homogeneous MoS_2 layers on plastic substrates for flexible visible-light photodetectors, *Adv. Mater.,* **28**(25), pp. 5025–5030.

81. Ma, Y., Liu, N., Li, L., Hu, X., Zou, Z., Wang, J., Luo, S., and Gao, Y. (2017). A highly flexible and sensitive piezoresistive sensor based on MXene with greatly changed interlayer distances, *Nat. Commun.,* **8**, pp. 1207.

82. Hirsch, A., Michaud, H. O., Gerratt, A. P., de Mulatier, S., and Lacour, S. P. (2016). Intrinsically stretchable biphasic (solid–liquid) thin metal films, *Adv. Mater.,* **28**(22), pp. 4507–4512.

83. Zhong, Q., Zhong, J., Cheng, X., Yao, X., Wang, B., Li, W., Wu, N., Liu, K., Hu, B., and Zhou, J. (2015), Paper-based active tactile sensor array, *Adv. Mater.,* **27**(44), pp. 7130–7136.

84. Huynh, T. P., Sonar, P., and Haick, H. (2017). Advanced materials for use in soft self-healing devices, *Adv. Mater.,* **29**(19), pp. 1604973.

85. Kim, J., Rim, Y. S., Chen, H. J., Cao, H. H., Nakatsuka, N., Hinton, H. L., Zhao, C. Z., Andrews, A. M., Yang, Y., and Weiss, P. S. (2015). Fabrication of high-performance ultrathin In_2O_3 film field-effect transistors and biosensors using chemical lift-off lithography, *ACS Nano,* **9**(4), pp. 4572–4582.

86. Lang, C., Fang, J., Shao, H., Ding, X., and Lin, T. (2016). High-sensitivity acoustic sensors from nanofibre webs, *Nat. Commun.,* **7**, pp. 11108.

87. Chen, Z., Wang, Z., Li, X., Lin, Y., Luo, N., Long, M., Zhao, N., and Xu, J. B. (2017). Flexible piezoelectric-induced pressure sensors for static measurements based on nanowires/graphene heterostructures, *ACS Nano,* **11**(5), pp. 4507–4513.

88. Pu, X., Liu, M. M., Chen, X. Y., Sun, J. M., Du, C. H., Zhang, Y., Zhai, J. Y., Hu, W. G., and Wang, Z. L. (2017). Ultrastretchable transparent triboelectric nanogenerator as electronic skin for biomechanical energy harvesting and tactile sensing, *Sci. Adv.,* **3**(5), pp. e1700015.

89. Li, T., Zou, J., Xing, F., Zhang, M., Cao, X., Wang, N., and Wang, Z. L. (2017). From dual-mode triboelectric nanogenerator to smart tactile sensor: A multiplexing design, *ACS Nano,* **11**(4), pp. 3950–3956.

90. Root, S. E., Savagatrup, S., Printz, A. D., Rodriquez, D., and Lipomi, D. J. (2017). Mechanical properties of organic semiconductors for stretchable, highly flexible, and mechanically robust electronics, *Chem. Rev.,* **117**(9), pp. 6467–6499.

91. Oh, J. Y., Kim, S., Baik, H. K., and Jeong, U. (2016). Conducting polymer dough for deformable electronics, *Adv. Mater.,* **28**(22), pp. 4455–4461.

92. Yao, Y., Dong, H., and Hu, W. (2016). Charge transport in organic and polymeric semiconductors for flexible and stretchable devices, *Adv. Mater.,* **28**(22), pp. 4513–4523.

93. Wang, L., Chen, D., Jiang, K., and Shen, G. (2017). New insights and perspectives into biological materials for flexible electronics, *Chem. Soc. Rev.,* **46**(22), pp. 6764–6815.

94. Montero, de Espinosa, L., Meesorn, W., Moatsou, D., and Weder, C. (2017). Bioinspired polymer systems with stimuli-responsive mechanical properties, *Chem. Rev.,* **117**(20), pp. 12851–12892.

95. Wang, L., Jackman, J. A., Tan, E.-L., Park, J. H., Potroz, M. G., Hwang, E. T., and Cho, N. J. (2017). High-performance flexible electronic skin sensor incorporating natural microcapsule actuators, *Nano Energy,* **36,** pp. 38–45.

96. Wang, Q., Jian, M. Q., Wang, C. Y., and Zhang, Y. Y. (2017). Carbonized silk nanofiber membrane for transparent and sensitive electronic skin, *Adv. Funct. Mater.,* **27**. pp. 1605657.

97. Zhu, B., Wang, H., Leow, W. R., Cai, Y., Loh, X. J., Han, M. Y., and Chen, X. (2016). Silk fibroin for flexible electronic devices, *Adv. Mater.,* **28**(22), pp. 4250–4265.

98. Irimia-Vladu, M. (2014). “Green” electronics, biodegradable and biocompatible materials and devices for sustainable future, *Chem. Soc. Rev.,* **43**(2), pp. 588–610.

99. Boutry, C. M., Nguyen, A., Lawal, Q. O., Chortos, A., Rondeau-Gagne, S., and Bao, Z. (2015). A sensitive and biodegradable pressure sensor array for cardiovascular monitoring, *Adv. Mater.,* **27**(43), pp. 6954–6961.

100. Curry, E. J., Ke, K., Chorsi, M. T., Wrobel, K. S., Miller III, A. N., Patel, A., Kim, I., Feng, J. L., Yue, L. X., Wu, Q., Kuo, C.-L., Lo, K. W.-H., Laurencin, C. T., Ilies, H., Purohit, P. K., and Nguyen, T. D. (2018). Biodegradable piezoelectric force sensor, *Proc. Natl. Acad. Sci. USA*, **115**(5), pp. 909–914.

101. Benight, S. J., Wang, C., Tok, J. B. H., and Bao, Z. (2013). Stretchable and self-healing polymers and devices for electronic skin, *Prog. Polym. Sci.*, **38**(12), pp. 1961–1977.

102. Tee, B. C., Wang, C., Allen, R., and Bao, Z. (2012). An electrically and mechanically self-healing composite with pressure- and flexion-sensitive properties for electronic skin applications, *Nat. Nanotechnol.*, **7**(12), pp. 825–832.

103. Hou, C., Huang, T., Wang, H., Yu, H., Zhang, Q., and Li, Y. (2013). A strong and stretchable self-healing film with self-activated pressure sensitivity for potential artificial skin applications, *Sci. Rep.*, **3**, pp. 3138.

104. Phadke, A., Zhang, C., Arman, B., Hsu, C. C., Mashelkar, R. A., Lele, A. K., Tauber, M. J., Arya, G., and Varghese, S. (2012). Rapid self-healing hydrogels, *Proc. Natl. Acad. Sci. USA*, **109**(12), pp. 4383–4388.

105. Lin, S., Yuk, H., Zhang, T., Parada, G. A., Koo, H., Yu, C., and Zhao, X. (2016). Stretchable hydrogel electronics and devices, *Adv. Mater.*, **28**(22), pp. 4497–4505.

106. Keplinger, C., Sun, J. Y., Foo, C. C., Rothemund, P., Whitesides, G. M., and Suo, Z. G. (2013). Stretchable, transparent, ionic conductors, *Science*, **341**(6149), pp. 984–987.

107. Wang, X. W., Gu, Y., Xiong, Z. P., Cui, Z., and Zhang, T. (2013). Silk-molded flexible, ultrasensitive, and highly stable electronic skin for monitoring human physiological signals, *Adv. Mater.*, **26**, pp. 1336–1342.

108. Gerratt, A. P., Michaud, H. O., and Lacour, S. P. (2015). Elastomeric electronic skin for prosthetic tactile sensation, *Adv. Funct. Mater.*, **25**(15), pp. 2287–2295.

109. Liu, W. H., Zhang, C., Lin, H. S., Qu, W. M., Li, X., and Wang, X. L. (2017). Texture and sliding motion sensation with a triboelectric-nanogenerator transducer, *Sensors Actuators A*, **256**, pp. 89–94.

110. Chun, S., Hong, A., Choi, Y., Ha, C., and Park, W. (2016). A tactile sensor using a conductive graphene-sponge composite, *Nanoscale*, **8**, pp. 9185–9192.

111. Park, J., Kim, M., Lee, Y., Lee, H. S., and Ko, H. (2015). Fingertip skin-inspired microstructured ferroelectric skins discriminate static/dynamic pressure and temperature stimuli, *Sci. Adv.,* **1**, pp. 1500661.

112. Tao, L. Q., Zhang, K. N., Tian, H., Liu, Y., Wang, D. Y., Chen, Y. Q., Yang, Y., and Ren, T. L. (2017). Graphene-paper pressure sensor for detecting human motions, *ACS Nano,* **11**(9), pp. 8790–8795.

113. Zhao, J. J., Han, S., Yang, Y., Fu, R. P., Ming, Y., Lu, C., Liu, H., Gu, H. W., and Chen, W. (2017). Passive and space-discriminative ionic sensors based on durable nanocomposite electrodes toward sign language recognition, *ACS Nano*, **11**, pp. 8590–8599.

114. Li, L. L., Bai, Y. Y., Li, L. L., Wang, S. Q., and Zhang, T. (2017). A superhydrophobic smart coating for flexible and wearable sensing electronics, *Adv. Mater.,* **29**, pp. 1702517

115. Khan, Y., Ostfeld, A. E., Lochner, C. M., Pierre, A., and Arias, A. C. (2016). Monitoring of vital signs with flexible and wearable medical devices, *Adv. Mater.*, **28**(22), pp. 4373–4395.

116. Schwartz, G., Tee, B. C.-K., Mei, J. G., Appleton, A. L., Kim, D. H., Wang, H. L., and Bao, Z. N. (2013). Flexible polymer transistors with high pressure sensitivity for application in electronic skin and health monitoring, *Nat. Commun.,* **4**, pp. 1859.

117. Choi, S., Lee, H., Ghaffari, R., Hyeon, T., and Kim, D. H. (2016). Recent advances in flexible and stretchable bio-electronic devices integrated with nanomaterials, *Adv. Mater.*, **28**(22), pp. 4203–4218.

118. Pang, C., Koo, J. H., Nguyen, A., Caves, J. M., Kim, M. G., Chortos, A., Kim, K., Wang, P. J., Tok, J. B.-H., and Bao, Z. (2015). Highly skin-conformal microhairy sensor for pulse signal amplification, *Adv. Mater.,* **27**, pp. 634–640.

119. Kim, J., Kim, M., Lee, M. S., Kim, K., Ji, S., Kim, Y. T., Park, J., Na, K., Bae, K. H., Kyun, K. H., Bien, F., Young, L. C., and Park, J. U. (2017). Wearable smart sensor systems integrated on soft contact lenses for wireless ocular diagnostics, *Nat. Commun.,* **8**, pp. 14997.

120. Song, Y. M., Xie, Y. Z., Malyarchuk, V., Xiao, J. L., Jung, I., Choi, K.-J., Liu, Z. J., Park, H., Lu, C. F., Kim, R.-H., Li, R., Crozier, K. B., Huang, Y. G., and Rogers, J. A. (2013). Digital cameras with designs inspired by the arthropod eye, *Nature*, **497**, pp. 95–99.

121. Henning, A., Swaminathan, N., Godkin, A., Shalev, G., Amit, I., and Rosenwaks, Y. (2015). Tunable diameter electrostatically formed nanowire for high sensitivity gas sensing, *Nano Res.,* **8**(7), pp. 2206–2215.

122. Gu, Y., Wang, X. W., Gu, W., Wu, Y. J., Li, T., and Zhang, T. (2017). Flexible electronic eardrum, *Nano Res.,* **10**(8), pp. 2683–2691.

123. Trung, T. Q. and Lee, N. E. (2016). Flexible and stretchable physical sensor integrated platforms for wearable human-activity monitoring and personal healthcare, *Adv. Mater.,* **28**(22), pp. 4338–4372.

124. Li, T., Li, L. H., Sun, H. W., Xu, Y., Wang, X. W., Luo, H., Liu, Z., and Zhang, T. (2017). Porous ionic membrane based flexible humidity sensor and its multifunctional applications, *Adv. Sci.,* **4**, pp.1600404.

125. Kim, J., Campbell, A. S., and Wang, J. (2018). Wearable non-invasive epidermal glucose sensors: A review, *Talanta,* **177**, pp. 163–170.

126. Wang, S. Q., Wu, Y. J., Gu, Y., Li, T., Luo, H., Li, L. H., Bai, Y. Y., Li, L.L. Liu, L., Cao, Y. D., Ding, H. Y., and Zhang, T. (2017). Wearable sweatband sensor platform based on gold nanodendrite array as efficient solid contact of ion-selective electrode, *Anal. Chem.,* **89**, pp. 10224–10231.

127. Gao, W., Emaminejad, S., Nyein, H. Y. Y., Challa, S., Chen, K., Peck, A., Fahad, H. M., Ota, H., Shiraki, H., Kiriya, D., Lien, D.-H., Brooks, G. A., Davis, R. W., and Javey, A. (2016). Fully integrated wearable sensor arrays for multiplexed in situ perspiration analysis, *Nature,* **529**, pp. 509–514.

Chapter 8

Liquid Metal–Enabled Functional Flexible and Stretchable Electronics

Xuelin Wang[a] **and Jing Liu**[a,b]
[a]*Department of Biomedical Engineering, School of Medicine, Room B201, Medical Science Building, Tsinghua University, Haidian District, Beijing 100084, China*
[b]*Beijing Key Lab of CryoBiomedical Engineering and Key Lab of Cryogenics, Technical Institute of Physics and Chemistry, Chinese Academy of Sciences, Haidian District, Beijing 100190, China*
jliubme@tsinghua.edu.cn

This chapter briefly presents an overview of material properties, design strategies, and typical applications of liquid metal–enabled (LM-enabled) flexible and stretchable electronics. In particular, it focuses on the core manufacturing material: room-temperature liquid metal (RTLM), currently mainly represented by gallium and its alloys with the properties of excellent resistivity, enormous bendability, low adhesion, and large surface tension. In addition, a series of recently developed printing technologies spanning from planar electronics printing (direct writing, mechanical printing, mask-based method, high-resolution nanoimprinting,

Flexible and Stretchable Electronics: Materials, Designs, and Devices
Edited by Run-Wei Li and Gang Liu

ISBN 978-981-4800-46-4 (Hardcover), 978-0-429-05890-5 (eBook)
www.jennystanford.com

microfluidic and subtractive manufacturing, etc.) to 3D printing is comprehensively introduced. The latest developments in liquid metal (LM) soft device applications have been discussed, which include but are not limited to strain sensors, stretchable coils, e-skin, wearable bioelectronics, and conformable bioelectronics. Finally, the chapter interprets the drawbacks and challenges of LM material properties and flexible fabrication technologies, and sets goals to achieve LM intelligent flexible future.

8.1 Introduction

In the recent years, room-temperature liquid metals (RTLMs), especially gallium-based RTLMs, have attracted great attention in the field of flexible and stretchable electronics owing to their excellent electrical conductivity, fluidity, and compliance [1]. RTLM, just as its name suggests, is a kind of metal that remains in liquid phase from room temperature up to 2000°C. Unlike mercury, a well-known toxic RTLM, gallium-based RTLMs remain nontoxic and benign biocompatible with a negligible vapor pressure and have made important progress in both laboratory and industrial applications, especially in the biomedical area, such as being used as the material of bone cement, as a vascular contrast agent, and in tumor therapy [2–4]. Compared to conventional rigid silicon wafer material and new emerging conductive inorganic and organic materials, like gold, silver, copper, carbon nanotube (CNT), graphene, etc. [5–9], gallium-based RTLMs can avoid the mismatch between soft wearables and biological tissues due to their intrinsic flexibility and liquidity [10–13] and are widely utilized in stretchable wireless sensor [14–16], e-skin [17, 18], wearable bioelectronics [19–21], and soft robots [22, 23]. Besides, the rapid and efficient fabrication technologies of these RTLM soft circuits, such as microchannel [15, 24], mask-based printing [25, 26], and direct writing [27], are much simpler and cheaper than that of conventional sputter coating or plasma-enhanced chemical vapor deposition, which make RTLMs a broad application prospect in stretchable and soft electronics.

Gallium-based RTLMs preponderate in making soft electronic devices with high deformation and mechanical tunability. A plenty of papers in this field have reported multifarious flexible electronics, including self-healing wires [28], stretchable wireless power transfer [14], soft loudspeaker [29], nerve connection [30], mechanical exoskeleton with both solidness and softness [31], implantable bioelectrodes [12], etc., all of which are enabled by multifunctional RTLM manufacturing technologies. Accordingly, RTLMs can be widely used in electronics, materials, and biomedical areas.

Along with the development of material science and technology, LM-enabled flexible electronics are quickly shaping the field of flexible electronic circuits and allied machines, serving as a basic way of quickly making functional devices. In this chapter, we are dedicated to integrating liquid metal (LM) alloys with flexible electronics and focus our attention on interpreting the recent advancements in LM flexible and stretchable electronics. In particular, we review the typical soft properties of RTLMs in flexible electronics and the fabrication methods for making various stretchable devices. Finally, we conclude with the potential applications of such alloys, such as monitoring sensors, functional coils, conformable electrode, e-skin, and wearable bioelectronics.

8.2 Materials and Properties of Gallium-Based RTLMs

8.2.1 Compositions of Gallium-Based RTLM Alloys

Generally speaking, alloy elements of low melting point include gallium, bismuth, lead, tin, cadmium, and indium. Among them, EGaIn ($GaIn_{24.5}$: 75.5 wt% gallium and 24.5 wt% indium) and Galinstan ($Ga_{68.5}In_{21.5}Sn_{10}$: 68.5 wt% gallium, 21.5 wt% indium and 10 wt% tin) are commonly used. From Ref. [32], one can get information on the main physical properties of frequently used LM alloys, such as composition, melting point, density, resistivity, and thermal conductivity. For the same composition, a different proportion may lead to quite varied properties in the alloys. Sometimes, even a small change in mass ratio would cause strong variation in the material behaviors. Table 8.1 compares the basic physical properties (melting point, viscosity, surface tension, electrical conductivity, etc.) between typical gallium-based RTLMs and water. The low melting point of −19°C (Table 8.1) makes Galinstan an excellent candidate in low temperature environment. If loaded with certain microelements, the alloy may also display additional properties accordingly. Therefore, one can change the ratio of chemical materials of alloys or add some microelements to adjust the melting point and other properties to fulfill various specific needs. These low-melting LMs can preserve liquid state at room temperature, which is zero stiffness and almost infinite stretchability [12, 33]. More importantly, LMs own high electroconductivity (EGaIn: 3.4×10^6 S/m) [10], which is far higher than those of non-metallic materials (carbon: 1.8×10^3 S/m, CNT: 5.03×10^3 S/m, PEDOT:PSS: 8.25×10^3 S/m) [24, 34]. The natures of metallicity and fluidity make LMs especially suitable for manufacturing flexible circuits on various soft substrates [35, 36]. LMs in the following refer to gallium-based RTLMs.

Table 8.1 Comparison of the basic physical properties between typical gallium-based RTLMs and water

	Ga	$GaIn_{24.5}$	$Ga_{68.5}In_{21.5}Sn_{10}$	$Ga_{67}In_{25}Sn_{13}Zn_1$	Water
Melting point (°C)	29.8	15.5	−19	7.6	0
Boiling point (°C)	2204	2000	1300	900	100
Density (10^3 kg/m)	6.08	6.28	6.36	6.5	1.0
Viscosity (10^{-7} m^2/s)	3.24	2.7	2.98	0.71	11.2
Surface tension (N/m)	0.72	0.624	0.533	0.5	0.07
Electrical conductivity (10^6 S/m)	3.7	3.4	3.1	2.8	10^{-10}
Thermal conductivity (W/m°C)	29.4	42.2	44.8	48.2	0.55

8.2.2 Basic Properties of LMs in Flexible Electronics

Excellent electroconductivity: When stretching or bending an LM wire, its resistance would change correspondingly. The resistance of an LM wire is $R = \rho \cdot L/A$, where R is resistance, ρ is electrical conductivity (EGaIn: 3.4×10^6 S/m), L is the length of the deformed LM wire, and A is its cross-sectional area. The relationship of volume V and length L is $A = V/L$. Once the LM wire is encapsulated, the volume keeps constant even under stretching, that is, $V = V_0$. Therefore, the relationship between the tensile elongation and resistance variation of the resistive sensor can be calculated by Eq. (8.1), i.e.

$$\frac{R}{R_0} = \frac{\frac{\rho \cdot L}{A}}{\frac{\rho \cdot L_0}{A_0}} = \frac{V_0}{V}\left(\frac{L}{L_0}\right)^2 = \left(\frac{L}{L_0}\right)^2 \quad (8.1)$$

The resistance variation has a linear relationship with the square of the tensile length before and after stretching [37]. In Fig. 8.1A, the measured data of R/R_0 completely coincides with the fitting parabolic curve when stretching the LM strain sensor from 0% to 190% [81]. When some LM wires are connected in parallel, we can get the resistance value by Ohm's law, $R = \left(\sum_{i=1}^{n} \frac{1}{R_i}\right)^{-1}$, where n is the number of wires connected in parallel. Ref. [39] demonstrates that the measured and predicted resistance values show a high degree of agreement with the number of wires changing. Figure 8.1B shows the resistance variation of LM lines with different width [82]. Additionally, when bending the LM wire at −180°, −90°, 0°, 90°, 180°, respectively, the resistance variation is very little (Fig. 8.1C) [40]. Furthermore, the fatigue test over 1000 cycles straining illustrates the resistance stability of the LM resistor line (Fig. 8.1D) [41]. All of them provide theoretical and practical support for the development of LM flexible and stretchable electronics.

Superior flexibility: From Fig. 8.2A, we can know that LM has the lowest Young's modulus value than rigid materials (polyimide, copper, graphene, etc.), elastomers (PDMS (polydimethylsiloxane), skin, etc.), and even gels (brain tissue, muscle, etc.) [83]. Due to its intrinsic fluidity and softness, the electrical circuits fabricated by LM can be strained, bent, twisted, and compressed largely depending on the flexibility of substrates. Encapsulated by PDMS and Ecoflex, LM is designed as a soft artificial arm with the sensing ability under different bending states (Fig. 8.2B) [84]. Besides, printing LM on luminous Ecoflex silicon rubber, we can arbitrarily stretch or bend it (Fig. 8.2C) [81]. The LM fiber can maintain metallic conductivity when highly strained up to ~700% with ultra-stretchable property [43]. All the bending and stretching tests suggest that LM wire has reliable flexibility when employed in flexible electronics.

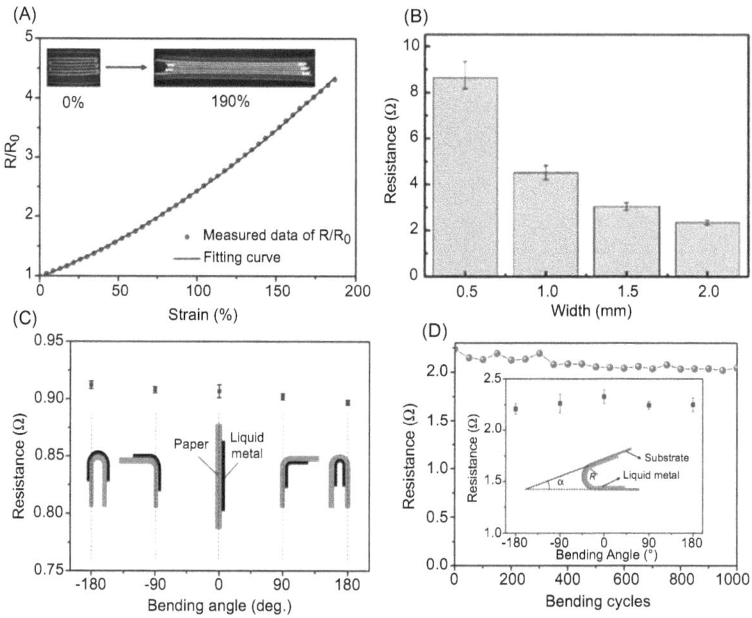

Figure 8.1 The relationship between strain and resistance of LM strain sensor from 0% to 190%. Reprinted with permission from Ref. [81], Copyright 2018, IEEE. (B) The resistances of LM lines with different width. Reprinted with permission from Ref. [82], Copyright 2018, John Wiley & Sons. (C) Resistance values of liquid metal line at five different bending angles. Reprinted with permission from Ref. [40], Copyright 2013, Springer Nature. (D) Resistance values of liquid metal line at different bending angles and different bending cycles. Reprinted with permission from Ref. [41], Copyright 2015, The Royal Society of Chemistry.

Tunable adhesion: Adhesion is the attraction among different molecules, such as the adhesion between particle and substrate. Based on the Young–Dupre equation [44], one has

$$W_{SL} = \gamma_L (1 + \cos \theta) \quad (8.2)$$

where W_{SL} is the adhesion work of the LM drop, γ_L is its surface tension, and θ is the contact angle between the droplet and substrate. The smaller the droplet, the larger the γ_L and the smaller the θ, hence the greater the W_{SL}, representing better adhesion. In other words, one can regulate the adhesion of LM ink on a substrate by controlling its oxide layer [45]. In spray-printing process, the surface of LM droplets generates a layer of oxide (Ga_2O_3/Ga_2O), which can greatly enhance its substrate adhesion [46]. In some cases, strong adhesion to a majority of substrates is undesirable during manufacturing micro-precision circuits [47].

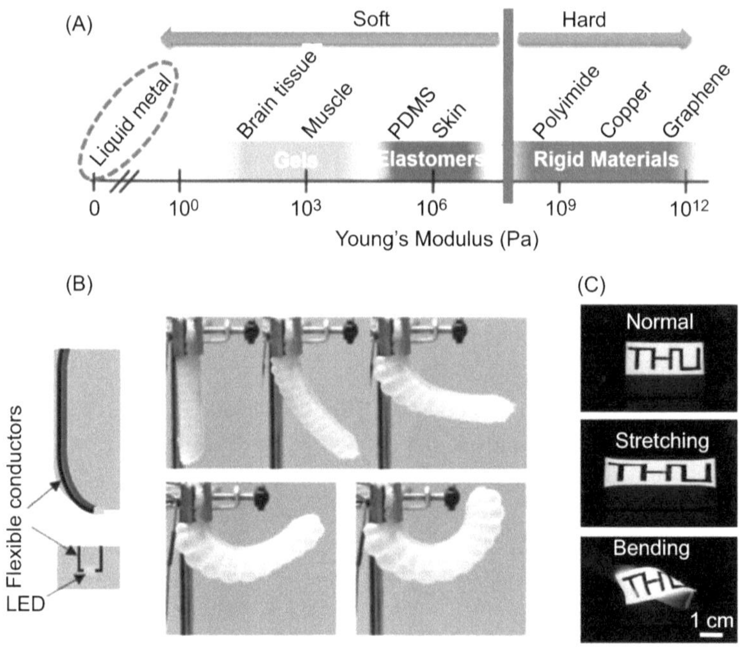

Figure 8.2 Superior flexibility of LM materials as flexible electronics. (A) The comparation of Young's modulus among liquid metal and common soft and rigid materials [83]. (B) Nano-Fe GaIn LM circuit with LED for smart flexible pneumatic actuator under different bending state. Reprinted with permission from Ref. [84], Copyright 2018, IOP Publishing. (C) Straining and bending displays of "THU" printed by LM on luminous substrate at night. Reprinted with permission from Ref. [81], Copyright 2018, IEEE.

Prominent wettability: Wettability is the ability of liquid to spread on a solid surface. The contact angle usually expresses the wettability: The smaller the contact angle, the better the wettability. The oxide skin outside LM can improve the wettability between metal material and substrate [46]. When increasing the oxygen content, the resistivity of LM decreases while the viscosity and the wettability increase. For this reason, we can control the thickness of the oxide layer and then the wettability of LM wire on soft substrate to obtain a superior performance of flexible printed circuit. Additionally, Gao et al. have disclosed the excellent wettability of LM $GaIn_{10}$ on various substrate materials with different surface roughness and material properties [48, 49].

8.3 Design and Fabrication of LM Flexible Electronics

8.3.1 Planar Electronics Printing

The invention of an excellent printing method has a huge impact on human life, such as the invention of movable-type printing method by Sheng Bi, a brilliant inventor in ancient China, with great spread of knowledge and culture. Nowadays, due to its high surface tension and good conductivity, LM has attracted great attention in printing circuit. We have proposed the concept of DREAM ink: direct writing or printing of electronics based on alloy and metal ink [32], and so far, many printing technologies have been applied to fabricate plenty of flexible electronic devices [48, 50, 51].

Direct writing: Because of its gravity and adhesion to substrates, LM can be directly installed in the refill as a kind of ink to write conductive texts or lines [52, 53]. Zheng et al. developed the first ever LM roller-ball pen (LMRP), which can write conductive lines or electronic devices on soft plates with the tip diameters ranging from 200 μm to 1000 μm [27]. Similarly, Boley et al. illustrated a direct-writing system, using LM to fabricate small-scale stretchable electronics [38]. Guo et al. reported a one-step liquid metal transfer printing method, by which LM could be directly and simply fabricated on wide range of substrates (Fig. 8.3A) [82]. Besides, increasing the oxidation degree or mixing it with other micro-/nanoparticles can largely improve the adhesion of LM on multifarious soft substrates, which guarantees the direct-writing or painting methods [54, 55].

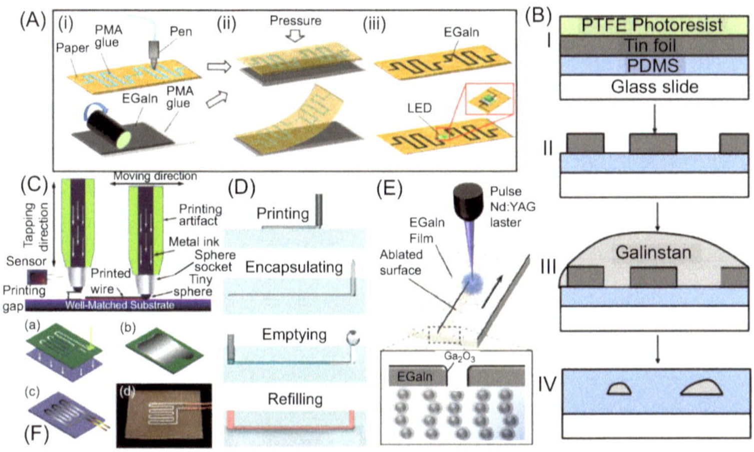

Figure 8.3 Planar electronics printing methods for fabricating LM flexible electronics. (A) Schematic illustration of LMs-based transfer printing method. Reprinted with permission from Ref. [82], Copyright 2018, John Wiley & Sons. (B) The masked deposition process of gallium–indium alloys for liquid-embedded elastomer conductors (modified from Ref. [26]). (C) The tapping mode mechanical printing method. Reprinted with permission from Ref. [57], Copyright 2014, Springer Nature. (D) Fabrication process of LM microchannels by direct-writing. Reprinted with permission from Ref. [24], Copyright 2017, The Royal Society of Chemistry. (E) Subtractive manufacturing of LM soft-matter electronics by UV laser. Reprinted with permission from Ref. [59], Copyright 2017, American Chemical Society. (F) Mask-based microcontact printing process to fabricate LM soft-matter circuit. Reprinted with permission from Ref. [60], Copyright 2013, American Chemical Society.

Mechanical printing: Compared to conventional direct-writing method, the mechanical printing method can realize digital control, which makes the design and fabrication process of LM flexible electronic circuits more accurate, smart, and rapid. Via a digital control system, a versatile desktop LM printer has been designed for manufacturing plane circuit or even 3D metal object [40]. Another common mechanical printing method is tapping mode printing [57]. From Fig. 8.3D, it is seen that when printing, the printer head falls to contact the base and then rises a certain height over the printing substrate to avoid the printed line. In the process of printing, the printing head bead rotates due to the basal friction, which makes the LM ink in the pen container run downwards. One can use the tapping mode method to manufacture various printed circuit boards (PCBs), antennas, and so on. In addition, other mechanical printing methods to print personal electronic circuits using LM include inkjet printing and coelectrospinning [61, 62].

Mask-based printing: Mask-based printings are the most commonly used methods to manufacture multifunctional LM flexible electronics in the laboratory. Kramer et al. provided a masked deposition method to fabricate LM hyperelastic electronic circuits (Fig. 8.3C) [26]. First, tin foil is patterned on the PDMS surface with photolithography (Fig. 8.3C(I)). After some time, a microchannel on the elastomer surface is formed (Fig. 8.3C(II)). And then Galinstan is flooded on the surface (Fig. 8.3C(III)). After removing excess Galinstan with a thin film, we cool the Galinstan in a freezer and encapsulate it by coating with an elastomer to fabricate the liquid-embedded elastomers (Fig. 8.3C(IV)). Besides, Gozen et al. used the masked deposition method to prepare microscale high-density soft-matter electronics [39]. Wang et al. proposed rapid fabrication of flexible functional circuits based on LM dual-trans printing, which can be applied to quickly fabricate a flexible functional electronic device fitting to any complex surface shape [19]. Additionally, widely utilized microcontact printing and spray-printing methods both need advanced customized masks. Unlike spray printing via airbrush to spray LM microdroplets onto the mask, microcontact printing directly paints LM on the mask and the specific circuit can be obtained after removing the mask (Fig. 8.3G) [60].

High-resolution nanoimprinting: Unlike the relatively large LM droplets (spray-printing process), one can prepare LM nanoparticles to print microscale or nanoscale wires. John et al. produced small features through mechanically sintering gallium–indium nanoparticles to realize high-resolution nanoimprint lithography (Fig. 8.3B) [56]. One micrometer coalesced EGaIn line arrays are formed by reducing the size of the sintering tool. EGaIn nanoparticles disperses on the surface of a nitrile glove via inkjet printing, and fabricated functional elastomer glove with the LM micro-wire arrays represents excellent stretchability when holding a tennis ball.

Microfluidic: Microchannels can be fabricated by lithographic method or subtractive method [63, 64]. Injecting LM into prepared elastomeric microchannels [33], one can manufacture functional soft microfluidic components such as pumps, electrodes, sensors, antennas, heaters, and coolers (Fig. 8.3E) [24, 58]. Due to its fluidity, the injected LMs in flexible and stretchable microchannels can maintain electrical continuity during deformation, which provides huge potential for highly integrated microfluidic systems.

Subtractive manufacturing: Unlike additive techniques as mentioned earlier, the subtractive technique selectively removes metal from the elastomer to pattern LM films. Figure 8.3F illustrates the subtractive manufacturing of LM soft-matter electronics by UV laser [59]. The oxide layer Ga_2O_3 is necessary for the patterned LM to hold its shape after laser ablation. Subtractive method provides a considerable method to produce accurate LM-enabled deformable electronics.

8.3.2 3D Printing

Three-dimensional printing is a technology for constructing an object by means of printing layer by layer; in other words, it is a kind of typical additive manufacturing technology. Different from conventional subtractive technology, additive manufacturing is a top-down, layer-by-layer manufacturing process. It can not only avoid corrosion and contribute to environmental protection, but also save a lot of raw materials. This significantly reduces environmental pollution and the cost of materials. However, subtractive manufacturing technology is more accurate than addictive technology, which can be utilized in machine finishing. Liquid phase 3D printing is the most commonly used additive mode and is achieved in a liquid environment (water, ethanol, kerosene, electrolyte solution, etc.). The melting point of LM must be lower than that of the liquid environment to ensure the printed item to run (Fig. 8.4D) [65]. Similarly, the suspension 3D printing method can print LM into arbitrary macroscopic 3D structures under supporting hydrogel (Fig. 8.4A) [66]. Besides, Fassler et al. have applied the common freeze-casting method to manufacture various LM-based 3D structures (Fig. 8.4B) [68]. Tang et al. tested the formability of Cu-GaIn LM amalgams for 3D freestanding LM sandcastles at room temperature [55]. LM can also be employed as an injectable 3D bioelectrode to measure electrocardiograph (ECG) or electroencephalogram (EEG) [12, 18, 69].

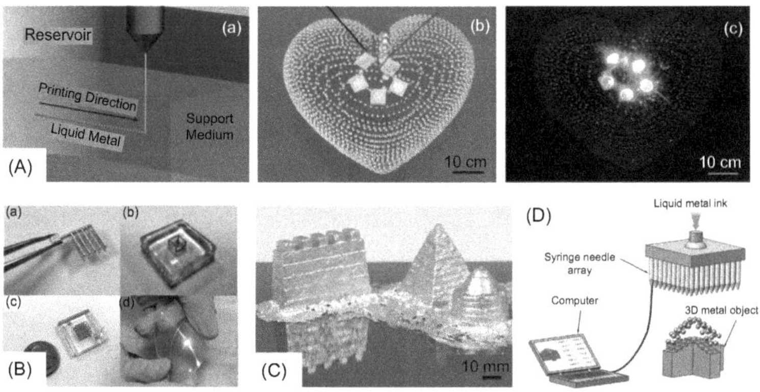

Figure 8.4 LM-enabled 3D printing. (A) Suspension 3D printing of LM into self-healing hydrogel: (a) schematic for the suspension 3D printing method, (b) 3D structural electronics, (c) embedding LEDs on the 3D printing electronics are lighted. Reprinted with permission from Ref. [66], Copyright 2017, John Wiley & Sons. (B) The electronic circuits and devices created by freeze casting. Reprinted with permission from Ref. [68], Copyright 2013, The Royal Society of Chemistry. (C) Freestanding "sandcastles" made directly by molding. Reprinted with permission from Ref. [55], Copyright 2017, American Chemical Society. (D) Liquid phase 3D printing method. Reprinted with permission from Ref. [65], Copyright 2014, Science China Press and Springer-Verlag Berlin Heidelberg.

8.4 Applications: LM Soft Devices

The flexibility and electrical conductivity of RTLM make it play an important role in soft electronics. As a kind of new fashioned and recently used soft material, LM has been widely utilized in skin electronics for health monitoring and detecting. Besides skin electronics, LM shows great potential in self-healing, reconfigurable circuits, soft diodes, and memristors [36]. Figure 8.5 presents some typical applications of LM-enabled soft devices, such as strain sensors

[70], stretchable coils [71], e-skin and wearable bioelectronics [17, 19], conformable bioelectrodes [72], etc. Both industrial production and experimental studies enlarge its applications in multifunctional flexible and stretchable devices.

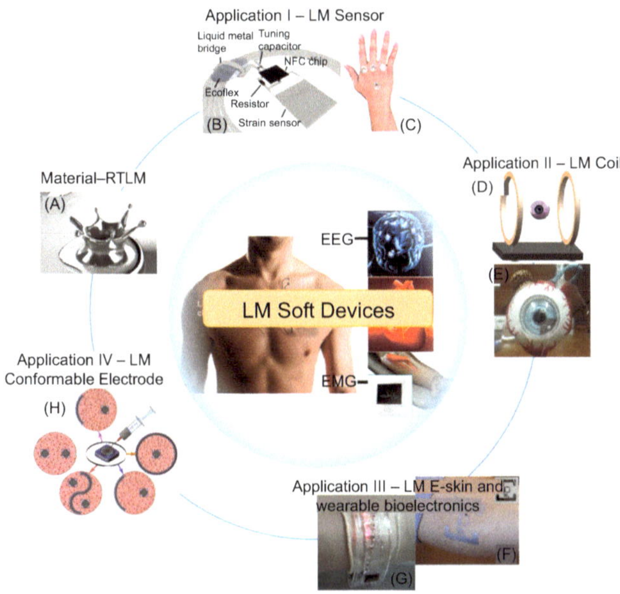

Figure 8.5 Typical applications of LM soft devices: sensor, coil, e-skin and wearable bioelectronics, and conformable electrode. (A) RTLM material (https://cn.bing.com/images/). (B) The liquid GaInSn strain sensor device and (C) its applications on human finger motion monitoring with multi-site sensing capabilities. Figures (B) and (C) reprinted from Ref. [70] under Creative Commons license. (D) The geometric model of LM induction coil, and (E) LM coil embedded in PDMS is put on the artificial eyeball for eye movement recording. Figures (D) and (E) reprinted with permission from Ref. [71], Copyright 2018, IEEE. (F) LM tag on human skin. Reprinted with permission from Ref. [17], Copyright 2012, The Royal Society of Chemistry. (G) LM wearable wristband. Reprinted with permission from Ref. [19], Copyright 2015, John Wiley & Sons. (H) LM-conformable electrode with five different shapes. Reprinted with permission from Ref. [72], Copyright 2017, Elsevier.

8.4.1 LM Sensor

Because of the rapid manufacturing and resistance variation under deformation, printing LM on stretchable elastomer to prepare a strain sensor has been widely used in soft electronics. From Section 8.2.1, the resistance of LM wires produces corresponding change along with its deformation under stretching and bending. Accordingly, many soft sensors using LM have been designed, such as the square-shaped sensor. When stretching the sensor along axial direction, the length of the sensor increases and the cross-sectional area decreases, and hence the resistance increases. Jeong et al. have combined LM strain sensor with wireless antenna to design an LM-integrated near-field communication (NFC) device for human multi-site motion sensing [70]. Figure 8.6A is the three-layered structure of LM sensor device, including top PDMS, bottom PDMS, and middle functional LM circuit. Among these, LM strain sensor is the core to detect the movement accompanied by the variation in electrical signals (Figs. 8.6B and 8.6C). Moreover, Matsuzaki et al. have designed a wearable data glove with local strain sensing using GaInSn electrodes [20]. Meanwhile, Wang et al. designed a kind of finger monitoring glove, which can judge the degree of finger bending according to the change in resistance at different monitoring points (Fig. 8.6D) [81]. Attaching the ultra-stretchable LM sensor to human joints, we can design an activity-monitoring rehabilitation system for real-time detection and long-term comfortable wearing [73, 74].

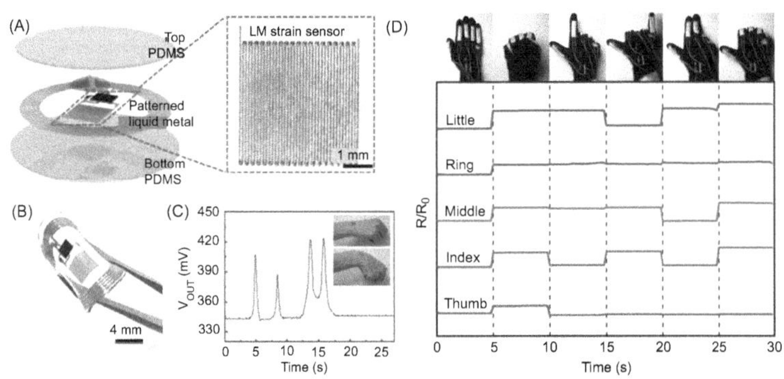

Figure 8.6 Various LM soft sensors. (A) The three-layered structure of LM sensor device. (B) Bending the device. (C) The electrical signals change under the device wireless monitoring of wrist motion. Figures (A), (B), and (C) reprinted from Ref. [70] under Creative Commons license. (D) Resistance changes of the LM sensors onto the glove at six different hand positions for real-time monitoring of finger movements. Reprinted with permission from Ref. [81], Copyright 2018, IEEE.

8.4.2 LM Coil

With the improvement in people's living standard, the design of coil systems is more and more focused on comfortable sensation to replace traditional stiff coils, such as copper wire coil. Zhao et al. have designed an innovative LM electronic coil for eye movement tracking (Figs. 8.7A and 8.7B) [71]. Experiments show that the results measured by LM coil accord well with that by classical copper coil during eye movement. Meanwhile, Fig. 8.7C fully showcases the excellent flexibility of LM coil. Due to its deformation capability and electromagnetic interaction between the magnet and LM coil, the LM coil is used as a highly efficient electromagnetic actuator at the soft fishtail (Fig. 8.7D) [75]. LM can also fabricate a stretchable loudspeaker to apply audio frequency electric signal (Fig. 8.7E) [29]. More importantly, an LM coil can be applied for wireless power transfer device. Figure 8.7F shows an LM-coil-powered device working well under bending condition [76]. Figure 8.7G displays an LM-coil-powered pulse wave sensor device placed on the wrist monitoring human pulse wave signals.

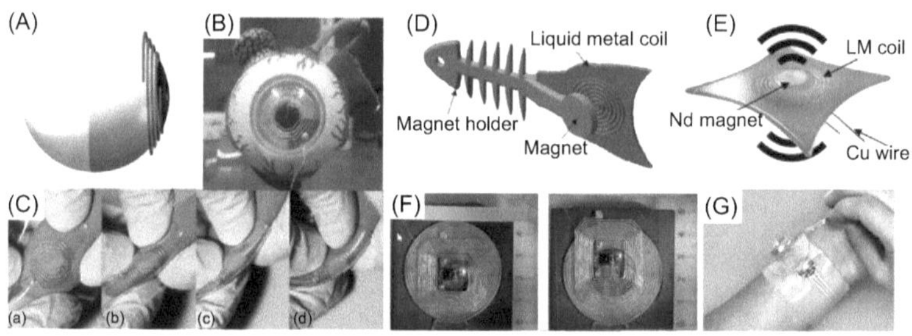

Figure 8.7 LM-enabled stretchable coils. (A) The geometric model of LM induction coil. (B) The LM coil embedded in PDMS is put on the artificial eyeball for eye movement recording. (C) Bending the LM coil to show its flexibility. Figures (A), (B), and (C) reprinted with permission from Ref. [71], Copyright 2018, IEEE. (D) The LM coil is used as an electromagnetic actuator at the soft fishtail. Reprinted with permission from Ref. [75], Copyright 2017, Science China Press and Springer-Verlag GmbH Germany. (E) The LM stretchable loudspeaker. Reprinted from Ref. [29] under Creative Commons license, Copyright 2015, Springer Nature. (F) A working stretchable wireless power transfer device. Reprinted from Ref. [85] under Creative Commons license, Copyright 2015, Springer Nature. (G) The pulse wave sensor with wireless power transfer coil is put on the wrist. Reprinted with permission from Ref. [86], Copyright 2018, John Wiley & Sons.

8.4.3 LM e-Skin and Wearable Bioelectronics

With high stretchability and conductivity, LM has formed ultra-flexible e-skin and wearable electronics. Combining with specific sensor or chip, multifunctional soft devices are produced. Attaching the device onto soft skin, stable physiological signals (such as EEG, ECG, sweat, temperature, pulse, and gait) are detected even in human activities. Intelligent electronic skin network (Fig. 8.8A) based on LM flexible electronics is expected to come true in the future with the functions of sensing, monitoring, and treating. Owing to the favorable adhesion of LM to skin, LM can be directly printed on the skin surface by the spraying method (Fig. 8.8B) [37]. Furthermore, LM can be directly written on the palm, and the soft skin circuit can work well once powered on (Figs. 8.8C and 8.8D) [18]. Besides, Wu et al. have designed a stretchable ultra-high frequency radio frequency identity tag printed on human skin and demonstrated that liquid alloy microfluidic electronics can be applied to electronic skin (Fig. 8.5F) [17]. Most recently, LM soft connector with excellently flexible PDMS substrate is developed for neuromuscular stimulation (Fig. 8.8E) [77]. As shown in Fig. 8.8F, the implanted LM electrodes are applied to the sural nerve and tibial nerve of a dead bullfrog.

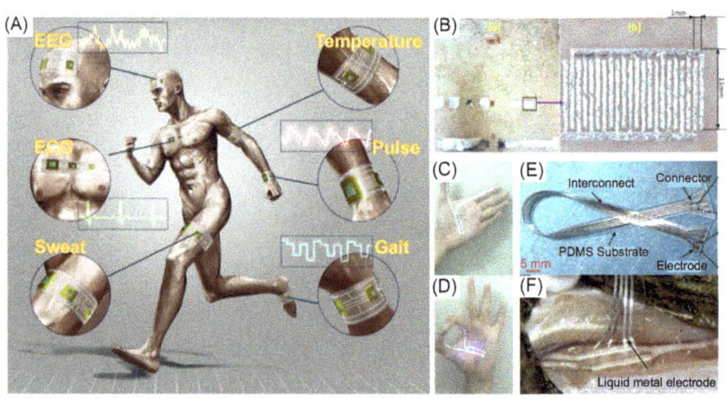

Figure 8.8 LM e-skin and wearable bioelectronics. (A) The intelligent electronic skin network based on LM e-skin and wearable bioelectronics. (B) Geometric graphics of LM e-skin components by spray printing. Reprinted with permission from Ref. [37], Copyright 2014, The Royal Society of Chemistry. (C) The LM skin circuit written on the palm. (D) The soft circuit works once the power is on. Figures (C) and (D) reprinted from Ref. [18] under Creative Commons license. (E) The optical image of LM electronical array. (F) The implanted LM electrodes for the bullfrog sural nerve and tibial nerve. Figures (E) and (F) reprinted with permission from Ref. [77], Copyright 2017, IOP Publishing.

8.4.4 LM-Conformable Electronics

Along with the development of smart health care, non-invasion and precision in medical monitoring and treatment are the mainstream trends, and LM-conformable electrodes can meet the requirements. LM can adapt to complex tissue shapes to achieve the desired biological performance (Fig. 8.9A) [72]. Meanwhile, LM can be shaped to arbitrary electrode shapes by certain molds (Fig. 8.9B). Due to conformability to any complex surface, LM soft electrodes are used in tumor therapy. By comparing the antitumor efficacy between LM electrode and Pt electrode, we know that LM electrode is more effective in inhibiting tumor growth than Pt electrode. LM electrode can also be used in implanted electrode and epidermal electrode for electrical stimulation or physiological signal monitoring [12, 78]. The utilization of LM-conformable electrode presents better signal stability and certain reduction in noise than conventional Ag/AgCl electrodes [78].

8.4.5 Other Applications

In addition to the aforementioned uses, there are some other applications of LM in soft electronic devices. Through the direct-writing method, Li et al. drew an LM thermocouple on paper (composed of Ga and EGaIn), which not only has temperature-sensing ability but also bending capacity [50]. So et al. have developed a kind of soft-matter diode, which is a new type of flexible functional equipment component with LM electrodes [36]. Because the sandwich layers of EGaIn electrodes are conductive forward from Ga_xO_y to Ga while nonconductive backward from Ga to Ga_xO_y, the soft-matter diode has the function of limiting current and can pass the current through one way. Additionally, Krupenkin et al. have modeled a footwear-embedded microfluidic energy harvester, which is a wearable mechanical-to-electrical energy convertor based on LM microdroplets [21]. Li et al. have demonstrated the self-healing capability of LM wire; they showed in an experiment that when an LM circuit is cut off, once reconnected, it can work well without additional repair [45].

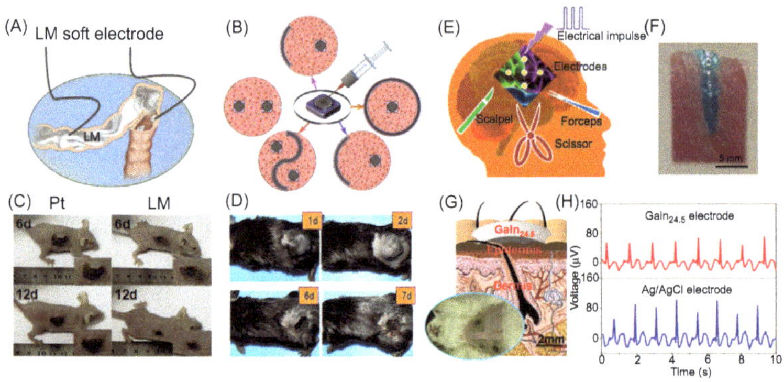

Figure 8.9 LM-conformable electronics. (A) The conceptual illustration of LM soft electrode. (B) LM-conformable electrode with five different shapes. (C) The comparison of antitumor efficacy of electrochemical therapy between LM electrode and rigid Pt electrode. Figures (A), (B), and (C) reprinted with permission from Ref. [72], Copyright 2017, Elsevier. (D) Electrical stimulation treatment of malignant melanoma tumor on C57BL/6 mice based on LM spray printing. Reprinted with permission from Ref. [67], Copyright 2016, the authors. (E) Conventional surgery in implanting electrode. (F) The cross-sectional photograph of LM injectable electrode in biological tissue. Figures (E) and (F) reprinted with permission from Ref. [12], Copyright 2013, Springer Nature. (G) Structural diagram of LM soft electrode on skin surface; inset is the LM electrodes on rabbit chest. (H) ECG signals, respectively, measured by LM-conformable electrodes and classical Ag/AgCl electrodes. Figures (G) and (H) reprinted with permission from Ref. [78], Copyright 2018, IOP Publishing.

8.5 Discussion and Conclusion

As new-generation soft materials for flexible and stretchable electronics, LMs possess outstanding versatility and specific merits of fluidity, conductivity, metallicity, and electricity properties. They thus offer tremendous opportunities for making future soft electronic devices. However, there are still challenges in preparing, processing, and utilizing LM alloys. For example, the low adhesion to many substrates, easy oxidation, and high surface tension of liquid alloys still restrict the realization of high-precision, stable circuits. To solve such key issues, we can mix LMs with various metal or non-metal materials. Therefore, new hybrid functional LM materials can be prepared with desirable properties, such as tunable and enhanced electrical, mechanical, and chemical properties. These recently emerging LM amalgams enable LMs to well adapt to practical situations and display many unconventional functions. For example, Tang et al. mixed LM with Cu particles to significantly enhance its electroconductivity and adhesion for directly writing soft electronics on papers and PVC substrates [55]. The method of doping LM with other particles provides more opportunities of LM in soft devices field.

LMs have been utilized to fabricate multifunctional flexible electronic components such as sensors, actuators, wearable bioelectronics, electrical skin, etc. The use of various stretchable substrates significantly enriches the application field of LM flexible electronics, and one can thus expand related practices from human beings to other organisms such as animals and plants. Imaginably, someone prints LM flexible circuits with features of energy harvesting and light emission on leaves to capture the energy of swinging leaves to supply light at night.

Various kinds of LM printing technologies are enabling efficient circuit fabrication. At this stage, a series of technological problems still need to be tackled, such as precise and reliable connection between the printed LM circuits with the traditional integrated circuits. High-precision printing is critical to achieve the LM integrated circuits. Regarding the specific fabrication, hybrid flexible printing with planar printing and 3D printing are worth trying. Furthermore, cooling of complex LM circuits should be carefully monitored to avoid damage from the "thermal barrier." Meanwhile, encapsulation is often a necessity for making end-user devices. For instance, long-lasting LM skin electronics mainly depend on packaging technology. PDMS and silicon rubber are often applied to package LM flexible circuits. However, for various substrates in different environments, one needs to consider diverse encapsulation technologies to make the circuit more secure in order to achieve the desired functionality. For applications in organisms, the nontoxicity of packaging materials especially needs to be considered. Sometimes, LM's movement and shape shifting under the electric field can produce more self-configurational operations of the circuits [79, 80]. These unconventional, dynamically changing or self-destructive circuits are also worth pursuing.

In summary, LM-enabled flexible electronics break through the limitations of conventional flexible electronics. The process of circuit manufacturing is straightforward, fast, stable, and relatively compatible with existing integrated circuit technology. LMs allow us to make high-quality products such as wearable devices, electronic skin, medical implants, flexible display, and solar panels that will meet the coming demands. However, the existing flexible LM printing technology is a combination of LM soft circuit and rigid components or chips, so not completely flexible electronics in the larger sense. Therefore, this kind of semi-flexible circuit may partially restrict its application. A basic significance of LM flexible electronics lies in the realization of personalized customization of any flexible functional devices on any surface, which can significantly expand the application of flexible electronics. By overcoming the problems in obtaining a high rate of good product, self-repair of damaged circuits, long service life, low environmental pollution, and waste recycling, LM flexible electronics can accelerate our pace toward a soft future.

Acknowledgments

This work is partially supported by the NSFC Key Project under Grant No. 91748206, the Ministry of Higher Education Equipment Development Fund, Dean's Research Funding, and the Frontier Project of the Chinese Academy of Science.

References

1. Wang, X. and Liu, J. (2016). Recent advancements in liquid metal flexible printed electronics: Properties, technologies, and applications, *Micromachines*, **7**, 206.
2. Yi, L., Jin, C., Wang, L., and Liu, J. (2014). Liquid–solid phase transition alloy as reversible and rapid molding bone cement, *Biomaterials*, **35**, pp. 9789–9801.
3. Yi, L. and Liu, J. (2017). Liquid metal biomaterials: A newly emerging area to tackle modern biomedical challenges, *International Materials Reviews*, **62**, pp. 415–440.

4. Wang, Q., Yu, Y., Pan, K., and Liu, J. (2014). Liquid metal angiography for mega contrast X-ray visualization of vascular network in reconstructing in-vitro organ anatomy, *IEEE Trans. Biomed. Eng.*, **61**, pp. 2161–2166.

5. Jung, Y. H., Seo, J.-H., and Zhang, H. (2017). Radio-frequency flexible and stretchable electronics: The need, challenges and opportunities, *Micro- Nanotechnology Sensors, Systems, and Applications IX*, 10194: 101941C.

6. Shin, G., Gomez, A. M., Al-Hasani, R., Jeong, Y. R., Kim, J., Xie, Z., Banks, A., Lee, S., Han, S. Y., Yoo, C. J., and Rogers, J. A. (2017). Flexible near-field wireless optoelectronics as subdermal implants for broad applications in optogenetics, *Neuron*, **93**, pp. 509–521.

7. Xu, R., Lee, J. W., Pan, T., Ma, S., Wang, J., Han, H., Ma, Y., Rogers, J. A., and Huang, Y. (2017). Designing thin, ultrastretchable electronics with stacked circuits and elastomeric encapsulation materials, *Adv. Funct. Mater.*, **27**, 1604545.

8. Kim, B. H., Kim, J.-H., Persano, L., Hwang, S.-W., Lee, S., and Rogers, J. A. (2017). Dry transient electronic systems by use of materials that sublime, *Adv. Funct. Mater.*, **27**, 1606008.

9. Hirsch, A., Michaud, H. O., Gerratt, A. P., Mulatier, S., and Lacour, S. P. (2016). Intrinsically stretchable biphasic (solid–liquid) thin metal films, *Adv. Mater.*, **28**, pp. 4507–4512.

10. Gao, Y., Li, H., and Liu, J. (2013). Directly writing resistor, inductor and capacitor to composite functional circuits: A super-simple way for alternative electronics, *PLoS One*, **8**, e69761.

11. Jeong, S. H., Hjort, K., and Wu, Z. (2014). Tape transfer printing of a liquid metal alloy for stretchable RF electronics, *Sensors*, **14**, pp. 16311–16321.

12. Jin, C., Zhang, J., Li, X., Yang, X., and Liu, J. (2013). Injectable 3-D fabrication of medical electronics at the target biological tissues, *Sci. Rep.*, **3**, 3442.

13. Li, H. and Liu, J. (2011). Revolutionizing heat transport enhancement with liquid metals: Proposal of a new industry of water-free heat exchangers, *Front. Energy*, **5**, pp. 20–42.

14. Jeong, S. H., Hjort, K., and Wu, Z. (2015). Tape transfer atomization patterning of liquid alloys for microfluidic stretchable wireless power transfer, *Sci. Rep.*, **5**, 8419.

15. Cheng, S. and Wu, Z. (2012). Microfluidic electronics, *Lab Chip*, **12**, pp. 2782–2791.

16. Cheng, S. and Wu, Z. (2011). A microfluidic, reversibly stretchable, large-area wireless strain sensor, *Adv. Funct. Mater.*, **21**, pp. 2282–2290.

17. Jeong, S. H., Hagman, A., Hjort, K., Jobs, M., Sundqvist, J., and Wu, Z. (2012). Liquid alloy printing of microfluidic stretchable electronics, *Lab Chip*, **12**, pp. 4657–4664.

18. Yu, Y., Zhang, J., and Liu, J. (2013). Biomedical implementation of liquid metal ink as drawable ECG electrode and skin circuit, *PLoS One*, **8**, e58771.

19. Wang, Q., Yu, Y., Yang, J., and Liu, J. (2015). Fast fabrication of flexible functional circuits based on liquid metal dual-trans printing, *Adv. Mater.*, **27**, pp. 7109–7116.

20. Matsuzaki, R. and Tabayashi, K. (2015). Highly stretchable, global, and distributed local strain sensing line using GaInSn electrodes for wearable electronics, *Adv. Funct. Mater.*, **25**, pp. 3806–3813.

21. Krupenkin, T. and Taylor, J. A. (2011). Reverse electrowetting as a new approach to high-power energy harvesting, *Nat. Commun.*, **2**, 448.

22. Sheng, L., Zhang, J., and Liu, J. (2014). Diverse transformations of liquid metals between different morphologies, *Adv. Mater.*, **26**, pp. 6036–6042.

23. Zhang, J., Yao, Y., Sheng, L., and Liu, J. (2015). Self-fueled biomimetic liquid metal mollusk, *Adv. Mater.*, **27**, pp. 2648–2655.

24. Khoshmanesh, K., Tang, S., Zhu, J. Y., Schaefer, S., Mitchell, A., Kalantar-zadeh, K., and Dickey, M. D. (2017). Liquid metal enabled microfluidics, *Lab Chip*, **17**, pp. 974–993.

25. Tabatabai, A., Fassler, A., Usiak, C., and Majidi, C. (2013). Liquid-phase gallium–indium alloy electronics with microcontact printing, *Langmuir*, **29**, pp. 6194–6200.

26. Kramer, R. K., Majidi, C., and Wood, R. J. (2013). Masked deposition of gallium–indium alloys for liquid-embedded elastomer conductors, *Adv. Funct. Mater*, **23**, pp. 5292–5296.

27. Zheng, Y., Zhang, Q., and Liu, J. (2013). Pervasive liquid metal based direct writing electronics with roller-ball pen, *AIP Adv.*, **3**, pp. 6459–6463.

28. Li, G., Wu, X., and Lee, D.-W. (2016). A galinstan-based inkjet printing system for highly stretchable electronics with self-healing capability, *Lab Chip*, **16**, pp. 1366–1373.

29. Jin, S. W., Park, J., Hong, S. Y., Park, H., Jeong, Y. R., Park, J., Lee, S.-S., and Ha, J. S. (2015). Stretchable loudspeaker using liquid metal microchannel, *Sci. Rep.*, **5**, 11695.

30. Liu, F., Yu, Y., Yi, L., and Liu, J. (2016). Liquid metal as reconnection agent for peripheral nerve injury, *Sci. Bull.*, **61**, pp. 939–947.

31. Deng, Y. and Liu, J. (2014). Flexible mechanical joint as human exoskeleton using low-melting-point alloy, *J. Med. Devices*, **8**, 044506.

32. Zhang, Q., Zheng, Y., and Liu, J. (2012). Direct writing of electronics based on alloy and metal ink (DREAM Ink): A newly emerging area and its impact on energy, environment and health sciences, *Front. Energy*, **6**, pp. 311–340.

33. Dickey, M. D. (2017). Stretchable and soft electronics using liquid metals, *Adv. Mater.*, 1606425.

34. Wang, L. and Liu, J. (2014). Research advancement of liquid metal printed electronics ink, *Imaging Sci. Photochem.*, **32**, pp. 382–392.

35. Hammock, M. L., Chortos, A., Tee, B. C. K., Tok, J. B. H., and Bao, Z. (2013). 25th anniversary article: The evolution of electronic skin (e-skin): A brief history, design considerations, and recent progress, *Adv. Mater.*, **25**, pp. 5997–6038.

36. So, J.-H., Koo, H.-J., Dickey, M. D., and Velev, O. D. (2012). Ionic current rectification in soft-matter diodes with liquid-metal electrodes, *Adv. Funct. Mater.*, **22**, pp. 625–631.

37. Guo, C., Yu, Y., and Liu, J. (2014). Rapidly patterning conductive components on skin substrates as physiological testing devices via liquid metal spraying and pre-designed mask, *J. Mater. Chem. B*, **2**, pp. 5739–5745.

38. Boley, J. W., White, E. L., Chiu, G. T. C., and Kramer, R. K. (2014). Direct writing of gallium–indium alloy for stretchable electronics, *Adv. Funct. Mater.*, **24**, pp. 3501–3507.

39. Gozen, B. A., Tabatabai, A., Ozdoganlar, O. B., and Majidi, C. (2014). High-density soft-matter electronics with micron-scale line width, *Adv. Mater.*, **26**, pp. 5211–5216.

40. Zheng, Y., He, Z., Gao, Y., and Liu, J. (2013). Direct desktop printed-circuits-on-paper flexible electronics, *Sci. Rep.*, **3**, 1786.

41. Wang, L. and Liu, J. (2015). Pressured liquid metal screen printing for rapid manufacture of high-resolution electronic patterns, *RSC Adv.*, **5**, pp. 57686–57691.

42. Kubo, M., Li, X., Kim, C., Hashimoto, M., Wiley, B. J., Ham, D., and Whitesides, G. M. (2010). Stretchable microfluidic radiofrequency antennas, *Adv. Mater.*, **22**, pp. 2749–2752.

43. Zhu, S., So, J.-H., Mays, R., Desai, S., Barnes, W. R., Pourdeyhimi, B., and Dickey, M. D. (2013). Ultrastretchable fiber with metallic conductivity using a liquid metal alloy core, *Adv. Funct. Mater.*, **23**, pp. 2308–2314.

44. Schrader, M. E. (1995). Young-dupre revisited, *Langmuir*, **11**, pp. 3585–3589.

45. Liu, Y., Gao, M., Mei, S., Han, Y., and Liu, J. (2013). Ultra-compliant liquid metal electrodes with in-plane self-healing capability for dielectric elastomer actuators, *Appl. Phys. Lett.*, **103**, 064101.

46. Kim, D., Thissen, P., Viner, G., Lee, D.-W., Choi, W., Chabal, Y. J., and Lee, J.-B. (2013). Recovery of nonwetting characteristics by surface modification of gallium-based liquid metal droplets using hydrochloric acid vapor, *ACS Appl. Mater. Interfaces*, **5**, pp. 179–185.

47. Doudrick, K., Liu, S., Mutunga, E. M., Klein, K. L., Damle, V., Varanasi, K. K., and Rykaczewski, K. (2014). Different shades of oxide: From nanoscale wetting mechanisms to contact printing of gallium-based liquid metals, *Langmuir*, **30**, pp. 6867–6877.

48. Gao, Y., Li, H., and Liu, J. (2012). Direct writing of flexible electronics through room temperature liquid metal ink, *PLoS ONE*, **7**, e45485.

49. Zhang, Q., Gao, Y., and Liu, J. (2014). Atomized spraying of liquid metal droplets on desired substrate surfaces as a generalized way for ubiquitous printed electronics, *Appl. Phys. A*, **116**, pp. 1091–1097.

50. Li, H., Yang, Y., and Liu, J. (2012). Printable tiny thermocouple by liquid metal gallium and its matching metal, *Appl. Phys. Lett.*, **101**, 073511.

51. Liu, J. and Li, H. (2011). A liquid metal based printed circuit board and its fabrication method. China Patent 201110140156, 14 October 2011.

52. Wang, L. and Liu, J. (2014). Printing low-melting-point alloy ink to directly make a solidified circuit or functional device with a heating pen, *Proc. R. Soc. A: Math. Phys. Eng. Sci.*, **470**, 0609.

53. Yang, J. and Liu, J. (2014). Direct printing and assembly of FM radio at the user end via liquid metal printer, *Circ. World*, **40**, pp. 134–140.

54. Sheng, L., Teo, S., and Liu, J. (2016). Liquid-metal-painted stretchable capacitor sensors for wearable healthcare electronics, *J. Med. Biol. Eng.*, **36**, pp. 265–272.

55. Tang, J., Zhao, X., Li, J., Guo, R., Zhou, Y., and Liu, J. (2017). Gallium-based liquid metal amalgams: Transitional-state metallic mixtures (TransM^{2}ixes) with enhanced and tunable electrical, thermal, and

mechanical properties, *ACS Appl. Mater. Interfaces*, **9**, pp. 35977–35987.

56. Boley, J. W., White, E. L., and Kramer, R. K. (2015). Mechanically sintered gallium–indium nanoparticles, *Adv. Mater.*, **27**, pp. 2355–2360.
57. Zheng, Y., He, Z.-Z., Yang, J., and Liu, J. (2014). Personal electronics printing via tapping mode composite liquid metal ink delivery and adhesion mechanism, *Sci. Rep.*, **4**, pp. 101–104.
58. Je, J. and Lee, J. (2014). Design, fabrication, and characterization of liquid metal microheaters, *J. Microelectromech. Syst.*, **23**, pp. 1156–1163.
59. Lu, T., Markvicka, E. J., Jin, Y., and Majidi, C. (2017). Soft-matter printed circuit board with UV laser micropatterning, *ACS Appl. Mater. Interfaces*, **9**, pp. 22055–22062.
60. Tabatabai, A., Fassler, A., Usiak, C., and Majidi, C. (2013). Liquid-phase gallium–indium alloy electronics with microcontact printing, *Langmuir*, **29**, pp. 6194–6200.
61. Jeong, J.-A. and Kim, H.-K. (2010). Characteristics of inkjet-printed nano indium tin oxide particles for transparent conducting electrodes, *Curr. Appl. Phys.*, **10**, e105–e108.
62. Kim, D., Yoo, J. H., Lee, Y., Choi, W., Yoo, K., and Lee, J. B. J. (2014). Gallium-based liquid metal inkjet printing, *Proc. 27th IEEE Inter. Conf. MEMS*, pp. 967–970 (in USA).
63. Sia, S. K. and Whitesides, G. M. (2003). Microfluidic devices fabricated in poly(dimethylsiloxane) for biological studies, *Electrophoresis*, **24**, pp. 3563–3576.
64. Roberts, M. A., Rossier, J. S., Bercier, P., and Girault, H. (1997). UV laser machined polymer substrates for the development of microdiagnostic systems, *Anal. Chem.*, **69**, pp. 2035–2042.
65. Wang, L. and Liu, J. (2014). Liquid phase 3D printing for quickly manufacturing conductive metal objects with low melting point alloy ink, *Sci. China Technol. Sci.*, **57**, pp. 1721–1728.
66. Yu, Y., Liu, F., Zhang, R., and Liu, J. (2017). Suspension 3D printing of LM into self-healing hydrogel, *Adv. Mater. Technol.*, **2**, 1700173.
67. Li, J., Guo, C., Wang, Z., Gao, K., Shi, X., and Liu, J. (2016). Electrical stimulation towards melanoma therapy via liquid metal printed electronics on skin, *Clin. Trans. Med.*, **5**, 21.
68. Fassler, A. and Majidi, C. (2013). 3D structures of liquid-phase gain alloy embedded in PDMS with freeze casting, *Lab Chip*, **13**, pp. 4442–4450.

69. Liu, R., Yang, X., Jin, C., Fu, J., Chen, W., and Liu, J. (2013). Development of three-dimension microelectrode array for bioelectric measurement using the liquid metal-micromolding technique, *Appl. Phys. Lett.*, **103**, 193701.

70. Jeong, Y. R., Kim, J., Xie, Z., Xue, Y., Won, S. M., Lee, G., Jin, S. W., Hong, S. Y., Feng, X., Huang, Y., Rogers, J. A., and Ha, J. S. (2017). A skin-attachable, stretchable integrated system based on liquid GaInSn for wireless human motion monitoring with multi-site sensing capabilities, *NPG Asia Mater.*, **9**, e443.

71. Zhao, Z., Lin, J., Zhang, J., Yu, Y., Yuan, B., Fan, C.-C., Wang, L., and Liu, J. (2018). Liquid metal enabled flexible electronic system for eye movement tracking, *IEEE Sens. J.*, **18**, pp. 2592–2598.

72. Sun, X., Yuan, B., Rao, W., and Liu, J. (2017). Amorphous liquid metal electrodes enabled conformable electrochemical therapy of tumors, *Biomaterials*, **146**, pp. 156–167.

73. Kramer, R. K., Majidi, C., Sahai, R., and Wood, R. J. (2011). Soft curvature sensors for joint angle proprioception, *2011 IEEERSJ Int. Conf. Intelligent Robots and Systems (IROS)*, pp. 1919–1926 (in USA).

74. Menguc, Y., Park, Y.-L., Pei, H., Vogt, D., Aubin, P. M., Winchell, E., Fluke, L., Stirling, L., Wood, R. J., and Walsh, C. J. (2014). Wearable soft sensing suit for human gait measurement, *Int. J. Robot. Res.*, **33**, pp. 1748–1764.

75. Guo, R., Sheng, L., Guo, Y., and Liu, J. (2017). Liquid metal spiral coil enabled soft electromagnetic actuator, *Sci. China Tech. Sci.*, **60**, pp. 1–6.

76. Wu, Z., Hjort, K., and Jeong, S. H. (2015). Microfluidic stretchable radio-frequency devices, *P. IEEE*, **103**, pp. 1211–1225.

77. Guo, R. and Liu, J. (2017). Implantable liquid metal-based flexible neural microelectrode array and its application in recovering animal locomotion functions, *J. Micromech. Microeng.*, **27**, 104002.

78. Wang, X., Zhang, Y., Guo, R., Wang, H., Yuan, B., and Liu, J. (2018). Conformable liquid metal printed epidermal electronics for smart physiological monitoring and simulation treatment, *J. Micromech. Microeng.*, **28**, 034003.

79. Wang, L. and Liu, J. (2015). Ink spraying based liquid metal printed electronics for directly making smart home appliances, *ECS J. Solid State Sci. Technol.*, **4**, pp. 3057–3062.

80. Fassler, A. and Majidi, C. (2013). Soft-matter capacitors and inductors for hyperelastic strain sensing and stretchable electronics, *Smart Mater. Struct.*, **22**, pp. 955–966.

81. Wang, X., Guo, R., Yuan, B., Yao, Y., Wang, F., and Liu, J. (2018). Ni-doped liquid metal printed highly stretchable and conformable strain sensor for multifunctional human-motion monitoring, *40th Annual International Conference of the IEEE Engineering in Medicine and Biology Society, Hawaii, USA*.

82. Guo, R., Tang, J., Dong, S., Lin, J., Wang, H., Liu, J., and Rao, W. (2018). One-step liquid metal transfer printing: toward fabrication of flexible electronics on wide range of substrates, *Adv. Mater. Technol.*, 1800265.

83. Wang, X., Ren, Y., and Liu, J. (2018). Liquid metal enabled electrobiology: a new frontier to tackle disease challenges, *Micromachines*, **9**, p. 360.

84. Guo, R., Wang, H., Duan, M., Yu, W., Wang, X., and Liu, J. (2018). Stretchable electronics based on Nano-Fe GaIn amalgams for smart flexible pneumatic actuator, *Smart Mater. Struct.*, **27**, 085022 (10 pp).

85. Jeong, S. H., Hjort, K., and Wu, Z. (2015). Tape transfer atomization patterning of liquid alloys for microfluidic stretchable wireless power transfer, *Sci. Rep.*, **5**, 8419.

86. Guo, R., Wang, X., Chang, H., Yu, W., Liang, S., Rao, W., and Liu, J. (2018). Ni-GaIn amalgams enabled rapid and customizable fabrication of wearable and wireless healthcare electronics, *Adv. Eng. Mater.*, 201800054.

Chapter 9

Printing Technology for Fabrication of Flexible and Stretchable Electronics

Wei Yuan and Zheng Cui
Printable Electronics Research Centre, Suzhou Institute of Nanotech and Nano-bionics, Chinese Academy of Sciences, Suzhou 215123, China
zcui2009@sinano.ac.cn

Recently, flexible and stretchable electronics have experienced tremendous development due to their promising applications in fields such as wearable electronics, portable energy devices, flexible display, and human-skin sensors. In order to effectively fabricate flexible and stretchable electronics, a high-throughput, cost-saving, and eco-friendly manufacturing technology is required. Printing, an additive manufacturing process, can serve as an effective tool for the fabrication of various flexible and stretchable electronics. In this chapter, the recent advances in printing processes, printable inks, post-printing treatments, and the important applications of printed flexible and stretchable electronics are introduced. The remaining challenges are analyzed, and future prospects are speculated.

Flexible and Stretchable Electronics: Materials, Designs, and Devices
Edited by Run-Wei Li and Gang Liu

ISBN 978-981-4800-46-4 (Hardcover), 978-0-429-05890-5 (eBook)
www.jennystanford.com

9.1 Introduction

Flexible electronics is a new class of electronics whose entities are inherently bendable, rollable, or even foldable. A stretchable electronic system is not only flexible but also works under strained condition, which adds an extra conformability to its flexibility, enabling it to be attached to non-planar surfaces. Such freedom opens a vast spectrum of applications that are not possible by conventional silicon or printed circuit boards (PCBs) type of rigid electronics [1–3]. Although flexible circuits or electronic devices can be made in the same way as for rigid electronics, which involves thin-film deposition, lithographic patterning, and etching as shown in Fig. 9.1a, the processes are complex and expensive, not to mention that these processes are subtractive, involving material wastes and discharge of pollutants. Recently, fabrication of flexible and stretchable electronics by printing is gaining more attention due to the advances in various printable functional ink materials [4, 5]. Printing is an additive process, and functional materials can be deposited directly into device structures. Only a subsequent annealing/sintering process is needed, as shown in Fig. 9.1b [6–9]. Conventional fabrication, as shown in Fig. 9.1a, is not easy to perform on a flexible substrate. A rigid substrate is often required to carry the flexible film substrate. After all the fabrication processes, the film is detached from the carrier. For printing fabrication, electronics can be directly printed onto a flexible film. Even high-throughput, roll-to-roll printing is possible.

Most printing methods, such as screen printing, inkjet printing, or gravure printing, have been used to deposit metallic, semiconducting, insulating as well as packaging materials onto various flexible substrates, including plastic (PET, PI, PEN film), paper, and even textiles [10]. It is very much like color printing in a conventional printing press, where each color ink is printed sequentially, and several color layers are overlaid to form the final color print. For example, thin-film transistors (TFTs) have been made by printing source, drain and gate electrodes, as well as semiconductor and insulating layers. Because of the similarity, the printers for printing flexible electronics are not much different from conventional media printers. The only difference is the ink materials. The inks for printing electronics have conducting, semiconducting, and dielectric properties. They are electronic materials, not conventional dyes or pigments, which is the key for printing to be used for fabrication of electronic devices.

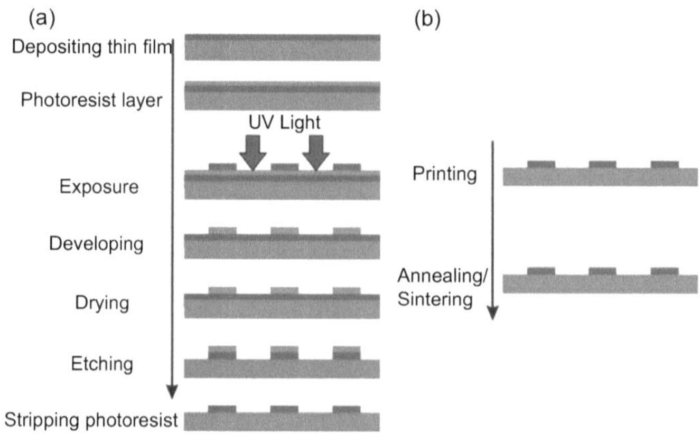

Figure 9.1 The traditional processes (a) and printing process (b) for making flexible electronics.

Recently, many studies have indicated that printing is becoming an effective technique to fabricate large-area, flexible, and stretchable electronics [11–13]. The authors' Printable Electronic Research Center (PERC) has been developing ink materials and printing processes for flexible and stretchable electronics since 2010. A variety of electronic systems made by printing have been demonstrated, including wearable sensors, smart packaging system, printed TFTs and solar cells, flexible multilayer circuit and transparent conductive film, as well as flexible display and smart textile, such as the examples shown in Fig. 9.2 [14]. In the course of printed electronics technology development at PERC, rich experience has been accumulated. A comprehensive book on printed electronics has been published elsewhere [5]. This chapter will focus on the technology for printing flexible and stretchable electronics only, based on the authors' own experience and other published work in literatures.

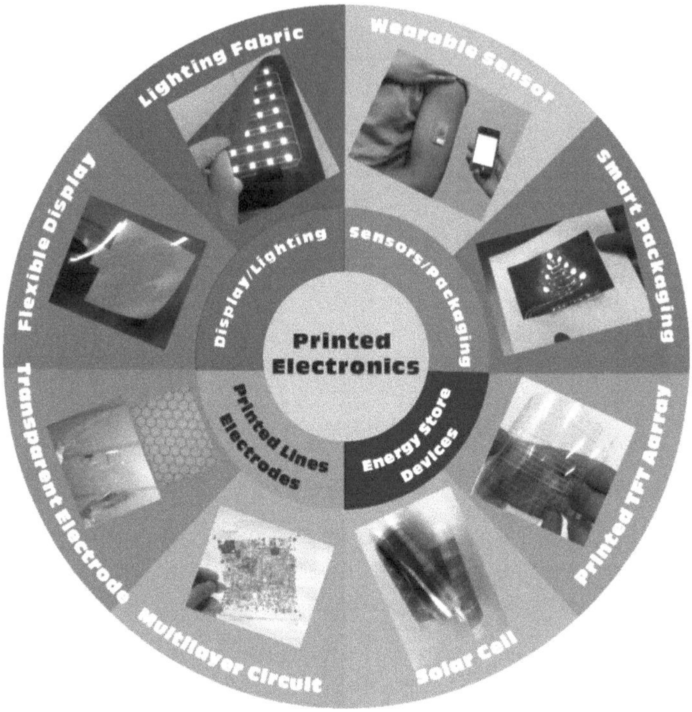

Figure 9.2 The various flexible devices produced by printing methods in PERC [14].

The chapter starts with discussion of various printing methods and their advantages and disadvantages for the fabrication of printed electronics. Then, ink formulations for inorganic nanomaterials, including metal nanoparticles, transparent conductive oxide, carbon nanomaterials, and semiconductor nanomaterials, are introduced. Furthermore, various post-printing treatments are presented in view of meeting the requirements of low-temperature sintering technique for plastic substrates. Finally, typical applications of printed flexible and stretchable electronics are demonstrated, including transistors, solar cells, light-emitting devices, and stretchable circuits.

9.2 Printing Process

According to whether a template is employed, printing can be divided into two categories. Printing without using a template is jet printing, which includes inkjet printing, aerosol-jet printing, and electrohydrodynamic (EHD)-jet printing. For printing that relies on a template is replicate printing, including screen printing, gravure printing, flexographic printing, and offset printing.

9.2.1 Jet Printing (Non-contact Printing)

Jet printing, also called non-contact printing or digital printing, is popular in research laboratories, because it deposits inks directly on substrate without the need for a pre-fabricated template and without any direct contact between equipment and substrate. The patterned deposition of electronic ink material is digitally controlled. Three common types of jet printing techniques are illustrated in Fig. 9.3.

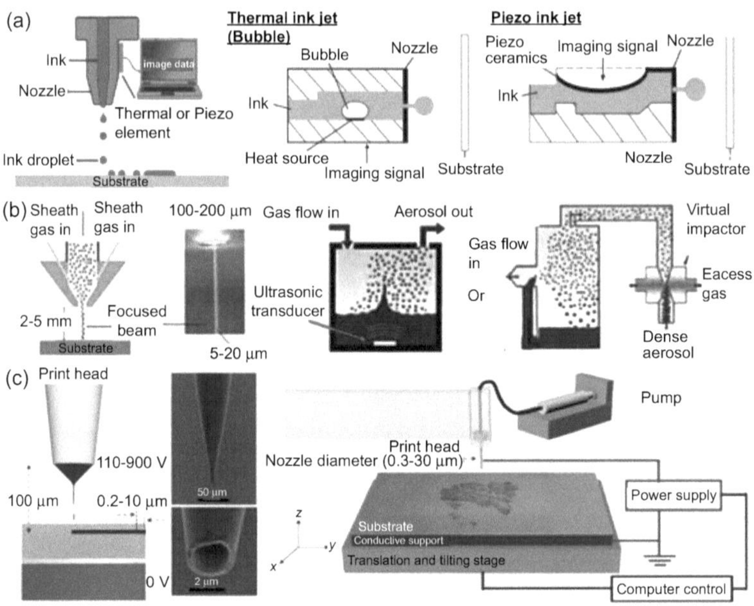

Figure 9.3 (a) Schematic illustration of inkjet printing (left) (reprinted with permission from Ref. [9], Copyright 2017, The Royal Society of Chemistry) and two different methods of propelling ink drops: thermal bubble propelling (middle) and piezoelectric propelling (right). (b) Schematic illustration of the principle of aerosol-jet printing. Reprinted with permission from Ref. [5], Copyright 2016, John Wiley & Sons. (c) Nozzle and substrate configuration for printing and SEM images of a gold-coated glass microcapillary nozzle (left) and schematic diagrams of a high-resolution e-jet printer (right). Reprinted with permission from Ref. [30], Copyright 2007, Springer Nature.

9.2.1.1 Inkjet printing

Inkjet printing, in principle, can be divided into two modes of operation: one is continuous inkjet (CIJ) mode and the other is drop on demand (DOD) inkjet mode. In a CIJ printer, ink droplets are produced and ejected continuously. The ink droplets are electrically charged, and their landing on substrate is controlled by an electrostatic field. Some droplets pass through the electrostatic field to form target pattern on substrate, while other droplets are deflected to an ink catcher for re-use. Compared to the CIJ printing model, a DOD printer only ejects ink droplets as required when a pulse of voltage is applied, without any drop deflecting and catching.

As shown in Fig. 9.3a, there are two different methods of propelling ink drops in an inkjet printer head, i.e., thermal bubble propelling and piezoelectric propelling [5, 9, 15]. In the thermal bubble-jet printing, a heating unit inside the ink reservoir creates an ink bubble, which forces a drop of ink out of the nozzle. In piezoelectric printing, a ceramic piezoelectric tile in the nozzle flexes into the ink reservoir in the printing head, forcing the ink drop out of the nozzle. In printed electronics, the most popular inkjet printers are based on piezoelectric DOD inkjet heads.

As an inexpensive tool, inkjet printing is most suitable for research laboratory for printed electronics [16]. Many parameters in inkjet printing, such as the volume, number, velocity, and direction of droplets, as well as the temperature of substrate, can be controlled precisely [17–19]. There is usually a camera equipped with an inkjet printer, which can observe the movement of ink droplets and study the interaction with substrates. Many types of organic and inorganic materials can be directly deposited onto the substrate by inkjet printing [20, 21]. Because the ink drops are digitally controlled, designs of printed patterns can be easily changed.

9.2.1.2 Aerosol-jet printing

Besides the conventional inkjet printing technique, aerosol-jet printing is also used in the fabrication of flexible electronics. Aerosol-jet printing is a new printing technology developed in the recent years, whose working principle is very different from inkjet printing. As shown in Fig. 9.3b, ink is first converted into a dense aerosol of tiny droplets between 1–5 microns in size pneumatically or ultrasonically in an atomizer. Then the ink aerosol is carried by an air stream to the nozzle, which is aerodynamically focused by a round compressed air sheath to form a fine and high-velocity aerosol stream spraying onto a substrate. The focused ink stream has a diameter as small as 5 μm [5, 22] and sprays continuously, with a shutter to stop the ink stream at non-patterned areas.

Aerosol-jet printing has many advantages over inkjet printing such as high resolution and suitability for a wide range of inks. The most important advantage is the high resolution generally down to 5 μm, which facilitates the printing of fine patterns [23, 24]. However, there are still some challenges when applying aerosol-jet printing to print electronics. The main challenge is that aerosol-jet printing relies on stable air flow, which means there will be a lagging effect when trying to adjust printing parameters. In addition, an aerosol-jet printing system is expensive than a desktop inkjet printer. So far, there is only one company (Optomec Inc.) providing commercial aerosol-jet printers [5, 22]. The multi-nozzle system of aerosol-jet printing is not as mature as that of inkjet printing, which makes the large-area printing much slower and cannot achieve high throughput.

9.2.1.3 Electrohydrodynamic-jet printing

For an inkjet printer, the printing resolution is directly related to the size of ink droplets, which is in turn related to the size of printing nozzle. It becomes more difficult to jet out inks if the nozzle is too small, because of the viscosity and surface tension of ink liquid. EHD-jet printing offers a solution to the limited resolution of the conventional inkjet printer systems by exploiting different means to generate the droplets [25]. EHD-jet printing is a technique that uses an electric field to create the ink flow through a fine nozzle onto a substrate [26–29]. Figure 9.3c shows the schematic setup of EHD-jet printing. The ink is fed to a nozzle by a syringe pump. When a strong electric field is applied between the nozzle and the opposing conducting holder, a stable cone-shaped droplet called "Taylor Cone" is formed on the tip of the nozzle. If the electric field force is larger than the surface tension force of the ink drop, nanodrops or nanofibers can be ejected from the cone to deposit on a substrate [30]. As reported, the EHD-jet printing technique can use 300 nm or even smaller diameter nozzle to achieve as low as 240 nm ultra-high-resolution printed patterns [30].

9.2.2 Replicate Printing (Impact Printing)

The aforementioned jet printing is also called digital printing, which means the patterns to be printed are in digital format. In contrast, replicate printing relies on a master or a template to print images or patterns. The printing process is basically a replication process, replicating the images or patterns from a master plate to other substrates [5, 9]. While digital images can be easily revised in inkjet printing, a master plate cannot be modified once it is made. Depending on the nature of a template, the replicate printing can be divided into screen printing, gravure printing, flexographic printing, and offset printing [15], as illustrated in Fig. 9.4.

9.2.2.1 Screen printing

Screen printing is a very simple, versatile, and inexpensive technology to print inks onto various rigid or flexible substrates [5, 9]. The printing is a push-through process using a porous mesh as the stencil, which is patterned so that ink material can only pass through some parts of the stencil by a squeegee press to deposit onto the substrate, as shown in Fig. 9.4a. The mesh is normally made from nylon fiber or metal wire in the range of 150–400 fibers per inch, and fixed by an aluminum frame. The inks used in screen printing are of high viscosity and usually called paste. The highly viscous paste ensures it will not leak onto the substrate without the squeegee pressed [15].

Compared with other printing methods, screen printing has two distinctive features: one is the printable paste, which can be formulated with micro-sized solid materials. There are various commercial pastes, including metal, carbon, and polymers, which can be printed to produce electrical components. The other is that the overall cost of screen printing is lower than other printing methods, both in equipment and in paste materials. However, although many functional devices have been fabricated by using the screen-printing method [31–33], the achievable resolution, usually in the range of 30–100 μm, is still too low for printing electronic devices of fine structures, such as source and drain electrodes in transistors.

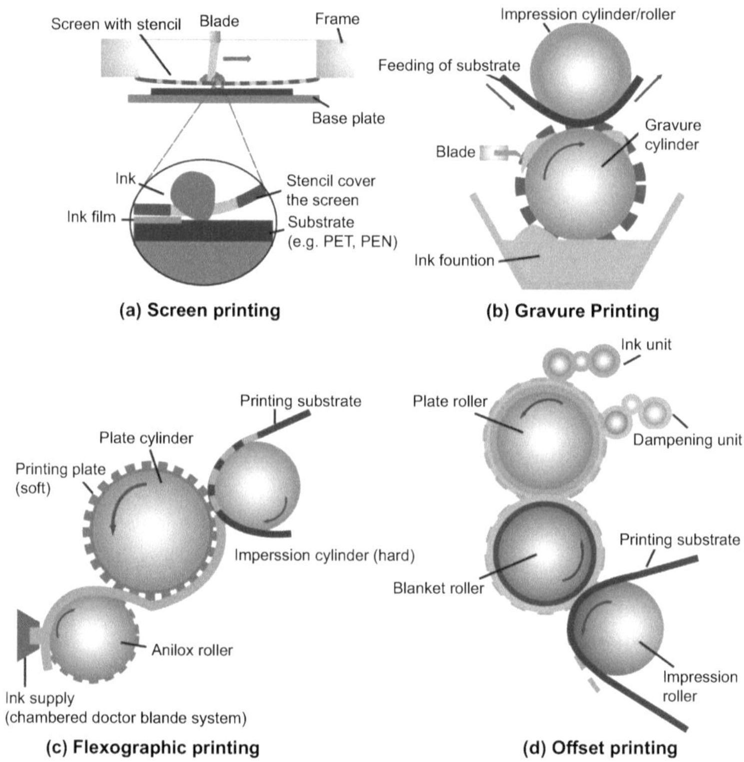

Figure 9.4 Schematic illustration of (a) screen printing, (b) gravure printing, (c) flexographic printing, and (d) offset printing.

9.2.2.2 Gravure printing

Gravure printing replicates patterns using a gravure roller. As shown in Fig. 9.4b, a gravure printer consists of a gravure roller, ink supply unit, a doctor blade, and an impression roller. The desired patterns are engraved onto the surface of the gravure roller in advance. The depth of gravure cell is 10–30 μm (maximum 50 μm), and the width of cell wall is 3–5 μm [5, 9, 15]. Before printing, the gravure roller is immersed in an ink tank, and excess inks are removed by a doctor blade. With the action of impression roller, the inks are transferred from the engraved area to the substrate due to the adhesive forces between the substrate and ink, and finally printed patterns are formed after drying.

Gravure printing allows good pattern reproduction; the metallic gravure roller usually has a long working life and can produce millions of prints without suffering pattern deterioration. However, because of the very high costs of gravure rollers, gravure printing can only be used in high-volume production for printed electronic devices. Recently, the gravure printing method has been employed to fabricate TFTs in the laboratory. Koo et al. used gravure printing to realize the fabrication of fully printed SWNT-based TFT arrays on PET film [34].

9.2.2.3 Flexographic printing

Flexographic printing uses an elastic plate or roller to replicate patterns. As shown in Fig. 9.4c, the pattern areas are raised above the non-pattern areas. Pattern areas on the soft plate are coated with a layer of functional inks by using an anilox roller. The anilox roller is very similar to a gravure roller except that the engraved cells cover all the surface of the roller. The depth of the cell in the anilox roller is constant, which results in a constant thickness of inks being transferred to the soft plate roller [5, 9]. Then the inks are transferred to substrate by pressures coming from the impression roller. It should be noted that because of the elastic plate roller, flexographic printing can print inks on various substrates, including rigid, flexible, or rough surface [35].

9.2.2.4 Offset printing

In contrast to gravure printing, inks in offset printing are first transferred onto a flexible blanket roller from a printing plate, and then transferred onto the substrate, as shown in Fig. 9.4d. Offset printing has a flat plate where pattern and non-pattern areas are distinguished by ink repellent and receptive properties. Inks only stick to the receptive area and are transferred to the substrate. Although offset printing is one of the most popular processes and has been widely used in traditional printing industry, reports on the manufacture of electronic devices are very rare so far. Recently, gravure offset printing is attracting much attention and has been used to print conductive components or devices [36, 37].

9.2.2.5 Roll-to-roll printing

Roll-to-roll (R2R) printing is a classic printing process. Patterns are designed onto a printing plate, which is often loaded onto a roller [5, 9, 15]. The patterns are printed on the surface of flexible substrates by the transfer of functional inks from the printing plate in R2R fashion. Generally, all the four printing processes shown in Fig. 9.4 can be done in R2R. R2R printing is considered to be a leading candidate for the manufacture of flexible electronic devices in the future due to its high-volume, large-scale, and low-cost characteristics [38, 39].

9.3 Printable Inks

Ink is the other crucial part in printed electronics. The development of low-cost functional inks with good printability is the key to realizing the manufacture of printed electronic devices for various applications. Printable ink is a complex mixture of ingredients combined in a specific formulation to possess specific characteristics, which are related to specific electronic properties. The core ingredient, which is usually a functional nanomaterial, performs the key function desired for printed electronic devices. Apart from the core ingredient, resins, carrier liquid, surfactants, and solvents are also added to make the ink printable and stick well to the substrate. Resin is often used as a binder to affix the core ingredient to the substrate. Carrier liquid is the medium for transferring the core ingredient and resin through the printing system. The surfactants and solvents are used to control the rheological properties of ink. In this section, many functional inks commonly used in printed flexible and stretchable electronics are discussed.

9.3.1 Metal Materials

Metals are widely used in electronic devices as connectors, electrodes, and antennas because of their high conductivity. However, metals cannot be dissolved in commonly used aqueous or organic solvents. To make metals printable, the conventional way is to mix metal particles or flakes with high-viscosity polymer solutions to form pastes [40–42]. The conductive pastes can provide good mechanical interconnection and excellent reliability but poor electrical conductivity with resistivity in the range of about 10^{-3} to 10^{-4} Ωcm, which is 2–3 orders higher than that of bulk metal (10^{-6} Ωcm).

Due to the progress in nanomaterials and surface modification techniques, metallic nanoparticles can be dispersed in an organic solvent or water with only a small amount of surfactants or lubricants and no resins added. They are normally called "inks," different from the aforementioned "paste." In the past few years, various metallic nanoparticles-based inks have been developed and used in printed electronics, including Au [43–45], Ag [46–48], Pd [49], Cu [50–53], Sn [54], Ni [55] nanoparticles and their alloys or composite nanoparticles [56, 57]. To be of practical importance, metal-based conductive inks should have high metal loading in the range of 20–80 wt% [7, 58]. Although the final conductivity of printed electrodes depends on several factors, including sintering methods (see next section) and the thickness of printed layer, high metal loading in ink formulation usually provides high conductivity in single-layer printing. Most methods of formulating inks with high metal loading are based on a two-step procedure. In the first step, the synthesized nanoparticles are separated by centrifugation, or by precipitation with alcohols (such as methanol, ethanol, isopropanol) or acetone, followed by washing the sediment with proper solvent to remove excess dispersion stabilizers. In the second step, the obtained nanoparticles are dispersed in a proper carrier liquid containing additives [7].

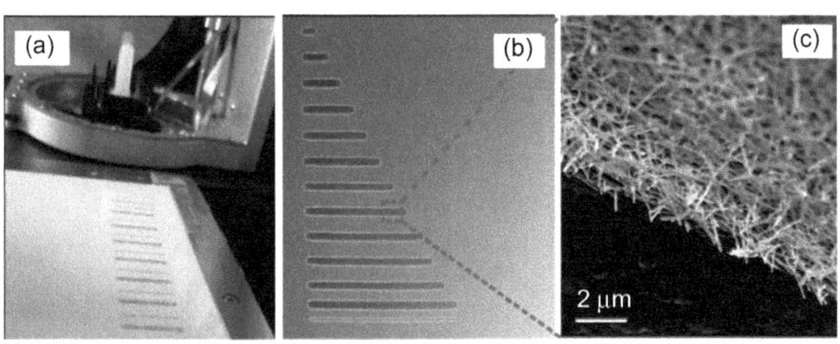

Figure 9.5 Inkjet printing silver nanowires: (a) Pictures of Dimatix materials printer 2800; (b) Printed Ag NWs lines of different lengths; (c) SEM image of the printed Ag NWs network. Reprinted with permission from Ref. [68], Copyright 2015, American Chemical Society.

Currently, Ag is the most reported metal because the electrical conductivity of Ag is the highest of all metals, its cost is lower than that of Au, and the air stability of Ag is better than that of Cu. Ag-based conductive inks with nanoparticles size in the range of 2–200 nm and metal loading in the range of 10–60 wt% are commercially available from a number of companies [7].

Besides the metal nanoparticles, nanowires-based inks have also been widely used to produce flexible transparent conductive films using drop casting [59–61], spray coating [62, 63], spin coating [64], rod coating [65, 66], and direct gravure printing [67]. Usually, the nanowire-based inks are not suitable for inkjet printing because the nanowires can aggregate easily and plug the printer nozzle due to its high aspect ratio. In order to pattern nanowire networks by the inkjet printing method, some nanowires of short length are chosen for ink preparation. For example, Coleman et al. deposited Ag NW networks with well-defined lines and patterns by the inkjet printing method, as shown in Fig. 9.5 [68]. The inks were made of Ag NWs shortened by ultrasonic with average length of only 8.1 μm and dispersed in isopropyl alcohol-diethylene glycol. The printed lines exhibited good electrical conductivity with a low sheet resistance of 8 Ω/sq (conductivity of 10^5 S/m). At the PERC, we also printed Ag NWs to form transparent conductive network as the top electrode for inverted semi-transparent poly(3-hexylthiophene):phenyl-C61–butyric acid methyl ester (P3HT:PC61BM) based OPV devices and achieved the highest power conversion efficiency of 2.71% with the printed semi-transparent top electrode [69].

The rheological property and wettability of an ink, such as viscosity and surface tension, are also important in order to obtain good quality and repeatable printed patterns. The rheological property is related not only to the solvent but also to the solid content, dispersant, nanoparticles size and shape. The contact angle of an ink droplet on substrate surface strongly depends on the surface tension of ink and the free energy of solid substrate surface. Therefore, it is necessary to add some agents into the ink formulation, including surfactant, wetting agent, reducing agent, anchoring agent, and defoaming agent.

9.3.2 Transparent Conducting Oxide Inks

Conventional transparent conducting oxide (TCO) materials include indium–tin oxide (ITO), ZnO doped with aluminum (AZO), gallium (GZO), and indium (IZO), SnO_2 doped with fluorine (FTO), CdO, Ga_2O_3, and In_2O_3. Traditionally, an ITO transparent electrode is often fabricated by depositing polycrystalline ITO thin films via sputtering techniques. Nowadays, researchers are looking for an alternative method to reduce the cost, improve material utilization, and improve the compatibility with various flexible substrates. Solution-based ITO deposition methods, especially printing techniques, have become a good candidate for practical fabrication of low-cost flexible electronic devices [70, 71].

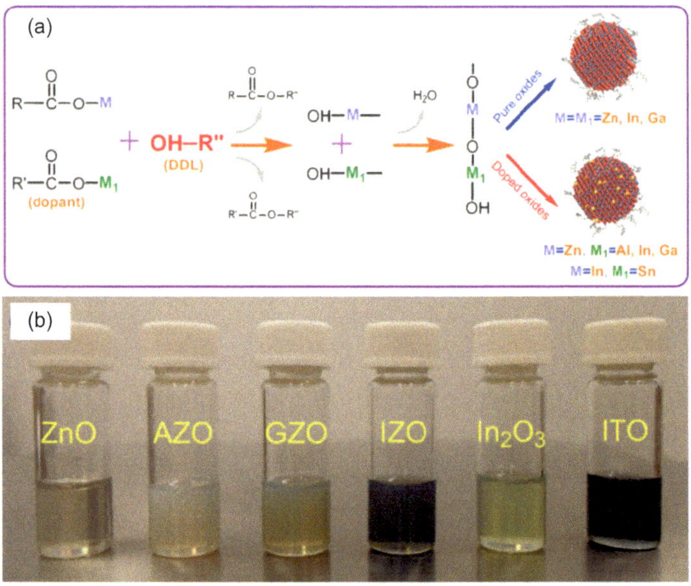

Figure 9.6 The proposed one-pot strategy: (a) Synthesis of both pure and doped TCO nanocrystals. (b) Photographs of a series of TCO nanocrystal inks dispersed in toluene with stabilities of more than 1 year. Reprinted with permission from Ref. [72], Copyright 2015, John Wiley & Sons.

In order to prepare TCO inks, the first step is to produce TCO nanoparticles. Recently, many methods have been used to synthesize TCO nanoparticles. The most convenient way is to decompose metal complexes in a wet-chemical reaction. Zeng et al.

developed a universal one-pot strategy to synthesize many kinds of TCO nanocrystal inks by simply heating a mixture of metal-organic precursors. The chemical reaction and synthesized TCO products are shown in Fig. 9.6. Transparent electrodes with sheet resistance of 110 Ω/sq and transmittance of 88% were achieved by solution-based techniques [72].

Though transparent electrodes made by printing TCO nanoparticle inks have been reported, they are still of limited practical application because of poor performance. For example, the sheet resistances of printed transparent electrodes are much higher than those by sputtering deposition method at a similar transparency [73]. Further improvements in synthesis, dispersion, and printing process for TCO nanoparticle inks are required.

9.3.3 Carbon Nanomaterials

Carbon nanomaterials exhibit excellent stability in various chemical and physical environments, such as in acid, at high temperature and pressure. In the past two decades, conductive carbon pastes have been widely used to print electrochemical electrodes for sensing applications because their price is relatively low, and their electrochemical performance is good [74]. Such conductive carbon pastes are often composed of graphite particles, polymer resin, and other additives, which are similar to the aforementioned conductive metal pastes [42].

In the recent years, carbon nanotubes (CNTs) have been adopted in printed devices because of their unique electrical properties [75]. The major challenge in formulating CNT inks is to obtain stable dispersion of non-aggregated nanotubes in a carrier liquid at low viscosity [76]. With the assistance of van der Walls adsorbed surfactants or covalently attached hydrophilic groups, improved dispersion has been achieved to meet the printability requirement [77–80]. By functionalizing single-wall carbon nanotubes (SWCNTs) with hydrophilic groups, such as carboxylic acid, amide, and poly(ethylene glycol), Vajtai et al. prepared stable water-based CNT inks and fabricated fine CNT conductive patterns by using inkjet printing [77]. Besides the dispersion issue, the length of CNTs should be limited to prevent clogging in the inkjet printer nozzles.

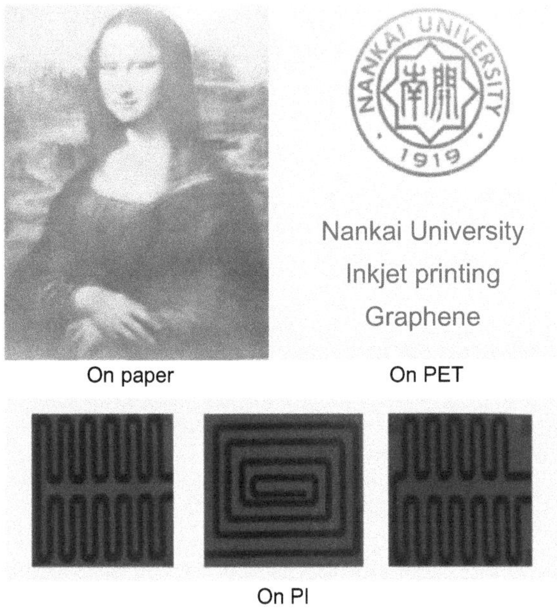

Figure 9.7 Graphene-based conducting inks for direct inkjet printing of flexible conductive patterns on various substrates. Reprinted with permission from Ref. [88], Copyright 2011, Springer Nature.

Graphene is a new type of 2D carbon material and has attracted increasing attention in both fundamental and applied research because of its remarkable physical and chemical properties [81]. Similar to CNTs, to obtain stable graphene dispersions suitable for ink formulation, various stabilizing agents are used, such as sodium dodecyl sulfate (SDS) [82], sodium dodecylbenzene sulfonate (NaDDBS) [83], sodium deoxycholate [84], and CTAB [85]. Hersam et al. prepared inkjet-printed graphene inks and achieved conductive features with a low resistivity of 4 mΩ cm after sintering at 250°C for 30 min. The samples show uniform morphology, compatibility with flexible substrates, and excellent tolerance to bending stresses [86]. Compared to the hydrophobic character of graphene, graphene oxide (GO) contains hydroxyl and carboxylic groups that make it easily dispersible in water [87, 88] and polar organic solvents. Preparation of stable GO dispersions does not require addition of stabilizing agents, which is an obvious advantage compared to graphene. Huang et al. printed various patterns on different flexible substrates, including paper, PET, and polyimide (PI), by using single-layered GO and few-layered GO aqueous solutions as inks. The printed patterns displayed high resolution, as shown in Fig. 9.7 [88].

Although more and more reports have been focused on printed carbon electronics, the conductivity of pure carbon nanomaterials is still very low, which puts constrains in some applications. In order to achieve high conductivity, some carbon/metal hybrid inks were reported, such as carbon–core/Ag–shell nanoparticles-based conductive ink [89] and Ag/reduced graphene oxide (RGO) composite conductive inkjet printing ink [90].

9.3.4 Semiconductor Nanomaterials

Semiconductors play an important role in electronic components, such as field-effect transistors. Typical inorganic semiconductor materials include Si, oxides of transition metals, and chalcogenides. For fabricating semiconductor thin films, the most commonly used technology is sputtering process. Solution-processing methods of inorganic semiconductor materials have also been employed to produce semiconductor devices [91]. Although inorganic semiconductors exhibit greater electric performance and excellent environmental stability, they often suffer from some drawbacks, such as inadequate dispersion and high sintering temperature, which is incompatible with most flexible plastic substrates. Recently, many semiconducting inorganic oxide nanomaterials are employed to fabricate transistors by printing methods, including ZnO, TiO_2, and perovskite [92, 93]. Wu et al. prepared ZnO nanoparticles–based inks and printed them as patterns by the screen-printing method, and finally a patterned ZnO nanowire array was achieved by hydrothermal method. The patterns of nanowires array could be easily controlled by the screen-printing plate [94].

In the past few years, many 2D semiconductors, especially the transition-metal dichalcogenides (TMDs) such as MoS_2 [95], $MoSe_2$ [96], and Bi_2Te_3 [97], have also been used to produce inks and printed onto different substrates to fabricate electronic devices. For example, Östling et al. demonstrated a reliable and efficient inkjet printing of multilayer MoS_2 nanosheets to form uniform patterns with a resolution of tens of micrometers [98]. The printing and patterning of high-quality 2D TMD nanocrystals have great potential for optoelectronic devices. However, there remain challenges such as how to precisely control the thickness and surface roughness of the printed layer.

CNTs, which have been discussed in carbon-nanomaterials-based ink in Section 9.3.3, are also promising candidates as high-mobility semiconductors. CNTs generally show both metallic and semiconducting behaviors depending on their chirality [99]. Recently, many methods have been developed to separate semiconducting nanotubes from metallic nanotubes [100–104], including density gradient centrifugation [100], gel chromatography [101], partition separation [102], and sorting by conjugated polymers [104]. At the PERC, a series of development work was carried out to separate semiconducting SWCNTs from commercial mixed as-grown SWCNTs [105–108]. Printable semiconductor SWCNT inks with purity better than 99% were made and TFTs with charge mobility over 40 cm^2/Vs have been fabricated by inkjet printing on flexible plastic substrate [108].

9.3.5 Reactive Inks

Recently, reactive inks or reactive printing methods are attracting much attention in printed electronics, especially in the fabrication of conductive metal electrodes [109, 110]. Reactive inks have become a powerful alternative to nanoparticles-based inks. The primary concept of reactive inkjet printing method consists of two processes, as shown in Fig. 9.8: First, the metal precursors are printed, and subsequently the reducing reactants are deposited at the same place. After the sintering process, a metal film is obtained directly on the surface of the substrates [111].

At the PERC, Su et al. synthesized an aluminum (Al) ink for screen printing of Al films. With the sintering temperature as low as 80°C for 30 s, the printed Al film with thickness of 50 nm showed a sheet resistance of 2.29 Ω/sq [112]. Lewis et al. prepared reactive Ag inks for printing highly conductive patterns. These inks are highly transparent and do not contain any nanoparticles. Upon sintering at 90°C, the printed electrodes exhibited an electrical conductivity equivalent to that of bulk Ag [113].

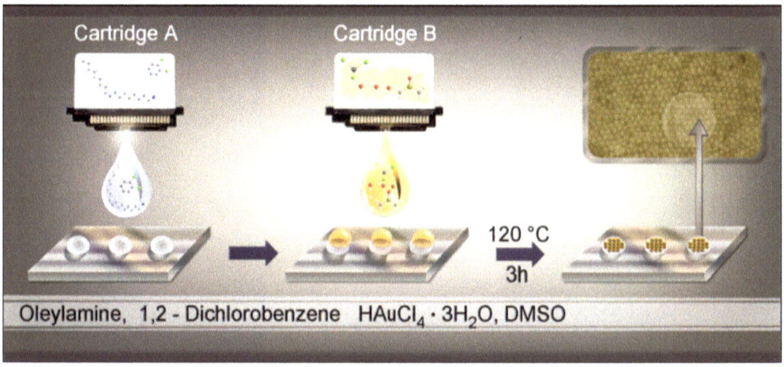

Figure 9.8 Schematic illustration of the in situ synthesis of self-assembled Au nanoparticles through the reactive inkjet printing method. Reprinted with permission from Ref. [111], Copyright 2014, John Wiley & Sons.

9.3.6 Stretchable Inks

Stretchable electronics or elastic electronics are constructed by building or embedding electronic circuits in/on elastomeric materials such as PDMS or polyurethane [114]. In order to fabricate stretchable circuits by printing method, the most important technology is elastic ink. Someya et al. reported a series of development work for stretchable conductive inks [115–118]. A printable stretchable ink with high conductivity and mechanical durability was fabricated by uniformly dispersing commercial Ag flacks in fluorinated rubber with the addition of a small amount nonionic surfactant, as shown in Fig. 9.9. The conductivity of as-obtained elastic conductor was up to 738 S/cm before stretching and remained 182 S/cm at a strain of 215%. The stretchable ink could directly be printed on various elastic substrates (PDMS, PU, Fabrics) by screen printing with fine pattering at a resolution of 50 μm [116]. At the PERC, we also reported a printing strategy to fabricate stretchable electrodes of Ag NWs/PDMS composite [119–123]. The printed electrode has a stable conductivity in a large range of tensile strain (0–50%) due to the buckling structure of Ag NWs/PDMS composite layer [119, 124]. Recently, a new type of elastic conductive paste has been developed at the PERC, which can be directly printed onto elastic textiles.

It should be noted that high stretchability will inevitably lead to increase in resistance and even pattern cracks, which are detrimental to electric performance. Therefore, in order to keep the printed circuit working properly under strained conditions, it is imperative to develop new stretchable inks, which are of stable conductivity under strain.

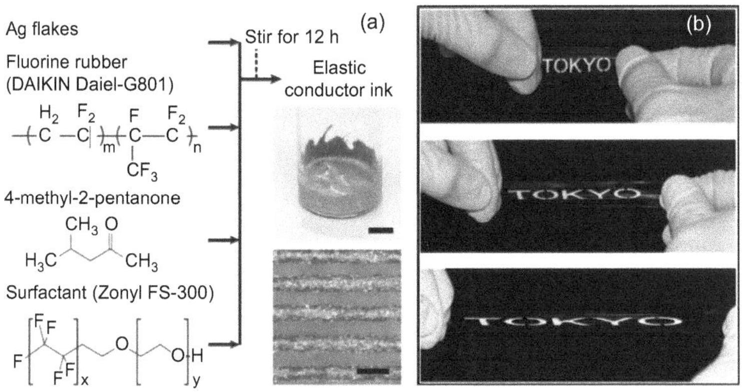

Figure 9.9 Highly stretchable conductive paste. (a) Fabrication process of elastic conductor ink. Upper picture, elastic conductor ink; lower picture, printed elastic conductor with high resolution. (b) Printed elastic conductor and demonstration of stretchability. Reprinted from Ref. [116] under Creative Commons license.

9.4 Post-printing Process

Besides printing processes and printable inks, post-printing treatment is also very important for manufacturing printed electronic devices. In the example of metallic nanoparticles-based inks, there are always some organic stabilizing agents and nonvolatile components remaining between the particles even after drying, which prevent the formation of continuous metallic contact between particles and result in low electrical conductivity. Therefore, a sintering process is required for particles-based printed patterns. Sintering is a process of welding particles together and usually requires the removal of organic materials, such as by decomposition. The main sintering process methods include thermal sintering, photonic sintering, laser sintering, plasma sintering, microwave sintering, and electrical sintering techniques. In this section, these sintering processes are introduced, respectively.

9.4.1 Thermal Sintering

Thermal treatment is the conventional method for metallic nanoparticles sintering, which is commonly performed in an oven, or in a hot air flow tunnel, or on a hotplate. Thermal sintering depends greatly on the size of metallic nanoparticles, the volatilization and decomposition temperature of organic additives, as well as the boiling point of solvent. Figure 9.10a shows the sintering mechanism during heating process. As the temperature increases, the thin organic insulating materials wrapped around nanoparticles are decomposed, which results in the formation of conductive necks between adjacent nanoparticles [9]. Extending the thermal sintering time is also beneficial to improve the conductivity of the printed patterns. Usually, heating at 200–350°C for 10–60 min is required to remove the organic insulating materials from the surface of the nanoparticles and to obtain sintered metal film with resistivity close to that of bulk metal [7]. Figure 9.10b shows the typical changes in the morphology of a printed silver layer after heating at various temperatures for 10 min. The critical changes in morphology with the formation of a continuous interconnection between silver nanoparticles are clearly seen at 260°C. At 320°C, the printed layer is well sintered and yields conductivity half of the bulk silver [125]. Thermal sintering with high temperature is routinely used for conductive patterns on glass or silicon substrates. However, for flexible electronics, in which the substrates are made of organic polymers, the sintering temperature has to be below 200°C. Lowering the sintering temperature can be achieved by properly tailoring the composition of ink, mainly by using stabilizers that decompose at low temperature [126, 127].

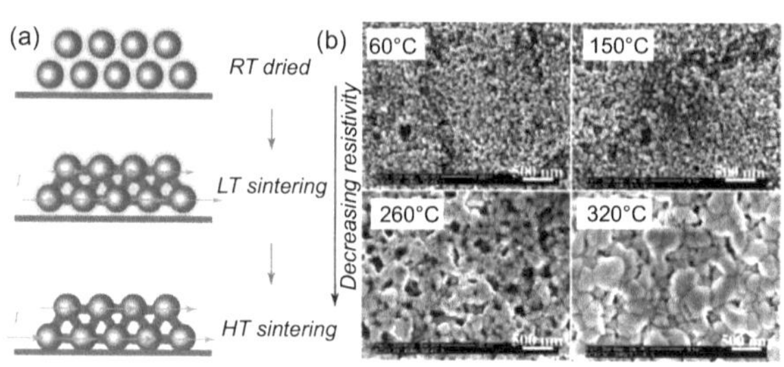

Figure 9.10 (a) Schematic illustration of thermal sintering process; RT, LT, and HT denote room temperature, low temperature, and hot temperature. Reprinted with permission from Ref. [9], Copyright 2017, The Royal Society of Chemistry. (b) SEM images of Ag nanoparticles sintered at different temperature. Reprinted with permission from Ref. [125], Copyright 2005, John Wiley & Sons.

9.4.2 Photonic Sintering

Photonic sintering is a low-temperature and selective sintering technology, which delivers energy from photon sources to the targeted material to be sintered [128]. Photonic sintering can sinter nanoparticles-based inks on various substrates and avoid destruction of the polymeric substrates. Photonic sintering technology includes laser sintering, pulsed/flashing light sintering, and infrared (IR) sintering.

Selective laser sintering utilizes laser as the energy source to sinter the metallic nanoparticles-based inks [129, 130]. The energy of focused laser beam absorbed by metallic nanoparticles generates instantaneous heat around it, which leads to the decomposition and evaporation of organic components and results in rapid sintering, as shown in Fig. 9.11a [9]. The laser source can be integrated to a scanning system to sinter a blanket of ink. Followed by the removal of ink, which was not exposed to the radiation (unsintered metal particles), metallic patterns can be selectively formed by scanned laser instead of printing. For example, Ko et al. reported a facile approach to fabricate a metallic grid transparent conductor using selective laser sintering of Ag nanoparticle ink. The metallic grid transparent conductor demonstrated high transmittance (>85%) and low sheet resistance (30 Ω/sq), which could be used to fabricate on flexible substrates with large area [131]. The scanning speed and laser power are crucial to the sintering results. For example, a higher power can result in excessive heating, which may lead to fast decomposition and evaporation of organic additives and cause delamination of the sintered layer [9].

Compared to focused laser sintering, an intense pulsed light (IPL) with a broadband of wavelength from 200 to 2000 nm can achieve large-area, high-speed sintering under ambient conditions. Very short irradiation on a printed pattern can produce instantaneous sintering at room temperature. IPL sintering is exclusively used to cure Cu nanoparticles-based inks, because this sintering process can instantly remove the oxide shells of Cu nanoparticles and generate a continuous conductive Cu film in a few milliseconds without damaging the underlying polymeric substrates [132–134]. At the PERC, Cu nanoparticles were synthesized and an IPL based on xenon lamp was used to sinter printed patterns of Cu nanoparticles [134].

The synthesized Cu nanoparticles have much thinner oxide shells, as shown in Fig. 9.11b, which significantly reduce the sintering energy. The sintered Cu films have low resistivity of 28 μΩcm on PEN substrate and 44 μΩcm on PET substrate. The sintered Cu film also showed strong adhesion to plastic substrate because the sintered Cu nanoparticles were embedded into the substrate after sintering, as shown in Fig. 9.11c, which results in the adhesion strength of 5B by the ASTM D3359 international standard. IPL technology can achieve large-area sintering, while a focused laser beam can only sinter a small region, which is more suitable for sintering fine conductive patterns.

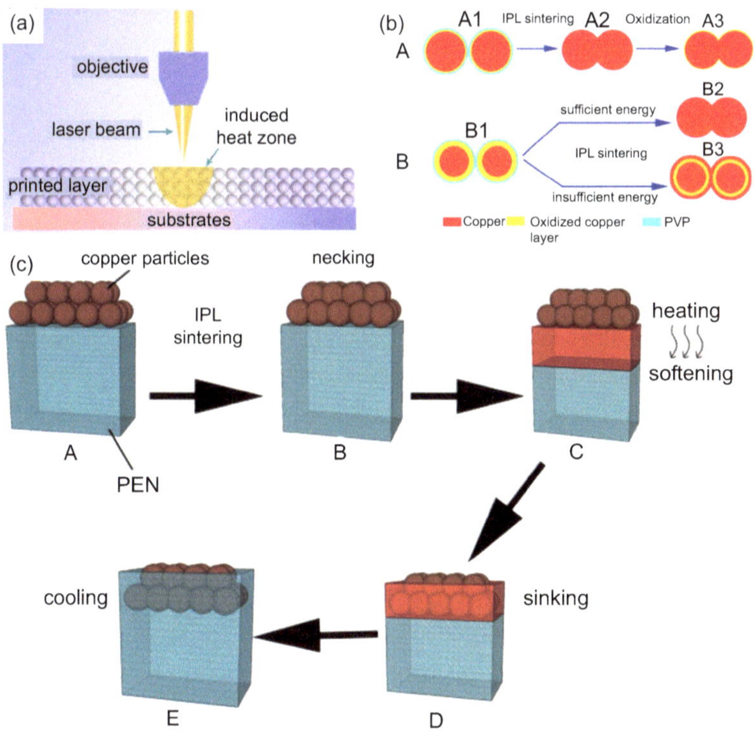

Figure 9.11 (a) Schematic illustration of laser sintering principle: A focused beam locally heats the printed nanoparticles layer. Reprinted with permission from Ref. [9], Copyright 2017, The Royal Society of Chemistry. (b) Schematic illustration of the difference between copper nanoparticles with thin and thick oxide layer under IPL treatment. (c) Schematic illustration of copper film on low-Tg substrate during the IPL-sintering process Figures (b) and (c) reprinted with permission from Ref. [134], Copyright 2017, IOP Publishing.

IR sintering is a conventional and low-cost method to obtain highly conductive printed patterns [135, 136]. Baumann et al. reported ultra-fast and highly efficient IR drying and sintering of inkjet-printed Ag layers on flexible polymeric substrate. Both drying and sintering can take place with IR post-processing in a fraction of seconds, which is suitable for R2R large-scale production [136]. However, the drawback of IR sintering technique is that the underlying plastic substrates are easily damaged because the polymers also have high absorbance in the IR region of the electromagnetic spectrum. Therefore, IR sintering techniques are mostly suitable for pre-curing with short time or curing printed patterns based on flexible papers, because papers have highly diffusive reflectance, relatively high thermal stability, and low thermal conductivity features.

9.4.3 Plasma, Microwave, and Electrical Sintering

Besides photonic sintering, plasma sintering, microwave sintering, and electrical sintering are also used in curing nanoparticles-based inks for manufacturing printed electronics [137–139]. Plasma exhibits good performance in the densification of particulate nanomaterials, which makes it suitable for the sintering of metallic nanoparticles-based inks. Plasma sintering is usually performed by exposing printed patterns to low-pressure Argon (Ar) plasma. Schubert et al. reported a rapid plasma sintering process of inkjet-printed Ag nanoparticle ink, and the sintered pattern has a conductivity of 11.4% of bulk Ag within 1 min of plasma exposure and a final conductivity up to 40% of bulk Ag after 60 min of exposure [139]. In order to avoid the sophisticated equipment for low-pressure plasma sintering, Ar plasma sintering at atmospheric pressure and room temperature was developed [140, 141]. Jung et al. demonstrated the printed Cu pattern, which was sintered by atmospheric-pressure plasma, as shown in Fig. 9.12a. The sintered Cu patterns exhibited conductivity of 21.06 μΩcm, which is around 12.61 times that of bulk Cu [140]. This approach enables sintering of patterns printed onto plastic substrates and can be utilized in R2R processes.

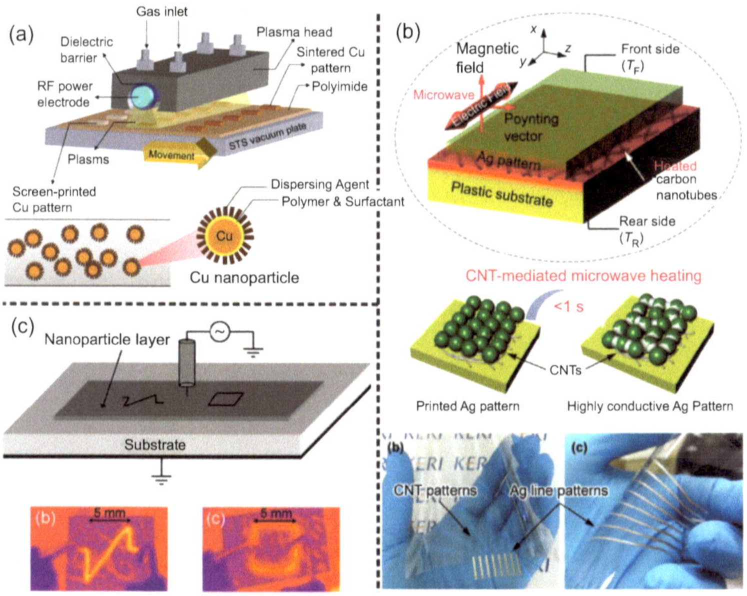

Figure 9.12 (a) Schematic illustration of the atmospheric-pressure plasma sintering system for sintering Cu nanopaste. Reprinted with permission from Ref. [140], Copyright 2013, Elsevier. (b) Schematic illustration of CNT-mediated microwave sintering (up), and the pictures of high-conductivity Ag patterns within a short sintering time by rapid and selective microwave heating of CNTs (down). Reprinted with permission from Ref. [145], Copyright 2009, The Royal Society of Chemistry. (c) Schematic illustration of contact-less AC sintering between a probe above the NPs layer and a ground plate beneath the printing substrate (up) and the corresponding infrared images (down). Reprinted with permission from Ref. [146], Copyright 2009, IOP Publishing.

Metals can also be sintered quickly by microwave radiation [142, 143]. In the microwave sintering process, the metallic nanoparticles absorb the electromagnetic energy and convert into heat, which can decompose the organic agents. For example, treatment of silver patterns on PI by a microwave reactor at constant power (300 W) for 240 s resulted in metallic tracks with conductivity 20 times lower than that of bulk silver [144]. Lee et al. reported a metal-on-CNT method to achieve effective sintering for printed Ag patterns, as shown in Fig. 9.12b. The Ag nanoparticles could be sintered rapidly (1 s) via heat transfer from CNTs by selective microwave heating of CNTs, and the rapid curing process does not damage the underlying heat-sensitive flexible substrates [145]. Though microwave sintering technology can achieve fast curing of printed inks, the biggest problem is the uniformity of sintering for large-area sample, which is extremely difficult to control.

Electrical sintering is another fast sintering technique. Sintering is done by applying a voltage over the printed pattern, and resistive heating is generated by an electric current flow in the metallic layer, as shown in Fig. 9.12c [146]. This method requires that the printed patterns should be slightly conductive before sintering. For example, Oh et al. reported that when the initial resistance is too high, electrical sintering will not be achieved and all the printed lines were burned and blown away when the input voltage was larger than 40 V. When the initial resistance is lower than ~100 kΩ and the input voltage is in the range of 20 V to 40 V, electrical sintering showed good result and there is no damage to the underlying flexible substrate [147]. The advantage of electrical sintering is its short sintering time (typically <100 ms), while the disadvantage is that it cannot be applied to sinter large-area R2R printed patterns.

9.5 Applications

Applications of printing flexible and stretchable electronics are discussed in this section, which include the fabrications and properties of flexible transparent conductive film (TCFs), printed TFTs, solar cells, organic light-emitting diodes (OLEDs), and stretchable electronic systems.

9.5.1 Transparent Conductive Films

The market for TCFs has grown rapidly in the past decade due to the wide proliferation of touchscreens in mobile electronic devices, such as mobile phones, pad computers, and smart wearable devices. Currently, more than 97% of TCFs rely on a transparent conductive material called indium–tin oxide (ITO). Although ITO is of low resistance, down to 10 Ω/sq, at about 90% optical transparency in visible wavelength, indium is a scarce mineral resource on earth and deposition of ITO relies on vacuum sputtering, which is a costly process. Moreover, the inherent brittleness of ITO imposes severe limitation on its utilization in flexible or stretchable electronics [7]. In addition, for touch control application in display, there is a need for patterning the ITO layer, which is usually made by chemical etching. Therefore, great efforts are being made in recent years to find alternatives to ITO. The materials and technologies that have been investigated include conductive polymers [148], CNTs [149], graphene [150], metal mesh (grid) [151], metal NWs [152], as well as printing processes instead of vacuum sputtering. An ideal replacement should be inexpensive, flexible, and simple to manufacture in high volume.

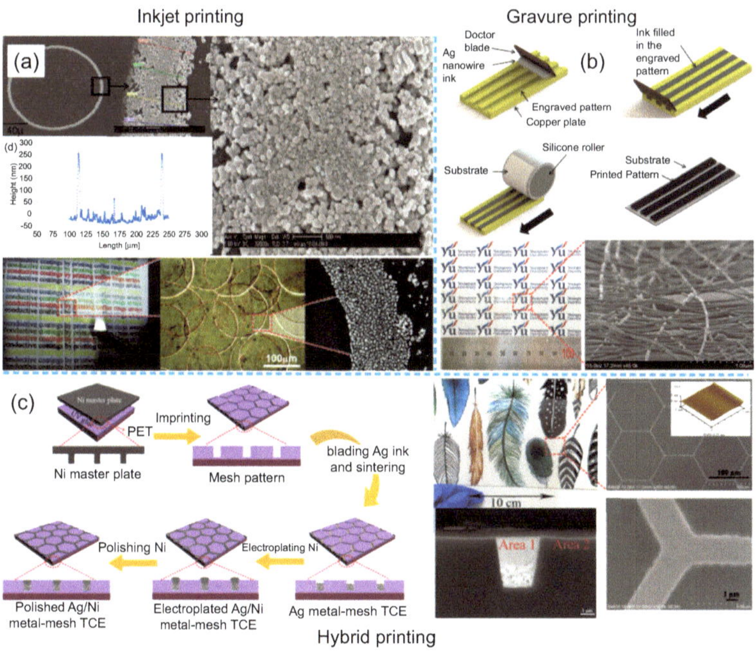

Figure 9.13 TCFs produced by various methods. (a) TCFs fabricated by inkjet printing and utilizing the coffee ring effect. Up: SEM images of the whole printed ring and the rim, and the height profile graph of ring. Down: Transparent conductive film fabricated by the ring arrays. Reprinted with permission from Ref. [153], Copyright 2009, American Chemical Society. (b) TCTs produced by gravure printing. Up: Schematic illustration of gravure printing method. Down: Optical picture and SEM image of obtained TCF. Reprinted with permission from Ref. [67], Copyright 2015, Elsevier. (c) TCTs produced by hybrid printing. Left: Schematic illustration of hybrid printing method. Right: Optical picture and SEM images (surface and cross section) of obtained TCF. Reprinted with permission from Ref. [151], Copyright 2017, American Chemical Society.

Magdassi et al. reported a concept utilizing the "coffee ring effect" in inkjet printing of Ag nanoparticles to fabricate ring-like metallic grids as transparent conductive patterns, as shown in Fig. 9.13a. The width of individual rings is less than 10 μm and less than 300 nm in height. Transparency up to 95% and sheet resistance down to 4 Ω/sq have been achieved with such an approach [153]. Coated metal NWs, especially Ag NWs, are another attractive option as TCFs. They are fabricated by various coating techniques, such as drop casting [59–61], spray coating [62, 63], rod coating [65, 66], spin coating [64], vacuum filtration followed by transfer printing [154], and printing [69]. Kim et al. used gravure printing to fabricate large-area patterned transparent and uniform Ag NW lines on flexible substrate. Continuous conductive patterns with transparency of 95% and sheet resistance of 28 Ω/sq were fabricated by printing Ag NW ink at printing speeds of 4–12 cm/s and a printing pressure of 174 kPa, as shown in Fig. 9.13b. The printed transparent conductive patterns are lines with width of 450 μm and spacing of 100 μm [67].

Besides the conventional printing methods, a new hybrid printing process has been successfully applied in the fabrication of flexible TCFs at the PERC. The hybrid printing approach uses a master template that has all the patterns needed and imprints into a polymeric layer. Conductive material such as Ag nanoparticles ink is then filled into the imprinted trenches and sintered to become embedded conductive silver tracks, as shown in Fig 9.13c. High-resolution patterns (less than 3 μm line width for the mesh) can be made on the master template, which ensures the making of high-resolution embedded silver lines on a substrate. TCFs made by the hybrid printing method have superior surface conductivity with sheet resistance less than 1 Ω/sq at 88% of transparency, which is superior to any other TCFs [151]. The big advantage of hybrid printing method is that the manufacturing of TCFs can be done in one continuous process in a R2R manner. High-throughput and high-volume manufacturing have been realized by the touch panel manufacturer O-film Co. Ltd. in China [5].

9.5.2 Printed TFTs

Transistors are the key components of all electronics. Making transistors by printing has been a dream since organic electronics research started 30 years ago [5]. Although organic polymer materials are inherently soluble and can be made into ink form, with the first all-printed transistor produced in 2000 [21], printed organic transistors remain as an interest of academic research and are not taken up by industry because of the low charge mobility of organic semiconductor. Printed TFTs have a long way to go before becoming mature and usable in practical electronic circuits. Apart from poor performances of printable semiconductor inks, printing as a manufacturing technique is inherently inferior to a conventional micro-fabrication process in terms of resolution, accuracy, structure complexity, and quality [5]. Consequently, the performances of TFTs made by printing are impossible to match those made by conventional processes such as silicon integrated circuits. However, progress in this field is rapid in recent years, particularly in the area of semiconductor ink development. Not only have the organic semiconductors advanced a great deal, the emergence of inorganic nanomaterials such as CNTs and metal oxides enables much better printable semiconductor inks to be made. High-performance printed TFTS have been achieved, including printed SWCNT TFTs [105–108, 155], metal oxide TFTs [156], and graphene TFTs [87]. At the PERC, considerable development work has been carried out to separate semiconducting SWCNTs from commercial mixed as-grown SWCNTs [105–108]. Top-gate TFT array was made by inkjet printing of semiconducting SWCNTs as active layer and Ag tracks as top electrodes, as shown in Fig. 9.14a (the printing fabrication process) and in Fig. 9.14b (the printed flexible TFT circuit array). The mobility of ~46.2 cm^2/VS, $I_{on/off}$ ratio of 10^5, and subthreshold swing of 109 mV per decade for the printed SWCNT TFTs were achieved [108]. Fully printed SWCNT TFTs were also demonstrated with mobility up to 7.67 cm^2/VS, on/off ratio of $10^4 \sim 10^5$, minimal hysteresis, and a low operation voltage (<10 V) [157]. However, the uniformity of printed TFTs is still an issue. Without good uniformity of every printed TFT, construction of complex functional electronic circuits will not be possible.

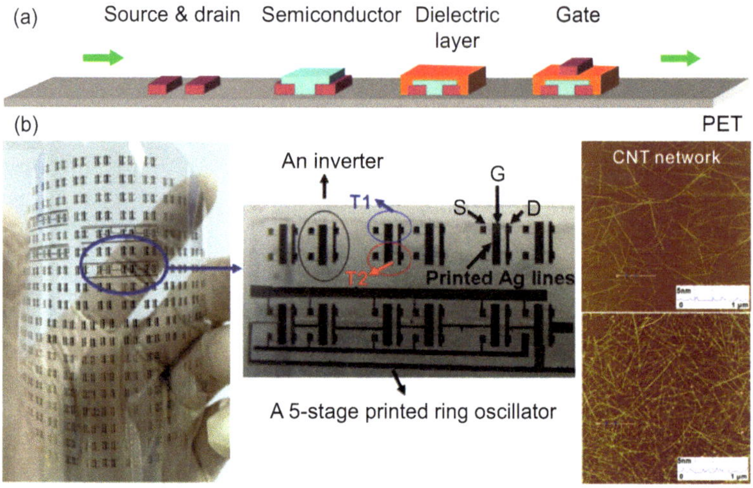

Figure 9.14 (a) Schematic illustration of the printing process for fabrication of TFT . Reprinted with permission from Ref. [9], Copyright 2017, The Royal Society of Chemistry. (b) Left: The optical picture of printed top-gate SWCNT TFT arrays on PET substrate. Middle: the enlarged images of printed top-gate TFTs. Right: AFM images of printed CNT networks with different density. Reprinted with permission from Ref. [108], Copyright 2009, The Royal Society of Chemistry.

9.5.3 Printed Solar Cells

A solar cell is a device that converts the energy of light directly into electricity by the photovoltaic effect. It is known that organic photovoltaic (OPV) has lower efficiency than silicon PV, which means OPV can only compete in the portable energy source market with its light weight and flexible features. Printing fabrication of OPV can lower its cost. Polymer solar cells fabricated by the screen-printing technique was reported in 2001 by Jabbour et al. [158]. In the report, an MDMO-PPV:PCBM-based organic solar cell was fabricated and power conversion efficiency of 4.3% was achieved when the solar cell was illuminated with a monochromatic light of 488 nm with a light intensity of 27 mW/cm^2. In 2009, Cho et al. compared the morphology of films prepared by screen printing with different organic solvents, including chloroform, toluene, chlorobenzene, and 1,2-dichlorobenzene, and demonstrated that thin film prepared from chloroform solution showed the lowest surface roughness.

The highest photoelectric conversion efficiency (PCE) of 4.23% was obtained by screen printing from chloroform [159]. Although the efficiency is rather low in these devices, the results confirmed that screen printing can be a viable technique to make organic solar cells.

Recently, many types of solar cells are being developed by inkjet printing. For example, Lund et al. successfully printed an ionic liquid electrolyte in dye-sensitized solar cells (DSSCs) as a part of a new cell fabrication sequence that eliminates drilling holes in the DSSC substrates. The as-obtained DSSCs exhibited 6% enhancement in overall PCE compared to the reference cells [160]. The inkjet printing of ionic liquid electrolyte did not only reduce the overall fabrication cost, but also remove one extra cell sealing step in traditional cell sealing process. Sirringhaus et al. demonstrated an all-inkjet-printed OPV with an average PCE of 2%, as shown in Fig. 9.15, and all the fabrication and measurement processes were carried out under ambient conditions [161]. In-depth study on PCDTBT:$PC_{70}BM$ layers revealed that the inkjet-printed blend layers exhibited similar nanoscale structure and excited state dynamics to spin-coated layers. Many reports indicate that inkjet printing is an attractive printing technology for the fabrication of photovoltaic cells on flexible substrates.

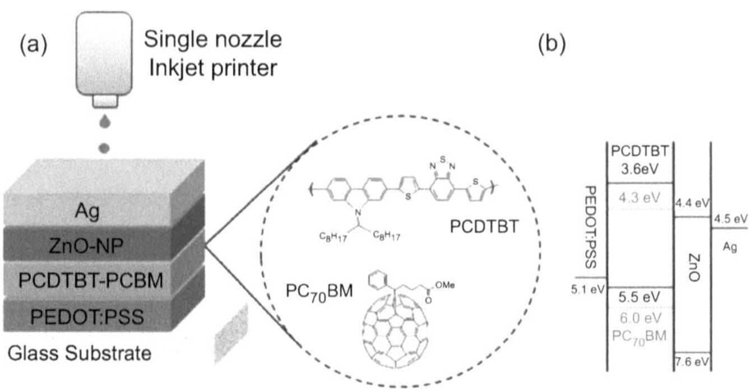

Figure 9.15 (a) Device structure of an inkjet-printed solar cell and (b) energy level diagram of its components. All the four layers are printed using a single nozzle inkjet printer, and all the fabrication processes, including printing, annealing, and measuring, have been performed in air at room temperature. Reprinted with permission from Ref. [161], Copyright 2014, John Wiley & Sons.

In the past few years, the most astonishing development in PV technology has been the newly emerged perovskite PV. The organic–

inorganic hybrid perovskite solar cells emerged as the third-generation solar cells, with a record efficiency of 22.1% in late 2017 [162]. The challenge for perovskite solar is to translate the lab-scale process to a large-area production process. Many efforts have been spent in this area, and great progresses have been made. For example, a fully printed perovskite solar cell was recently demonstrated, which was fabricated by large-scale R2R slot-die coating process. All layers, except the electrode, were deposited by slot-die coating, and the cells exhibited high PCE of 11.96% [163]. Though it is still very early to consider practical applications of perovskite solar cells, one would expect that low-cost perovskite solar cells fabricated by printing method will be on the market in the near future.

9.5.4 Printed OLEDs

OLED is another area where printing can play an important role. OLED lighting has many distinctive features compared to conventional lighting. It is a diffused light source instead of conventional point or line light source, which has no glare to human eyes. However, OLED lighting is facing brutal competition in a conventional lighting market where current light sources, especially the inorganic light-emitting diode (LED), are of much lower cost. Therefore, to lower the manufacturing cost of OLED lighting panels, one obvious option is to use printing instead of vacuum deposition process. Printing is naturally suitable for making OLEDs on flexible substrates. Research and development on printing OLED has been going on for many years. In order to make OLED by printing method, several key issues need to be solved. First, the OLED materials must be soluble, whether for polymers or for small molecules. Second, the ink must have good film-forming property. Third, the soluble ink must cause no damage to each functional layer. At the PERC, some cross-linking type bipolar host, electron transport materials, and even thermally activated delayed fluorescence (TADF) materials have been developed for printing OLED [164–166]. The cross-linking approach can facilitate the printing of OLEDs with multilayer architectures.

A wide range of coating/printing processes can be employed for making OLEDs, including spin coating or slot-die coating, inkjet printing, and screen printing. Inkjet printing can deposit ink material in precise locations, which is essential for making color pixels in OLED display [5]. Currently in OLED display panel manufacturing, red, green, and blue (RGB) organic light-emitting materials are deposited by vacuum evaporation through a fine metallic shadow mask, which has wasted 90% of the RGB materials. Using inkjet printing, the RGB materials can be deposited with only the amount needed and precisely at each pixel location. Panasonic has been developing the inkjet printing process for some years and demonstrated the world's first printed 55-inch OLED TV at the 2013 Consumer Electronics Show. AUO, a display panel manufacturer in Taiwan, demonstrated a 65-inch OLED TV made by inkjet printing at the 2014 Display Week conference [167]. China launched a national R&D program to develop printable OLED display in 2016, with participation of all major Chinese display panel manufacturers. It is expected that printing will play a major role in future manufacturing of flexible displays.

9.5.5 Printed Stretchable Circuits

Stretchable electronic systems may open many possible applications in wearable, implantable, and biocompatible electronic devices, and smart textiles [2, 114]. In the past few years, various strategies have been developed to enable the fabrication of stretchable electronic systems. The most reported method is transfer printing, in which serpentine structures were first made on rigid substrate by conventional micro-fabrication technique (see Fig. 9.1a) and then transferred to a prestretched elastic substrate. On releasing the strained substrate, the serpentine structures buckle, which enables stretchability of the electronic systems [2]. To avoid the complex and expensive fabrication process, many efforts have been made recently to fabricate stretchable electronics by using printing method [115–123]. Unfortunately, the traditional nanoparticles-based inks are not suitable for fabrication of stretchable circuit despite the necking formation between adjacent nanoparticles after sintering process, though some researchers have used inkjet-printed conductive patterns to make highly sensitive micro-deformation sensors [168].

The stability of resistance in stretching state is the most crucial factor for stretchable circuit. Currently, there are two approaches for fabricating stretchable electrodes by printing process. One is to print metal nanowires to obtain a 3D conductive network, which is then combined with an elastic polymer to form stretchable electrodes; the other is to print stretchable inks directly on elastic substrates to form stretchable electrodes. At the PERC, stretchable electrodes were fabricated by using the first method [119–123], and the multi-step fabrication process is shown in Fig. 9.16a. Complex patterns can be made easily by printing, and the produced stretchable electrodes have a stable conductivity in a large range of tensile strain (0–50%) due to the buckling structures of the Ag NWs/PDMS composite layer. A functional circuit integrated with a microprocessor chip, capacitors, resistors, and LED is also shown in Fig. 9.16a. With stretchable inks (see in Section 9.3.6), electronic circuits can be directly printed on various elastic substrates. Wang et al. demonstrated the full-printed stretchable electrochemical devices by screen printing, as shown in Fig. 9.16b. The printed device showed good performance under twisting, bending, and stretching conditions [169]. So far there are still many challenges to build complex functional stretchable

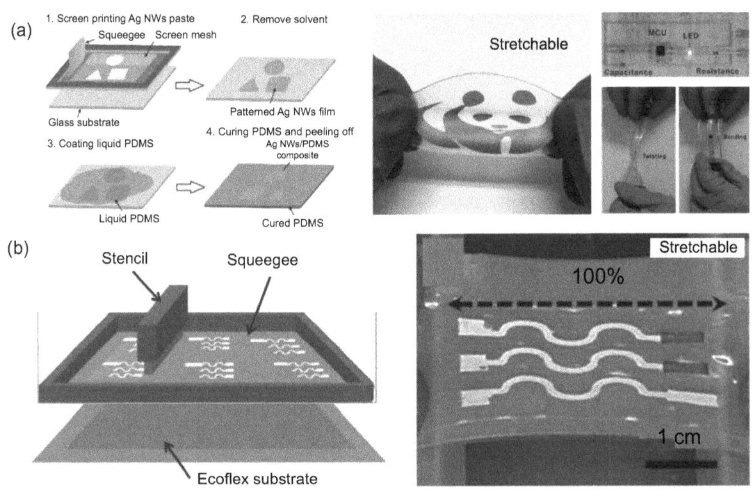

Figure 9.16 Stretchable conductive patterns fabricated by printing method. (a) Schematic illustration of the fabrication of stretchable electrodes by multi-step process (left), the printed panda pattern (middle), the simple system integrated with stiff chips under twisting and bending condition (right). Reprinted with permission from Ref. [119]. (b) Schematic illustration of screen-printed stretchable patterns by one-single process (left), the picture of printed stretchable device under 100% strain (right). Reprinted with permission from Ref. [169], Copyright 2015, John Wiley & Sons.

electronic systems, such as how to integrate commercial rigid ICs into a stretchable system and how to reduce the influence of strain-induced resistance change on its electronic functions. For future development of electronic textiles, printing stretchable inks seems to be a promising direction.

9.6 Summary

The chapter briefly reviewed the printing technologies that can be applied to the fabrication of flexible and stretchable electronics, including printing methods, pintable inks, and post-printing treatments. Printing is an attractive manufacturing technique due to its distinctive features of low-cost, large-area manufacturing capability, additive manufacturing with no particular requirements on substrate materials, green manufacturing with low energy consumption, and low discharge of pollutants. Printing particularly facilitates the manufacturing of electronics on flexible and stretchable substrates, which is essential to flexible and stretchable electronic systems. Although printing itself is an old technology, printed electronics is a new field. There are many challenges remaining to be solved before real applications to be widely available. The challenges include improving printing resolution, reducing surface roughness of printed layers, producing low-cost functional inks, exploring low-temperature sintering technique, designing high-speed R2R printing lines, and viable packaging and encapsulation methods. However, it is the authors' belief that printing will prevail as an electronic manufacturing technology that is complimentary to conventional silicon microelectronics manufacturing.

Acknowledgment

This work was supported by the National Science Foundation of China (Nos. 51603227), National Program on Key Basic Research Project (No. 2015CB351901), and the Strategic Priority Research Program of the Chinese Academy of Sciences (No. XDA09020201).

References

1. Suo, Z. (2012). Mechanics of stretchable electronics and soft machines, *MRS Bull.*, **37**, pp. 218–225.
2. Kim, D. H., Kim, Y. S., Wu, J., Liu, Z., Song, J., Kim, H. S., Huang, Y. Y., Hwang, K. C., and Rogers, J. A. (2009). Flexible electronics: Ultrathin silicon circuits with strain-isolation layers and mesh layouts for high-performance electronics on fabric, vinyl, leather, and paper, *Adv. Mater.*, **21**, pp. 3703–3707.
3. Su, Y. and Rogers, J. A. (2007). Inorganic semiconductors for flexible electronics, *Adv. Mater.*, **19**(15), pp. 1897–1916.
4. Minemawari, H., Yamada, T., Matsui, H., Tsutsumi, J., Haas, S., Chiba, R., Kumai, R., and Hasegawa, T. (2011). Inkjet printing of single-crystal films, *Nature*, **475**(7356), pp. 364–367.
5. Cui, Z., Qiu, S., Zhou, C., Chen, Z., Lin, J., Zhao, J., Ma, C., and Su, W. (2016). *Printed Electronics: Materials, Technologies and Applications* (Wiley).
6. Gao, M., Li, L., and Song, Y. (2017). Inkjet printing wearable electronic devices, *J. Mater. Chem. C*, **5**, pp. 2971–2993.
7. Kamyshny, A. and Magdassi, S. (2014). Conductive nanomaterials for printed electronics, *Small*, **10**(17), pp. 3515–3535.
8. Magliulo, M., Mulla, M. Y., Singh, M., Macchia, E., Tiwari, A., and Torsi, L., and Manoli, K. (2016). Printable and flexible electronics: From TFTs to bioelectronic devices, *J. Mater. Chem. C*, **3**, pp. 12347–12363.
9. Wu, W. (2017). Inorganic nanomaterials for printed electronics: A review, *Nanoscale*, **9**, pp. 7342–7372.
10. Hoeng, F., Denneulin, A., and Bras, J. (2016). Use of nanocellulose in printed electronics: A review, *Nanoscale*, **8**, pp. 13131–13154.
11. Khan, S. and Lorenzelli, L. (2017). Recent advances of conductive nanocomposites in printed and flexible electronics, *Smart Mater. Struct.*, **26**, pp. 083001.
12. Chen, K., Gao, W., Emaminejad, S., Kiriya, D., Ota, H., Nyein, H. Y. Y., Takei, K., and Javey, A. (2016). Carbon nanotubes: Printed carbon nanotube electronics and sensor systems, *Adv. Mater.*, **28**, pp. 4397–4414.
13. Parashkov, R., Becker, E., Riedl, T., Johannes, H. H., and Kowalsky, W. (2005). Large area electronics using printing methods, *Proc. IEEE*, **93**(7), pp. 1321–1329.
14. Website of PERC. http://www.perc-sinano.com/
15. Kipphan, H. (Ed.) (2001) *Handbook of Print Media* (Berlin: Springer).

16. Tekin, E., Smith, P. J., and Schubert, U. S. (2008). Inkjet printing as a deposition and patterning tool for polymers and inorganic particles, *Soft Matter.*, **4**(4), pp. 703–713.

17. Singh, M., Haverinen, H. M., Dhagat, P., and Jabbour, G. E. (2010). Inkjet printing-process and its applications, *Adv. Mater.*, **22**, pp. 673–685.

18. Derby, B. and Reis, N. (2003). Inkjet printing of highly loaded particulate suspensions, *MRS Bull.*, **28**(11), pp. 815–818.

19. Jang, D., Kim, D., and Moon, J. (2009). Influence of fluid physical properties on ink-jet printability, *Langmuir*, **25**(5), pp. 2629–2635.

20. Torrisi, F., Hasan, T., Wu, W., Sun, Z., Lombardo, A., Kulmala, T. S., Hsieh, G. W., Jung, S., Bonaccorso, F., Paul, P. J., Chu, D., and Ferrari, A. C. (2012). Inkjet-printed graphene electronics, *ACS Nano*, **6**(4), pp. 2992–3006.

21. Sirringhaus, H., Kawase, T., Friend, R. H., Shimoda, T., Inbasekaran, M., Wu, W., and Woo, E. P. (2000). High-resolution inkjet printing of all-polymer transistor circuits, *Science*, **290**, pp. 2123–2126.

22. Aersol jet printer. Available from: http://www.optomec.com.

23. Wang, K., Chang, Y. H., Zhang, C., and Wang, B. (2016). Conductive-on-demand: Tailorable polyimide/carbon nanotube nanocomposite thin film by dual-material aerosol jet printing, *Carbon*, **98**, pp. 397–403.

24. Hong, K., Kim, Y. H., Kim, S. H., Xie, W., Xu, W. D., Kim, C. H., and Frisbie, C. D. (2014). Aerosol jet printed, sub-2 V complementary circuits constructed from p- and n-type electrolyte gated transistors, *Adv. Mater.*, **26**, pp. 7032–7037.

25. Onses, M. S., Sutanto, E., Ferreira, P. M., Alleyne, A. G., and Rogers, J. A. (2015). Mechanisms, capabilities, and applications of high-resolution electrohydrodynamic jet printing, *Small*, **11**(34), pp. 4237–4266.

26. Seong, B., Yoo, H., Nguyen, V. D., Jang, Y., Ryu, C., and Byun, D. (2014). Metal-mesh based transparent electrode on a 3-d curved surface by electrohydrodynamic jet printing, *J. Micromech. Microeng.*, **24**(9), pp. 097002.

27. Shin, D. Y., Seo, J. Y., Tak, H., and Byun, D. (2015). Bimodally dispersed silver paste for the metallization of a crystalline silicon solar cell using electrohydrodynamic jet printing, *Sol. Energy Mater. Sol. Cells*, **136**, pp. 148–156.

28. Kim, B. H., Onses, M. S., Lim, J. B., Nam, S., Oh, N., Kim, H., Yu, K. J., Lee, J. W., Kang, S. K., Lee, C. H., Lee, J., Shin, J. H., Kim, N. H., Leal, C., Shim, M., and Rogers, J. A. (2015). High-resolution patterns of quantum dots formed by electrohydrodynamic jet printing for light-emitting diodes, *Nano Lett.*, **15**(2), pp. 969–973.

29. Shin, D. Y., Yoo, S. S., Song, H., Tak, H., and Byun, D. (2015). Electrostatic-force-assisted dispensing printing to construct high-aspect-ratio of 0.79 electrodes on a textured surface with improved adhesion and contact resistivity, *Sci. Rep.*, **5**, pp. 16704.

30. Park, J. U., Hardy, M., Kang, S. J., Barton, K., Adair, K., Mukhopadhyay, D. K., Lee, C. Y., Strano, M. S., Alleyne, A. G., Georgiadis, J. G., Ferreira, P. M., and Rogers, J. A. (2007). High-resolution electrohydrodynamic jet printing, *Nat. Mater.*, **6**(10), pp. 782–789.

31. Yao, W., Tian, Q., Liu, J., Wu, Z., Cui, S., Ding, J., Dai, Z., and Wu, W. (2016). Large-scale synthesis and screen printing of upconversion hexagonal-phase nayf4: yb3+, tm3+/er3+/eu3+ plates for security applications, *J. Mater. Chem. C*, **4**, pp. 6327–6335.

32. Sekitani, T., Takamiya, M., Noguchi, Y., Nakano, S., Kato, Y., Sakurai, T., and Someya, T. (2007). A large-area wireless power-transmission sheet using printed organic transistors and plastic mems switches, *Nat. Mater.*, **6**(6), pp. 413–417.

33. Kuo, H. P., Yang, C. F., Huang, A. N., Wu, C. T., and Pan, W. C. (2014). Preparation of the working electrode of dye-sensitized solar cells: Effects of screen printing parameters, *J. Taiwan Inst. Chem. Eng.*, **45**(5), pp. 2340–2345.

34. Koo, H., Lee, W., Choi, Y., Sun, J., Bak, J., Noh, J., Subramanian, V., Azuma, Y., Majima, Y., and Cho, G. (2015). Scalability of carbon-nanotube-based thin film transistors for flexible electronic devices manufactured using an all roll-to-roll gravure printing system, *Sci. Rep.*, **5**, pp. 14459.

35. Deganello, D., Cherry, J. A., Gethin, D. T., and Claypole, T. C. (2012). Impact of metered ink volume on reel-to-reel flexographic printed conductive networks for enhanced thin film conductivity, *Thin Solid Films*, **520**(6), pp. 2233–2237.

36. Izumi, K., Yoshida, Y., and Tokito, S. (2017). Soft blanket gravure printing technology for finely patterned conductive layers on three-dimensional or curved surfaces, *Jpn. J. Appl. Phys.*, **56**(5S2), pp. 05EA03.

37. Cheng, H. C., Chen, Y. W., Chen, W. H., Lu, S. T., and Lin, S. M. (2018). Assessing ink transfer performance of gravure-offset fine-line circuitry printing, *J. Electron. Mater.*, **47**, pp. 1832–1846.

38. Krebs, F. C., Fyenbo, J., and Jørgensen, M. (2010). Product integration of compact roll-to-roll processed polymer solar cell modules: Methods and manufacture using flexographic printing, slot-die coating and rotary screen printing, *J. Mater. Chem.*, **20**, pp. 8994–9001.

39. Noh, J., Yeom, D., Lim, C., Cha, H., Han, J., Kim, J., Park, Y., Subramanian, V., and Cho, G. (2010). Scalability of roll-to-roll gravure-printed electrodes on plastic foils, *IEEE Trans. Electron. Packag. Manuf.*, **33**(4), pp. 275–283.

40. Guo, G., Gan, W., Xiang, F., Zhang, J., Zhou, H., Liu, H., and Luo, J. (2011). Effect of dispersibility of silver powders in conductive paste on microstructure of screen-printed front contacts and electrical performance of crystalline silicon solar cells, *J. Mater. Sci.: Mater. Electron.*, **22**(5), pp. 527–530.

41. Kure, K., Kanda, T., Suzumori, K., and Wakimoto, S. (2008). Flexible displacement sensor using injected conductive paste, *Sens. Actuators, A*, **143**(2), pp. 272–278.

42. Yang, C., Wong, C. P., and Yuen, M. M. F. (2013). Printed electrically conductive composites: Conductive filler designs and surface engineering, *J. Mater. Chem. C*, **1**, pp. 4052–4069.

43. Chung, J., Ko, S., Bieri, N. R., Grigoropoulos, C. P., and Poulikakos, D. (2004). Conductor microstructures by laser curing of printed gold nanoparticle ink, *Appl. Phys. Lett.*, **84**(5), pp. 801–803.

44. Urban, A. S., Lutich, A. A., Stefani, F. D., and Feldmann, J. (2010). Laser printing single gold nanoparticles, *Nano Lett.*, 2010, pp. 4794–4798.

45. Bakhishev, T. and Subramanian, V. (2009). Investigation of gold nanoparticle inks for low-temperature lead-free packaging technology, *J. Electron. Mater.*, **38**(12), pp. 2720–2725.

46. Jong Lee, K., Jun, B. H., Choi, J., Lee, Y. I., Joung, J., and Oh, Y. S. (2007). Environmentally friendly synthesis of organic-soluble silver nanoparticles for printed electronics, *Nanotechnology*, **18**(33), pp. 335601.

47. Lee, H. H., Chou, K. S., and Huang, K. C. (2005). Inkjet printing of nanosized silver colloids, *Nanotechnology*, **16**(10), pp. 2436–2441.

48. Rama, V. K. R., Korada, V. A., Karthik, P. S., and Singh, S. P. (2015). Conductive silver inks and its applications in printed and flexible electronics, *RSC Adv.*, **5**, pp. 77760–77790.

49. Tseng, C. C., Chang, C. P., Sung, Y., Chen, Y. C., and Ger, M. D. (2009). A novel method to produce pd nanoparticle ink for ink-jet printing technology, *Colloids Surf. A*, **339**, pp. 206–210.

50. Venkata, A. K., Rao, R. V. K., Karthik, P. S., and Singh, S. P. (2015). Copper conductive inks: Synthesis and utilization in flexible electronics, *RSC Adv.*, **5**, pp. 63985–64030.

51. Tang, X. F., Yang, Z. G., and Wang, W. J. (2010). A simple way of preparing high-concentration and high-purity nano copper colloid for conductive ink in inkjet printing technology, *Colloids Surf. A*, **360**, pp. 99–104.

52. Jeong, S., Woo, K., Kim, D., Lim, S., Kim, J. S., Shin, H., Xia, Y., and Moon, J. (2010). Controlling the thickness of the surface oxide layer on Cu nanoparticles for the fabrication of conductive structures by ink-jet printing, *Adv. Funct. Mater.*, **18**, pp. 679–686.

53. Jeong, S., Song, H. C., Lee, W. W., Lee, S. S., Choi, Y., Son, W., Kim, E. D., Paik, C. H., Oh, S. H., and Ryu, B. H. (2011). Stable aqueous based Cu nanoparticle ink for printing well-defined highly conductive features on a plastic substrate, *Langmuir*, **27**, pp. 3144–3149.

54. Jo, Y. H., Jung, I., Choi, C. S., Kim, I., and Lee, H. M. (2011). Synthesis and characterization of low temperature Sn nanoparticles for the fabrication of highly conductive ink, *Nanotechnology*, **22**(22), pp. 225701.

55. Tseng, W. J. and Chen, C. N. (2006). Dispersion and rheology of nickel nanoparticle inks, *J. Mater. Sci.*, **41**(4), pp. 1213–1219.

56. Grouchko, M., Kamyshny, A., and Magdassi, S. (2009). Formation of air-stable copper–silver core–shell nanoparticles for inkjet printing, *J. Mater. Chem.*, **19**, pp. 3057–3062.

57. Kim, C. K., Lee, G. J., Lee, M. K., and Rhee, C. K. (2014). A novel method to prepare Cu@Ag core–shell nanoparticles for printed flexible electronics, *Powder Technol.*, **263**(5), pp. 1–6.

58. Magdassi, S. (2009). *The Chemistry of Inkjet Inks* (Singapore: World Scientific Publishing Co).

59. Lee, J. Y., Connor, S. T., Yi, C., and Peumans, P. (2008). Solution-processed metal nanowire mesh transparent electrodes, *Nano Lett.*, **8**(2), pp. 689–692.

60. Gaynor, W., Lee, J. Y., and Peumans, P. (2010). Fully solution-processed inverted polymer solar cells with laminated nanowire electrodes, *ACS Nano*, **4**(1), pp. 30–34.

61. Yu, Z., Zhang, Q., Li, L., Chen, Q., Niu, X., Liu, J., and Pei, Q. (2011). Highly flexible silver nanowire electrodes for shape-memory polymer light-emitting diodes, *Adv. Mater.*, **23**, pp. 664–668.

62. Madaria, A. R., Kumar, A., and Zhou, C. (2011). Large scale, highly conductive and patterned transparent films of silver nanowires on arbitrary substrates and their application in touch screens, *Nanotechnology*, **22**(24), pp. 245201.

63. Scardaci, V., Coull, R., Lyons, P. E., Rickard, D., and Coleman, J. N. (2011). Spray deposition of highly transparent, low-resistance networks of silver nanowires over large areas, *Small*, **7**(18), pp. 2621–2628.

64. Leem, D. S., Edwards, A., Faist, M., Nelson, J., Bradley, D. D., and de Mello, J. C. (2011). Efficient organic solar cells with solution-processed silver nanowire electrodes, *Adv. Mater.*, **23**(38), pp. 4371–4375.

65. Liu, C. H. and Yu, X. (2011). Silver nanowire-based transparent, flexible, and conductive thin film, *Nanoscale Res. Lett.*, **6**(1), pp. 75.

66. Hu, L., Kim, H. S., Lee, J. Y., Peumans, P., and Cui, Y. (2010). Scalable coating and properties of transparent, flexible, silver nanowire electrodes, *ACS Nano*, **4**(5), pp. 2955–2963.

67. Park, J. D., Lim, S., and Kim, H. (2015). Patterned silver nanowires using the gravure printing process for flexible applications, *Thin Solid Films*, **586**, pp. 70–75.

68. Finn, D. J., Lotya, M., and Coleman, J. N. (2015). Inkjet printing of silver nanowire networks, *ACS Appl. Mater. Interfaces.*, **7**(17), pp. 9254–9261.

69. Lu, H., Lin, J., Wu, N., Nie, S., Luo, Q., Ma, C. Q., and Cui, Z. (2015). Inkjet-printed silver nanowire network as top electrode for semi-transparent organic photovoltaic devices, *Appl. Phys. Lett.*, 106.093302.

70. Muramatsu, A., Kanie, K., Sasaki, T., and Nakaya, M. (2014). High performance ITO nanoparticles as nanoink for printing as a substitute process of sputtering, *MRS Proc.*, **1699**, mrss14-1699-ll06-04.

71. Dattoli, E. N. and Lu, W. (2011). ITO nanowires and nanoparticles for transparent films, *MRS Bull.*, **36**(10), pp. 782–788.

72. Song, J., Kulinich, S. A., Li, J., Liu, Y., and Zeng, H. (2015). A general one-pot strategy for the synthesis of high-performance transparent-conducting-oxide nanocrystal inks for all-solution-processed devices, *Angew. Chem. Int. Ed.*, **127**(2), pp. 472–476.

73. Hecht, D. S. and Kaner, R. B. (2011). Solution-processed transparent electrodes, *MRS Bull.*, **36**(10), pp. 749–755.

74. Wang, J., Tian, B., Nascimento, V. B., and Angnes, L. (1998). Performance of screen-printed carbon electrodes fabricated from different carbon inks, *Electrochimi. Acta.*, **43**(23), pp. 3459–3465.

75. De Volder, M. F., Tawfick, S. H., Baughman, R. H., and Hart, A. J. (2013). Carbon nanotubes: Present and future commercial applications, *Science*, **339**(6119), pp. 535–539.

76. Tortorich, R. P. and Choi, J. W. (2013). Inkjet printing of carbon nanotubes, *Nanomaterials*, **3**(3), pp. 453–468.

77. Gracia-Espino, E., Sala, G., Pino, F., Halonen, N., Luomahaara, J., Mäklin, J., Tóth, G., Kordás, K., Jantunen, H., Terrones, M., Helistö, P., Seppä, H., Ajayan, P. M., and Vajtai, R. (2010). Electrical transport and field-effect transistors using inkjet-printed SWCNT films having different functional side groups, *ACS Nano*, **4**(6), pp. 3318–3324.

78. Chen, P., Fu, Y., Aminirad, R., Wang, C., Zhang, J., Wang, K., Galatsis, K., and Zhou, C. (2011). Fully printed separated carbon nanotube thin film transistor circuits and its application in organic light emitting diode control, *Nano Lett.*, **11**(12), pp. 5301–5308.

79. Kordás, K., Mustonen, T., Tóth, G., Jantunen, H., Lajunen, M., Soldano, C., Taplapatra, S., Kar, S., Vajtai, R., and Ajayan, P. M. (2006). Inkjet printing of electrically conductive patterns of carbon nanotubes, *Small*, **2**(8–9), pp. 1021–1025.

80. Baechler, C., Gardin, S., Abuhimd, H., and Kovacs, G. (2016). Inkjet-printed multiwall carbon nanotube electrodes for dielectric elastomer actuators, *Smart Mater. Struct.*, **25**(5), 055009.

81. Geim, A. K. and Novoselov, K. S. (2007). The rise of graphene, *Nat. Mater.*, **6**(3), pp. 183–191.

82. Lee, C. L., Chen, C. H., and Chen, C. W. (2013). Graphene nanosheets as ink particles for inkjet printing on flexible board, *Chem. Eng. J.*, **230**, pp. 296–302.

83. Lotya, M., Hernandez, Y., King, P. J., Smith, R. J., Nicolosi, V., Karlsson, L. S., Blighe, F. M., De, M., Wang, Z., McGovern, I. T., Duesberg, G. S., and Coleman, J. N. (2009). Liquid phase production of graphene by exfoliation of graphite in surfactant/water solutions, *J. Am. Chem. Soc.*, **131**(10), pp. 3611–3620.

84. Hasan, T., Torrisi, F., Sun, Z., Popa, D., Nicolosi, V., Privitera, G., Bonaccorso, F., and Ferrari, A. C. (2010). Solution-phase exfoliation of graphite for ultrafast photonics, *Phys. Status Solidi B*, **247**(11–12), pp. 2953–2957.

85. Vadukumpully, S., Paul, J., and Valiyaveettil, S. (2009). Cationic surfactant mediated exfoliation of graphite into graphene flakes, *Carbon*, **47**(14), pp. 3288–3294.

86. Secor, E. B., Prabhumirashi, P. L., Puntambekar, K., Geier, M. L., and Hersam, M. C. (2013). Inkjet printing of high conductivity, flexible graphene patterns, *J. Phys. Chem. Lett.*, **4**(8), pp. 1347–1351.

87. Torrisi, F., Hasan, T., Wu, W., Sun, Z., Lombardo, A., Kulmala, T. S., Hsieh, G. W., Jung, S., Bonaccorso, F., Paul, P. J., Chu, D., and Ferrari, A. C. (2012). Inkjet-printed graphene electronics, *ACS Nano*, **6**(4), pp. 2992–3006.

88. Huang, L., Huang, Y., Liang, J., Wan, X., and Chen, Y. (2011). Graphene-based conducting inks for direct inkjet printing of flexible conductive patterns and their applications in electric circuits and chemical sensors, *Nano Res.*, **4**(7), pp. 675–684.

89. Wu, W., Yang, S., Zhang, S., Zhang, H., and Jiang, C. (2014). Fabrication, characterization and screen printing of conductive ink based on carbon@ Ag core–shell nanoparticles, *J. Colloid Interface Sci.*, **427**, pp. 15–19.

90. Zhang, W., Bi, E., Li, M., and Gao, L. (2016). Synthesis of Ag/RGO composite as effective conductive ink filler for flexible inkjet printing electronics, *Colloids Surf. A.*, **490**, pp. 232–240.

91. Lim, S. C., Koo, J. B., Park, C. W., Jung, S. W., Na, B. S., Lee, S. S., and Chu, H. Y. (2014). Device characteristics of inkjet-printed ZnO TFTs by solution process, *Jpn. J. Appl. Phys.*, **53**(5S3), 05HB10.

92. Rehman, M. M., Siddiqui, G. U., Gul, J. Z., Kim, S. W., Lim, J. H., and Choi, K. H. (2016). Resistive switching in all-printed, flexible and hybrid MoS 2-PVA nanocomposite based memristive device fabricated by reverse offset, *Sci. Rep.*, **6**, 36195.

93. Kim, A., Won, Y., Woo, K., Kim, C. H., and Moon, J. (2013). Highly transparent low resistance ZnO/Ag nanowire/ZnO composite electrode for thin film solar cells, *ACS Nano*, **7**(2), pp. 1081–1091.

94. Cui, S., Dai, Z., Tian, Q., Liu, J., Xiao, X., Jiang, C., Wu, W., and Roy, V. A. (2016). Wetting properties and SERS applications of ZnO/Ag nanowire arrays patterned by a screen-printing method, *J. Mater. Chem. C*, **4**(26), pp. 6371–6379.

95. Lim, S., Cho, B., Bae, J., Kim, A. R., Lee, K. H., Kim, S. H., Hahm, M. G., and Nam, J. (2016). Electrohydrodynamic printing for scalable MoS_2 flake coating: Application to gas sensing device, *Nanotechnology*, **27**(43), pp. 435501.

96. Rowley-Neale, S. J., Foster, C. W., Smith, G. C., Brownson, D. A., and Banks, C. E. (2017). Mass-producible 2D-MoSe2 bulk modified screen-printed electrodes provide significant electrocatalytic performances towards the hydrogen evolution reaction, *Sustainable Energy Fuels*, **1**(1), pp. 74–83.

97. Madan, D., Wang, Z., Chen, A., Juang, R. C., Keist, J., Wright, P. K., and Evans, J. W. (2012). Enhanced performance of dispenser printed MA n-type Bi_2Te_3 composite thermoelectric generators, *ACS Appl. Mater. Interfaces*, **4**(11), pp. 6117–6124.

98. Li, J., Naiini, M. M., Vaziri, S., Lemme, M. C., and Östling, M. (2014). Inkjet printing of MoS_2, *Adv. Funct. Mater.*, **24**(41), pp. 6524–6531.
99. Saito, R., Fujita, M., Dresselhaus, G., and Dresselhaus, U. M. (1992). Electronic structure of chiral graphene tubules, *Appl. Phys. Lett.*, **60**(18), pp. 2204–2206.
100. Arnold, M. S., Green, A. A., Hulvat, J. F., Stupp, S. I., and Hersam, M. C. (2006). Sorting carbon nanotubes by electronic structure using density differentiation, *Nat. Nanotechnol.*, **1**, pp. 60–65.
101. Liu, H., Nishide, D., Tanaka, T., and Kataura, H. (2011). Large-scale single-chirality separation of single-wall carbon nanotubes by simple gel chromatography, *Nat. Commun.*, **2**, pp. 309.
102. Khripin, C. Y., Fagan, J. A., and Zheng, M. (2013). Spontaneous partition of carbon nanotubes in polymer-modified aqueous phases, *J. Am. Chem. Soc.*, **135**(18), pp. 6822–6825.
103. Hu, Y., Chen, Y., Li, P., and Zhang, J. (2013). Sorting out semiconducting single-walled carbon nanotube arrays by washing off metallic tubes using SDS aqueous solution, *Small*, **9**(8), pp. 1306–1311.
104. Lee, H. W., Yoon, Y., Park, S., Oh, J. H., Hong, S., Liyanage, L. S., Wang, H., Morishita, S., Patil, N., Park, Y. J. , Park, J., Spakowitz, A., Galli, G., Gygi, F., Wong, P. H. S., Tok, J. B. H., Kim, J. M., and Bao, Z. (2011). Selective dispersion of high purity semiconducting single-walled carbon nanotubes with regioregular poly (3-alkylthiophene)s, *Nat. Commun.*, **2**, pp. 541.
105. Gao, W., Xu, W., Ye, J., Liu, T., Wang, J., Tan, H., Lin, Y., Tange, M., Sun, D., Wu, L., Okazaki, T., Yang, Y., Zhang, Z., Zhao, J., Cui, Z., and Ma, C. Q. (2017). Selective dispersion of large-diameter semiconducting carbon nanotubes by functionalized conjugated dendritic oligothiophenes for use in printed thin film transistors, *Adv. Funct. Mater.*, **27**(44), pp. 1703938.
106. Wang, C., Qian, L., Xu, W., Nie, S., Gu, W., Zhang, J., Zhao, J., Lin, J., Chen, Z., and Cui, Z. (2013). High performance thin film transistors based on regioregular poly (3-dodecylthiophene)-sorted large diameter semiconducting single-walled carbon nanotubes, *Nanoscale*, **5**(10), pp. 4156–4161.
107. Zhao, J., Gao, Y., Gu, W., Wang, C., Lin, J., Chen, Z., and Cui, Z. (2012). Fabrication and electrical properties of all-printed carbon nanotube thin film transistors on flexible substrates, *J. Mater. Chem.*, **22**(38), pp. 20747–20753.

108. Xu, W., Liu, Z., Zhao, J., Xu, W., Gu, W., Zhang, X., Qian, L., and Cui, Z. (2014). Flexible logic circuits based on top-gate thin film transistors with printed semiconductor carbon nanotubes and top electrodes, *Nanoscale*, **6**(24), pp. 14891–14897.

109. Li, D., Sutton, D., Burgess, A., Graham, D., and Calvert, P. D. (2009). Conductive copper and nickel lines via reactive inkjet printing, *J. Mater. Chem.*, **19**(22), pp. 3719–3724.

110. Smith, P. J. and Morrin, A. (2012). Reactive inkjet printing, *J. Mater. Chem.*, **22**(22), pp. 10965–10970.

111. Abulikemu, M., Da'as, E. H., Haverinen, H., Cha, D., Malik, M. A., and Jabbour, G. E. (2014). In situ synthesis of self-assembled gold nanoparticles on glass or silicon substrates through reactive inkjet printing, *Angew. Chem.*, **126**(2), pp. 430–433.

112. Fei, F., Zhuang, J., Wu, W., Song, M., Zhang, D., Li, S., Su, W., and Cui, Z. (2015). A printed aluminum cathode with low sintering temperature for organic light-emitting diodes, *RSC Adv.*, **5**(1), pp. 608–611.

113. Walker, S. B. and Lewis, J. A. (2012). Reactive silver inks for patterning high-conductivity features at mild temperatures, *J. Am. Chem. Soc.*, **134**(3), pp. 1419–1421.

114. Kim, D. H., Xiao, J., Song, J., Huang, Y., and Rogers, J. A. (2010). Stretchable, curvilinear electronics based on inorganic materials, *Adv. Mater.*, **22**(19), pp. 2108–2124.

115. Sekitani, T., Nakajima, H., Maeda, H., Fukushima, T., Aida, T., Hata, K., and Someya, T. (2009). Stretchable active-matrix organic light-emitting diode display using printable elastic conductors, *Nat. Mater.*, **8**(6), pp. 494–499.

116. Matsuhisa, N., Kaltenbrunner, M., Yokota, T., Jinno, H., Kuribara, K., Sekitani, T., and Someya, T. (2015). Printable elastic conductors with a high conductivity for electronic textile applications, *Nat. Commun.*, **6**, 7461.

117. Matsuhisa, N., Inoue, D., Zalar, P., Jin, H., Matsuba, Y., Itoh, A., Yokota, T., Hashizume, D., and Someya, T. (2017). Printable elastic conductors by in situ formation of silver nanoparticles from silver flakes, *Nat. Mater.*, **16**(8), pp. 834–840.

118. Jinno, H., Fukuda, K., Xu, X., Park, S., Suzuki, Y., Koizumi, M., Yokota, T., Osaka, I., Takimiya, K., and Someya, T. (2017). Stretchable and waterproof elastomer-coated organic photovoltaics for washable electronic textile applications, *Nat. Energy*, **2**, pp. 780–785.

119. Yuan, W., Wu, X., Gu, W., Lin, J., and Cui, Z. (2018). Printed stretchable circuit on soft elastic substrate for wearable application, *J. Semicond.*, **39**(1), 015002.

120. Yuan, W., Gu, W., Lin, J., and Cui, Z. (2017). Fabrication of stretchable circuit with high resolution line width and complex pattern via printing method on silicone substrate, *Sci. Technol. Rev.*, **35**(17), pp. 73–79.

121. Yuan, W., Lin, J., Gu, W. B., and Cui, Z. (2016). Preparation of flexible and stretchable circuit via printing method based on silver nanowires (in Chinese), *Sci. Sin-Phys. Mech. Astron.*, **46**, pp. 044611.

122. Yuan, W., Gu, W., Lin, J., and Cui, Z. (2016). Flexible and stretchable hybrid electronics systems for wearable applications, *Dig. Tech. Pap. Soc. Inf. Disp. Int. Symp.*, **47**(1), pp. 668–671.

123. Yuan, W., Gu, W., Lin, J., and Cui, Z. (2016). Printed flexible and stretchable hybrid electronic systems for wearable applications, *Electronic System-Integration Technology Conference* (pp. 1–4). IEEE.

124. Xu, F. and Zhu, Y. (2012). Highly conductive and stretchable silver nanowire conductors, *Adv. Mater.*, **24**(37), pp. 5117–5122.

125. Kamyshny, A., Ben-Moshe, M., Aviezer, S., and Magdassi, S. (2005). Ink-jet printing of metallic nanoparticles and microemulsions, *Macromol. Rapid Commun.*, **26**(4), pp. 281–288.

126. Huang, D., Liao, F., Molesa, S., Redinger, D., and Subramanian, V. (2003). Plastic-compatible low resistance printable gold nanoparticle conductors for flexible electronics, *J. Electrochem. Soc.*, **150**(7), pp. G412–G417.

127. Perelaer, J., de Laat, A. W. M., Hendriks, C. E., and Schubert, U. S. (2008). Inkjet-printed silver tracks: Low temperature curing and thermal stability investigation, *J. Mater. Chem.*, **18**(27), pp. 3209–3215.

128. MacNeill, W., Choi, C. H., Chang, C. H., and Malhotra, R. (2015). On the self-damping nature of densification in photonic sintering of nanoparticles, *Sci. Rep.*, **5**, 14845.

129. Ermak, O., Zenou, M., Toker, G. B., Ankri, J., Shacham-Diamand, Y., and Kotler, Z. (2016). Rapid laser sintering of metal nano-particles inks, *Nanotechnology*, **27**(38), 385201.

130. Zenou, M., Ermak, O., Saar, A., and Kotler, Z. (2013). Laser sintering of copper nanoparticles, *J. Phys. D: Appl. Phys.*, **47**(2), 025501.

131. Hong, S., Yeo, J., Kim, G., Kim, D., Lee, H., Kwon, J., Lee, H., and Ko, S. H. (2013). Nonvacuum, mask-less fabrication of a flexible metal grid transparent conductor by low-temperature selective laser sintering of nanoparticle ink, *ACS Nano*, **7**(6), pp. 5024–5031.

132. Hwang, H. J., Chung, W. H., and Kim, H. S. (2012). In situ monitoring of flash-light sintering of copper nanoparticle ink for printed electronics, *Nanotechnology*, **23**(48), 485205.

133. Ryu, J., Kim, H. S., and Hahn, H. T. (2011). Reactive sintering of copper nanoparticles using intense pulsed light for printed electronics, *J. Electron. Mater.*, **40**(1), pp. 42–50.

134. Wu, X., Shao, S., Chen, Z., and Cui, Z. (2017). Printed highly conductive Cu films with strong adhesion enabled by low-energy photonic sintering on low-Tg flexible plastic substrate, *Nanotechnology*, **28**(3), 035203.

135. Tobjörk, D., Aarnio, H., Pulkkinen, P., Bollström, R., Määttänen, A., Ihalainen, P., Mäkelä, T., Peltonen, J., Toivakka, M., Tenhu, H., and Österbacka, R. (2012). IR-sintering of ink-jet printed metal-nanoparticles on paper, *Thin Solid Films*, **520**(7), pp. 2949–2955.

136. Sowade, E., Kang, H., Mitra, K. Y., Weiß, O. J., Weber, J., and Baumann, R. R. (2015). Roll-to-roll infrared (IR) drying and sintering of an inkjet-printed silver nanoparticle ink within 1 second, *J. Mater. Chem. C.*, **3**(45), pp. 11815–11826.

137. Reinhold, I., Hendriks, C. E., Eckardt, R., Kranenburg, J. M., Perelaer, J., Baumann, R. R., and Schubert, U. S. (2009). Argon plasma sintering of inkjet-printed silver tracks on polymer substrates, *J. Mater. Chem.*, **19**(21), pp. 3384–3388.

138. Sairam, K., Sonber, J. K., Murthy, T. C., Subramanian, C., Fotedar, R. K., Nanekar, P., and Hubli, R. C. (2014). Influence of spark plasma sintering parameters on densification and mechanical properties of boron carbide, *Int. J. Refract. Hard Mater.*, **42**, pp. 185–192.

139. Wolf, F. M., Perelaer, J., Stumpf, S., Bollen, D., Kriebel, F., and Schubert, U. S. (2013). Rapid low-pressure plasma sintering of inkjet-printed silver nanoparticles for RFID antennas, *J. Mater. Res.*, **28**(9), pp. 1254–1261.

140. Kim, K. S., Bang, J. O., Choa, Y. H., and Jung, S. B. (2013). The characteristics of Cu nanopaste sintered by atmospheric-pressure plasma, *Microelectron. Eng.*, **107**, pp. 121–124.

141. Wünscher, S., Stumpf, S., Perelaer, J., and Schubert, U. S. (2014). Towards single-pass plasma sintering: Temperature influence of atmospheric pressure plasma sintering of silver nanoparticle ink, *J. Materi. Chem. C.*, **2**(9), pp. 1642–1649.

142. Oghbaei, M. and Mirzaee, O. (2010). Microwave versus conventional sintering: A review of fundamentals, advantages and applications, *J. Alloys Compd.*, **494**(1–2), pp. 175–189.

143. Vaseem, M., Lee, S. K., Kim, J. G., and Hahn, Y. B. (2016). Silver-ethanolamine-formate complex based transparent and stable ink: Electrical assessment with microwave plasma vs thermal sintering, *Chem. Eng. J.*, **306**, pp. 796–805.

144. Perelaer, J., de Gans, B. J., and Schubert, U. S. (2006). Ink-jet printing and microwave sintering of conductive silver tracks, *Adv. Mater.*, **18**(16), pp. 2101–2104.

145. Jung, S., Chun, S. J., Han, J. T., Woo, J. S., Shon, C. H., and Lee, G. W. (2016). Sub-second carbon-nanotube-mediated microwave sintering for high-conductivity silver patterns on plastic substrates, *Nanoscale*, **8**(9), pp. 5343–5349.

146. Allen, M. L., Aronniemi, M., Mattila, T., Alastalo, A., Ojanperä, K., Suhonen, M., and Seppä, H. (2008). Electrical sintering of nanoparticle structures, *Nanotechnology*, **19**(17), pp. 175201.

147. Jang, S., Dong, J. L., Lee, D., and Oh, J. H. (2013). Electrical sintering characteristics of inkjet-printed conductive Ag lines on a paper substrate, *Thin Solid Films*, **546**(6), pp. 157–161.

148. Li, L., Xin, Z., Zhao, W., Tang, X., Du, P., Hu, X., and Leng, X. (2010). Research of transparent conductive film prepared with polymer, *Mater. Sci. Forum*, 663–665, pp. 532–537.

149. Geng, H. Z., Kim, K. K., and Lee, Y. H. (2008). Recent progress in carbon nanotube-based flexible transparent conducting film, *Proc. SPIE*, **7037**(5), pp. 70370A.

150. Li, X., Zhu, Y., Cai, W., Borysiak, M., Han, B., Chen, D., Piner, R. D., Colombo, L., and Ruoff, R. S. (2009). Transfer of large-area graphene films for high-performance transparent conductive electrodes, *Nano Lett.*, **9**(12), pp. 4359–4363.

151. Chen, X., Guo, W., Xie, L., Wei, C., Zhuang, J., Su, W., and Cui, Z. (2017). Embedded Ag/Ni metal-mesh with low surface roughness as transparent conductive electrode for optoelectronic applications, *ACS Appl. Mater. Interfaces*, **9**, pp. 37048–37054.

152. Rathmell, A. R., Bergin, S. M., Hua, Y. L., Li, Z. Y., and Wiley, B. J. (2010). The growth mechanism of copper nanowires and their properties in flexible, transparent conducting films, *Adv. Mater.*, **22**(32), pp. 3558–3563.

153. Layani, M., Gruchko, M., Milo, O., Balberg, I., Azulay, D., and Magdassi, S. (2009). Transparent conductive coatings by printing coffee ring arrays obtained at room temperature, *ACS Nano*, **3**(11), pp. 3537–3542.

154. Lee, P., Lee, J., Lee, H., Yeo, J., Hong, S., Nam, K. H., Lee, D., Lee, S. S., and Ko, S. H. (2012). Flexible electronics: Highly stretchable and highly conductive metal electrode by very long metal nanowire percolation network, *Adv. Mater.*, **24**, pp. 3326–3332.

155. Homenick, C. M., James, R., Lopinski, G. P., Dunford, J. L., Sun, J., Park, H., Jung, Y., Cho, G., and Malenfant, P. R. L. (2016). Fully printed and encapsulated SWCNT-based thin film transistors via a combination of R2R gravure and inkjet printing, *ACS Appl. Mater. Interfaces*, **8**, pp. 27900–27910.

156. Lee, Y. G. and Choi, W. S. (2014). Electrohydrodynamic jet-printed zinc-tin oxide TFTs and their bias stability, *ACS Appl. Mater. Interfaces*, **6**, pp. 11167–11172.

157. Cao, X., Chen, H., Gu, X., Liu, B., Wang, W., Cao, Y., Wu, F., and Zhou, C. (2014). Screen printing as a scalable and low-cost approach for rigid and flexible thin-film transistors using separated carbon nanotubes, *ACS Nano*, **8**, pp. 12769–12776.

158. Shaheen, S. E., Radspinner, R., Peyghambarian, N., and Jabbour, G. E. (2001). Fabrication of bulk heterojunction plastic solar cells by screen printing, *Appl. Phys. Lett.*, **79**(18), pp. 2996–2998.

159. Zhang, B., Chae, H., and Cho, S. M. (2009). Screen-printed polymer: fullerene bulk-heterojunction solar cells, *Jpn. J. Appl. Phys.*, **48**(2), pp. 020208.

160. Hashmi, S. G., Ozkan, M., Halme, J., Misic, K. D., Zakeeruddin, S. M., Paltakari, J., Grätzel, M., and Lund, P. D. (2015). High performance dye-sensitized solar cells with inkjet-printed ionic liquid electrolyte, *Nano Energy*, **17**, pp. 206–215.

161. Jung, S., Sou, A., Banger, K., Ko, D. H., Chow, P. C. Y., McNeill, C. R., and Sirringhaus, H. (2014). All-inkjet-printed, all-air-processed solar cells, *Adv. Energy Mater.*, **4**, pp. 1400432.

162. Yang, W. S., Park, B. W., Jung, E. H., Jeon, N. J., Kim, Y. C., Lee, D. U., Shin, S. S., Seo, J. Kim, E. K., Noh, J. H., and Seok, S. I. (2017). Iodide management in formamidinium-lead-halide-based perovskite layers for efficient solar cells, *Science*, **356**, pp. 1376–1379.

163. Hwang, K., Jung, Y., Heo, Y., Scholes, F. H., Watkins, S. E., Subbiah, J., Jones, D. J., Kim, D. Y., and Vak, D. (2015). Toward large-scale roll-to-roll production of fully printed perovskite solar cells, *Adv. Mater.*, **27**(7), pp. 1241–1247.

164. Wei, C., Zhuang, J., Chen, Y., Zhang, D., Su, W., and Cui, Z. (2016). Highly air-stable electron-transport material for ink-jet-printed OLEDs, *Chem. Eur. J.*, **22**, pp. 16576–16585.

165. Xing, Z., Zhuang, J., Wei, C., Zhang, D., Xie, Z. Z., Xu, X. P., Ji, S., Tang, J., Su W., and Cui Z. (2017). Inkjet-printed quantum dot light-emitting diodes with an air-stable hole transport material, *ACS Appl. Mater. Interfaces*, **9**, pp. 16351–16359.

166. Wei, C., Zhuang, J., Zhang, D., Guo, W., Yang, D., Xie, Z., Tang, J., Su, W., Zeng, H., and Cui, Z. (2017). Pyridine-based electron-transport materials with high solubility, excellent film-forming ability, and wettability for inkjet-printed OLEDs, *ACS Appl. Mater. Interfaces*, **9**, pp. 38716–38727.

167. Chen, P. Y., Chen, C. L., Chen, C. C., Tsai, L., Ting, H. C., Lin, L. F., Chen, C. C., Chen, C. Y., Chang, L. H., Shih, T. H., Chen, Y. H., Huang, J. C., Lai, M. Y., Hsu, C. M., and Lin Y. (2014). Inch inkjet-printed organic light-emitting display panel with high degree of pixel uniformity, *Dig. Tech. Pap. Soc. Inf. Disp. Int. Symp.*, **45**, pp. 396–398.

168. Su, M., Li, F., Chen, S., Huang, Z., Qin, M., Li, W., Zhang, X., and Song, Y. (2016). Nanoparticle based curve arrays for multirecognition flexible electronics, *Adv. Mater.*, **28**, pp. 1369–1374.

169. Bandodkar, A. J., Nuñez-Flores, R., Jia, W., and Wang, J. (2015). All-printed stretchable electrochemical devices, *Adv. Mater.*, **27**(19), pp. 3060–3065.

Chapter 10

Mechanics and Control of Smart Flexible Structures

Guoyong Mao and Shaoxing Qu
Department of Engineering Mechanics, State Key Laboratory of Fluid Power & Mechatronic System, and Key Laboratory of Soft Machines and Smart Devices of Zhejiang Province, Zhejiang University, Hangzhou 310027, China
squ@zju.edu.cn

Recent advances in mechanics have provided new routes for the design and fabrication of smart flexible structures, which can offer the electrical properties of conventional, rigid wafer-based technologies but may also be bent, stretched, twisted, and even deformed arbitrarily. The mechanical design of these flexible structures has enabled the development of many new applications, such as wearable strain sensors, stretchable skin-like sensor matrices, stretchable power sources, and electronic eye cameras. In this chapter, we review the mechanical strategies employed in stretchable electronics as well as their design concepts, theory, simulation, and applications. Finally, we provide some perspectives on the remaining challenges and opportunities in the future of stretchable electronics.

Flexible and Stretchable Electronics: Materials, Designs, and Devices
Edited by Run-Wei Li and Gang Liu

ISBN 978-981-4800-46-4 (Hardcover), 978-0-429-05890-5 (eBook)
www.jennystanford.com

10.1 Introduction

The great demand for wearable devices suitable for daily life has promoted the development of fundamental enabling technologies related to stretchable electronics. Traditional electronics employ rigid wafer-based systems that cannot be stretched or bent.

Recently, the scope of stretchable electronics research has expanded to include more complex components and achieve better mechanical performance—not only to be stretched, but also compressed, twisted, and deformed into arbitrary shapes. These mechanical demands require that stretchable electronics be as reliable and perform as well as those of well-developed wafer-based systems [1].

There are two main approaches to achieving high-performance stretchable electronics. The first uses new structural layouts with conventional materials, whereas the other adopts new materials with traditional layouts [2, 3]. In this chapter, we introduce the approaches with new structural layouts via a mechanical analysis. With proper mechanical design, existing stretchable electronics can be implemented for epidermal health/wellness monitors [4], flexible surgical instruments [2], tunable antennas [5], and eyeball-like digital cameras [6].

Early work on stretchable electronics began roughly 30 years ago with bendable polymer transistors formed on sheets of plastics, which were used to develop a flexible display device [7–10]. The key progress of these technologies is the patterning technique used to produce ultrathin circuits. When a material is ultrathin, it becomes flexible. For example, a 100 nm thick ribbon experiences a maximum strain of 0.0005% when it is bent to a radial curvature of 1 cm [1]. Many flexible electronics harness such properties to deposit silicon ribbons on the surface of plastic sheets. Further, if the silicon ribbon lies in the neutral mechanical plane of the plastic sheet, it can be bent to a radius of approximately 150 μm with a maximum strain of about 0.1% [11, 12]. Further development of conductors such as nanotubes and nanowires has led to the continued improvement of flexible electronics: for instance, successful flexible displays, power sources, and monitors composed of thousands of active components.

By contrast, the biological materials are usually soft, elastic, curved, and stretchable. Therefore, flexible electronics cannot serve as wearable devices; stretchable electronics are required instead. Several strategies can be used to make silicon ribbons stretchable. The simplest way is to put the silicon ribbons on a prestretched elastomer substrate such as poly(dimethylsiloxane) (PDMS) and then release the whole device [13], thus obtaining a wavy silicon ribbon that can be stretched and compressed. Si/PDMS systems can achieve a strain ranging from 10% to 20%, which is much larger than the intrinsic fracture limit of the silicon (roughly 1%) [14].

Island–bridge design has been put forward for using traditional electronic layouts in stretchable electronics, especially with commercial chips and elements, in which the chips/elements are placed on discrete islands, which are rigid platforms [15–27]. The interconnections between the islands can be straight, serpentine, or fractal; their elastic stretchability can be as large as about 2000% with fractal interconnections [24]. Further, wavy membrane designs enable electronics to be stretchable in two directions [28, 29].

The requirements of 3D functional structures have inspired the development of 3D mesostructures based on origami and kirigami design. Guided by mechanics modeling, the mesostructures can be helices, toroids, flowers, scaffolds, frameworks, cages, and so on [30–33]. This technology provides new routines for developing complex 3D structures within a mesoscopic range, which can be harnessed in tunable spiral micro-inductors and near-field communication devices [30, 32].

In the following, we will give details of such mechanical strategies and their development, including wavy, island–bridge, and origami/kirigami designs [12, 34–37].

10.2 Wavy Designs

Wavy silicone nanoribbons are shown in Fig. 10.1a, and their fabrication is illustrated in Fig. 10.2. A flat ribbon is printed on a prestrained PDMS substrate; then, when the strain is released, the substrate shrinks, compressing the silicone ribbons into waves via mechanical buckling. Wavy ribbons can sustain finite stretching or compression.

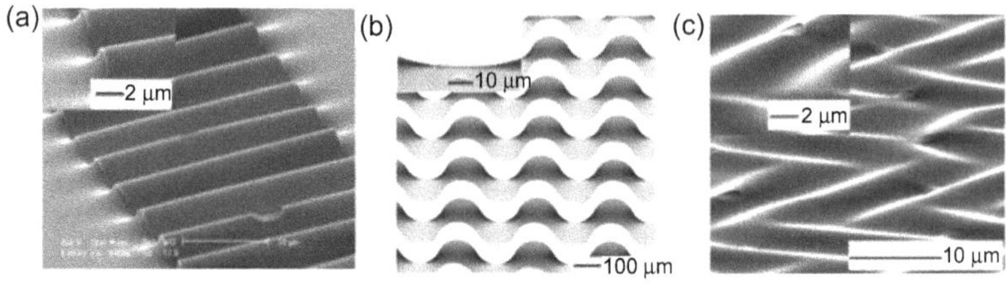

Figure 10.1 Scanning electron microscope (SEM) images of (a) stretchable nanoribbons, (b) partially boned ribbons, and (c) stretchable nanomembranes. Reprinted with permission from Ref. [38], Copyright 2009, AIP Publishing.

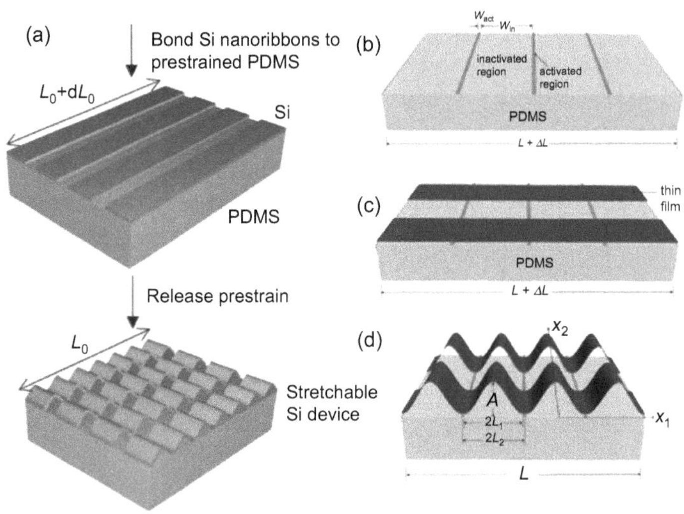

Figure 10.2 Schematic illustration of the process for fabricating wavy ribbons on a PDMS substrate. (a) The ribbons are fully bonded to the substrate. (b)–(d) The ribbons are partially bonded to the substrate. Reprinted with permission from Ref. [39], Copyright 2015, Elsevier.

10.2.1 Small Deformations of Wavy Ribbons

The mechanical response of wavy ribbons undergoing small deformations can be predicted precisely. The thin ribbon is considered an elastic nonlinear von Kármán beam, whereas the PDMS substrate is modeled as a semi-infinite solid because of its thickness (~mm) compared to that of the film (~μm). The total energy of the system consists of the bending energy and membrane energy in the film and the strain energy in the substrate.

The profile of wavy silicon ribbons can be described by the function $w = A\cos(2\pi/\lambda)$, where A is the amplitude and λ is the wavelength. A detailed derivation of the prediction of A and λ for small deformations can be obtained in Song's paper [39]; here, we simply give the results. When the strain is released, the amplitude is

$$\lambda = 2\pi h_{\mathrm{f}}\left(\frac{\bar{E}_{\mathrm{f}}}{\bar{E}_{\mathrm{s}}}\right)^{1/3}, \quad A = h_{\mathrm{f}}\sqrt{\frac{\varepsilon_{\mathrm{pre}}}{\varepsilon_{\mathrm{c}}} - 1}, \tag{10.1}$$

where h_{f} is the thickness of the film, $\varepsilon_{\mathrm{pre}}$ is the prestrain of the substrate, $\varepsilon_{\mathrm{c}} = 0.25(3\bar{E}_{\mathrm{s}}/\bar{E}_{\mathrm{f}})^{2/3}$ is the critical strain for buckling, and E_{f} is the modulus of the film, and E_{s} is that of the substrate. Then, $\bar{E}_{\mathrm{f}} = E_{\mathrm{f}}/(1-v_{\mathrm{f}}^2)$, where v_{f} is Poisson's ratio of the ribbon. Similarly, $\bar{E}_{\mathrm{s}} = E_{\mathrm{s}}/(1-v_{\mathrm{s}}^2)$ is Poisson's ratio of the substrate. Generally, $E_{\mathrm{f}} >> E_{\mathrm{s}}$: for Si/PDMS systems, E_{f} = 130 GPa and E_{s} = 1.8 MPa. Hence, critical strain ε_{c} is about 0.034% for Si/PDMS systems [39]. When the buckled system is subjected to applied strain $\varepsilon_{\mathrm{applied}}$, the amplitude and wavelength are

$$\lambda = 2\pi h_{\mathrm{f}}\left(\frac{\bar{E}_{\mathrm{f}}}{3\bar{E}_{\mathrm{s}}}\right)^{1/3}, \qquad A = h_{\mathrm{f}}\sqrt{\frac{\varepsilon_{\mathrm{pre}} - \varepsilon_{\mathrm{applied}}}{\varepsilon_{\mathrm{c}}} - 1} \tag{10.2}$$

and the peak strain in the ribbon is $\varepsilon_{\mathrm{peak}} = 2\sqrt{(\varepsilon_{\mathrm{pre}} - \varepsilon_{\mathrm{applied}})\varepsilon_{\mathrm{c}}}$ [39].

10.2.2 Large Deformations of Wavy Ribbons

In practice, when the initial strain is large, Eq. (10.1) does not produce predictions that are consistent with experimental observations; therefore, a theory considering finite deformation is required. Jiang et al. [14] and Song et al. [40] considered the finite deformation of the substrate and established a new theory for predicting the wavelength and amplitude of the ribbons that involves three extra factors in its analyses: the finite deformation of the ribbons and substrate, the nonlinear geometrical relationship of the substrate, and the nonlinear constitutive model of the substrate. Further, Chen and Song have shown that finite geometry changes are the main factor in the analysis [41]. By minimizing the total energy of the system, the wavelength and amplitude of the ribbons are

$$\lambda = \frac{2\pi h_f}{1+\varepsilon_{pre}}\left(\frac{\bar{E}_f}{3E_s}\right)^{1/3}, A = h_f\sqrt{\frac{\varepsilon_{pre}}{(1+\varepsilon_{pre})\varepsilon_c}-1}, \tag{10.3}$$

where the critical stress is $\varepsilon_c = 1/4(3\bar{E}_s/\bar{E}_f)^{2/3}$ and the peak strain in the buckled film/substrate system is

$$\varepsilon_{peak} \approx 2\sqrt{\frac{\varepsilon_{pre}}{(1+\varepsilon_{pre})}\varepsilon_c}. \tag{10.4}$$

Equation (10.3) and the finite deformation analysis [40] show that the finite strain and the constitutive model of the substrate do not affect the results and can be neglected.

When the buckled system is subjected to an applied strain, minimizing the total energy gives a ribbon wavelength and amplitude of

$$\lambda = \frac{2\pi h_f(1+\varepsilon_{applied})}{1+\varepsilon_{pre}}\left(\frac{\bar{E}_f}{3E_s}\right)^{1/3}, A = h_f\sqrt{\frac{\varepsilon_{pre}-\varepsilon_{applied}}{(1+\varepsilon_{pre})\varepsilon_c}-1}, \tag{10.5}$$

which is consistent with experimental observations [41]. Equation (10.5) shows that the amplitude decreases and the wavelength increases with increases in the applied strain. Because of the strain limit of the film, $\varepsilon_{fracture}$, the stretchability of the system is equal to $\varepsilon_{pre} + \varepsilon_{fracture} + \varepsilon_c$. Then, the peak strain of the ribbons is given by

$$\varepsilon_{peak} \approx 2\sqrt{\frac{\varepsilon_{pre}-\varepsilon_{applied}}{1+\varepsilon_{pre}}\varepsilon_c}. \tag{10.6}$$

10.2.3 Partially Boned Wavy Ribbons

The maximum strain that can be applied to fully boned ribbons is about 20%. Moreover, according to Eq. (10.5), the wavelength and amplitude are dependent on the moduli of the film and substrate and the thickness of the film. To overcome these disadvantages, partially boned wavy ribbons were developed by Sung et al. [42]. Figure 10.1b shows a tilted-view SEM image of buckled GaAs ribbons on a PDMS substrate, where ε_{pre} = 60%, W_{act} = 10 μm, and W_{in} = 400 μm. Experimental results with GaAs or Si film and a PDMS substrate demonstrate that such a method can produce electronics with a stretchability of up to 100%, a compressibility of up to 25%, and bendability with curvature radius as small as 5 mm [42]. The manufacture of partially boned wavy ribbons is shown in Fig. 10.2b. Jiang et al. provided a theory for predicting their wavelength and amplitude.

During manufacture, ribbons are bonded to a prestrained substrate (Fig. 10.2b, with strain $\varepsilon_{\text{pre}} = \Delta L/L$) at locations called activated regions (Fig. 10.2c). Between the activated areas, there are inactivated regions where there are only weak van der Waals interactions at the interface. The widths of the activated and inactivated areas are W_{act} and W_{in}, respectively. Finally, the substrate is released and the ribbons buckle (Fig. 10.2d). The buckling wavelength is $L_1 = W_{\text{in}}/(1 + \varepsilon_{\text{pre}})$, and the sum of the activated and inactivated regions after relaxation is $2L_2 = W_{\text{in}}/(1 + \varepsilon_{\text{pre}}) + W_{\text{act}}$ (Fig. 10.2d). The profile of the buckling film can be expressed as

$$w = \begin{cases} w_1 = \dfrac{1}{2}A\left(1+\cos\dfrac{\pi x_1}{L_1}\right), -L_1 < x_1 < L_1 \\ w_2 = 0, \qquad\qquad\qquad\qquad L_1 < |x_1| < L_2 \end{cases}. \tag{10.7}$$

The energy method is used to determine the buckling geometry, and the thin film is modeled as an elastic nonlinear von Kármán beam. As the strain of the film is negligibly small and the thickness of the film (~0.1 μm) is much smaller than any characteristic length (wavelength ~200 μm) or dimension of the film, the von Kármán approximation is always valid. The total energy of the system equals the sum of the bending and strain energy of the film and the energy in the substrate. By minimizing the total energy, the amplitude of the ribbons is

$$A = \frac{4}{\pi}\sqrt{L_1 L_2(\varepsilon_{\text{pre}} - \varepsilon_{\text{c}})},\ \varepsilon_{\text{pre}} > \varepsilon_{\text{c}}, \tag{10.8}$$

where the critical strain is $\varepsilon_{\text{c}} = h^2\pi^2/(12L_1^2)$. Critical strain ε_{c} is of the order of 10^{-6} with a buckling wavelength of ~200 μm and a film thickness of ~0.1 μm. Substituting the critical strain into Eq. (10.8) gives the approximate amplitude:

$$A \approx (4/\pi)\sqrt{L_1 L_2 \varepsilon_{\text{pre}}} = 2/\pi\sqrt{W_{\text{in}}[W_{\text{in}} + W_{\text{act}}(1+\varepsilon_{\text{pre}})]\varepsilon_{\text{pre}}}/1+\varepsilon_{\text{pre}}. \tag{10.9}$$

From Eq. (10.9), the amplitude is independent of the thin film's properties and is completely determined by the layout of interfacial patterns (W_{in} and W_{act}) and strain ε_{pre} [43]. With this approach, films made of different materials can form almost the same buckling profile, which has been illustrated by Sun et al., who examined Si and GaAs films [42].

Then, for the buckled system, the maximum strain in the thin film is the bending energy considering the small membrane strain in the electronics. The maximum strain in the thin film is

$$\varepsilon_{peak} = \frac{h}{2}\max\left(\frac{d^2w}{dx_1^2}\right) = \frac{h\pi}{L_1^2}\sqrt{L_1 L_2 \varepsilon_{pre}} \, . \tag{10.10}$$

As the activated region W_{act} is much smaller than the inactivated region W_{in}, Eq. (10.10) can be approximated by $\varepsilon_{peak} \approx (h\pi/L_1)\sqrt{\varepsilon_{pre}}$. For a 0.3 μm thick GaAs film buckled on a PDMS substrate with W_{act} = 10 μm W_{in} = 400 μm, and ε_{pre} = 60%, ε_{peak} is only 0.6% when the prestrain is 60%. Thus, the approach can significantly reduce the strain on the film and increase the system's stretchability [43].

10.2.4 Wavy Membranes

Ribbons with wave designs can be stretched in the direction of the waves. However, stretchable electronics may need to deform along more than one direction. Stretchable membranes, illustrated in Fig. 10.1c, were developed to solve such problems; their fabrication is shown in Fig. 10.3. Choi et al. and Song et al. conducted experiments and analyzed this approach [44, 45]. In these experiments, an Si nanomembrane was chemically bonded to a heated PDMS substrate subjected to isotropic expansion. Then, the PDMS substrate was cooled, forming a wavy nanomembrane as it shrank. Three patterns of wrinkled nanomembrane can be observed, including 1D periodic waves, a 2D checkerboard, and a 2D herringbone (Fig. 10.4).

Several numerical studies have examined film/substrate systems. For example, Huang et al. studied the nonlinear wrinkling of film/substrate systems with a spectral method [46, 47]. They found that the pattern of the system depends on the anisotropy of the membrane forces, which is consistent with the experimental observation in Ref. [44]. Of the 2D modes, the herringbone mode always has the lowest total energy that can be most easily observed in experiments [48–50]. The analytical solution is developed by Song et al. [44], which will be described as below. As in the previous section, the thin film is considered a von Kármán plate with finite rotation and the substrate is assumed to be a 3D elastic solid. The system can be analyzed as follows.

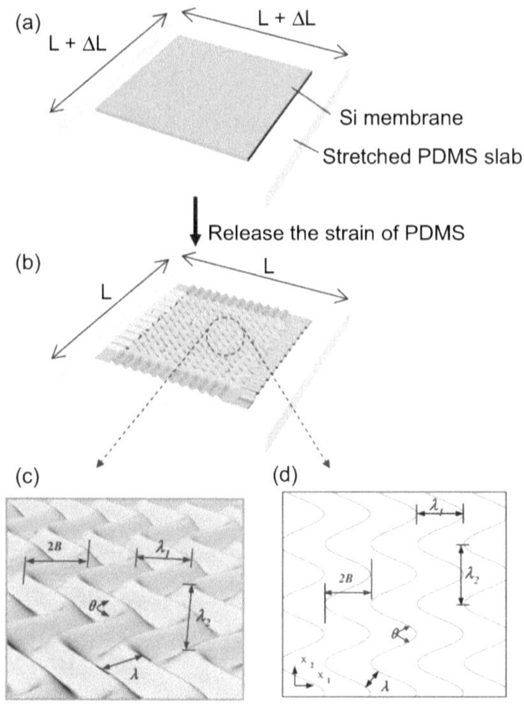

Figure 10.3 Schematic of the process for fabricating 2D wavy Si nanomembranes on a PDMS substrate. (a) The Si membrane is bonded on the stretched PDMS, (b) forming a 2D herringbone wave pattern when the PDMS is relaxed: (c) close-up and (d) top-down views of the herringbone mode. Reprinted with permission from Ref. [44], Copyright 2008, AIP Publishing.

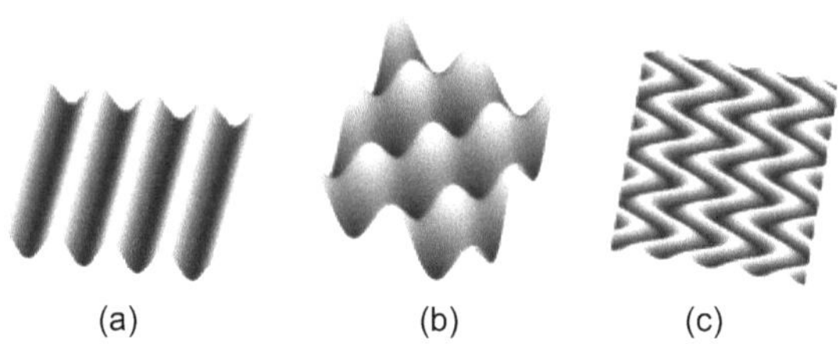

Figure 10.4 Schematic of different modes of buckling in a membrane/substrate system: (a) 1D, (b) checkerboard, and (c) herringbone. Reprinted with permission from Ref. [44], Copyright 2008, AIP Publishing.

The 1D mode is similar to a fully boned ribbon, but wider. The profile of the membrane can be described by

$$w = A\cos(2\pi/\lambda) = A\cos(kx), \tag{10.11}$$

where k is the wave number of the wrinkle. For equal biaxial prestrain, the wavelength and amplitude of the membrane are

$$\lambda = 2\pi h_{\mathrm{f}}\left(\frac{\bar{E}_{\mathrm{f}}}{3\bar{E}_{\mathrm{s}}}\right)^{1/3}, A = h_{\mathrm{f}}\sqrt{\frac{\varepsilon_{\mathrm{pre}}}{\varepsilon_{\mathrm{1D}}^{c}} - 1}, \tag{10.12}$$

where critical strain $\varepsilon_{\mathrm{1D}}^{c} = (3\bar{E}_{\mathrm{s}}/\bar{E}_{\mathrm{f}})^{2/3}/(4(1+\nu_{\mathrm{f}}))$. If the prestrain is not equal, the amplitude is modified by

$$A = h_{\mathrm{f}}\sqrt{\frac{\varepsilon_{11}^{\mathrm{pre}} + \nu_{\mathrm{f}}\varepsilon_{22}^{\mathrm{pre}}}{1/4(3\bar{E}_{\mathrm{s}}/\bar{E}_{\mathrm{f}})^{2/3}} - 1}, \tag{10.13}$$

where $\varepsilon_{11}^{\mathrm{pre}}$ and $\varepsilon_{22}^{\mathrm{pre}}$ are strains of the substrate in orthogonal directions.

The profile of the checkerboard mode (Fig. 10.4b) can be expressed as

$$w = A\cos(k_1x_1)\cos(k_2x_2), \tag{10.14}$$

Where k_1 and k_2 are wave numbers in two directions x_1 and x_2, respectively. In the checkerboard mode, the two wave numbers are equal (i.e., $k_1 = k_2$). By minimizing the total energy of the film/substrate system, the wave numbers and amplitude of the buckled membrane are

$$k_1 = k_2 = \frac{1}{\sqrt{2}h_{\mathrm{f}}}\left(\frac{3\bar{E}_{\mathrm{s}}}{\bar{E}_{\mathrm{f}}}\right)^{1/3}, A = h_{\mathrm{f}}\sqrt{\frac{8(\varepsilon_{\mathrm{pre}}/\varepsilon_{\mathrm{cb}}^{\mathrm{c}} - 1)}{(3-\nu_{\mathrm{f}})(1+\nu_{\mathrm{f}})}}, \tag{10.15}$$

where critical strain $\varepsilon_{\mathrm{cb}}^{\mathrm{c}} = (3\bar{E}_{\mathrm{s}}/\bar{E}_{\mathrm{f}})^{2/3}/(4(1+\nu_{\mathrm{f}}))$, the same as 1D mode.

The profile of herringbone mode can be expressed as

$$w = A\cos\{k_1[x_1 + B\cos(k_2x_2)]\} \tag{10.16}$$

or

$$w = \sum_{n=0}^{\infty}\left[A_{1n}\cos(k_1x_1)\cos(nk_2x_2) + B_{1n}\sin(k_1x_1)\cos(nk_2x_2)\right]. \tag{10.17}$$

The wavelength and amplitude of herringbone mode can be solved with the quasi-Newtonian and finite difference gradient methods in the IMSL (International Mathematics and Statistics Library) program [44]. For each k_2, there are multiple local minima for a range of A, B, and k_1. Then, the global minimum (with respect to A, B, and k_1) is obtained by comparing all local minima in the range.

10.3 Island–Bridge Designs

The island–bridge design was developed to increase the stretchability of electronics. In the following, we introduce these designs with three kinds of interconnects: straight, serpentine, and fractal.

10.3.1 Straight Interconnects

Lacour et al. developed elastic electronic surfaces by depositing the gold film on a prestretched PDMS substrate (10–25%). The gold films connect electronic elements distributed on rigid subcircuit islands over the PDMS substrate. When the substrate is relaxed, the gold film forms surface waves, like the wavy ribbons described previously [22]. Then, Kim et al. put entire ultrathin circuits near the neutral layer of a PDMS substrate [20], reducing the strain in the silicon to ~0.04% for an applied strain of ~10%. Kim et al. developed the interconnected-island mesh design shown in Fig. 10.5, which uses straight interconnections to connect metal-oxide-semi-conductor (CMOS) [18]. Straight interconnects are not boned to the prestretched substrate that will buckle when the substrate is released. The full system can accommodate even demanding configurations, such as a corkscrew twist with tight pitch and linear stretching (~140%) [18].

Song et al. established a mechanical model to explain the high stretchability of the island–bridge design with straight interconnects [21]. A straight interconnect is modeled as a beam with clamped ends.

Its out-of-plane displacement can be expressed as

$$w = \frac{A}{2}\left(1+\cos\left(\frac{2\pi x}{L_{\text{bridge}}}\right)\right) = \frac{A}{2}\left(1+\cos\left(\frac{2\pi X}{L^0_{\text{bridge}}}\right)\right) \tag{10.18}$$

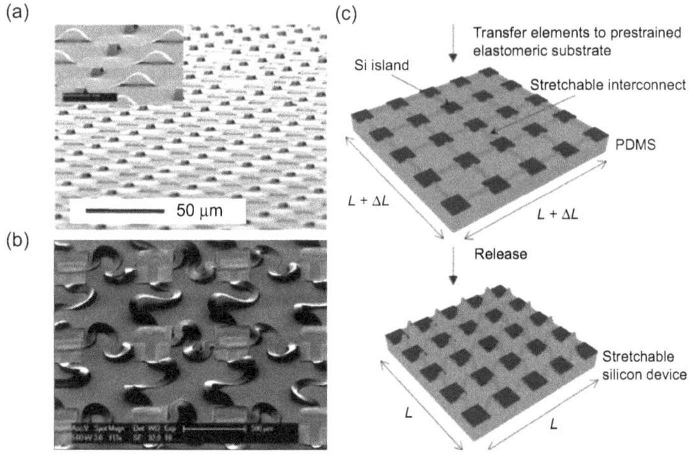

Figure 10.5 SEM images of island–bridge design with (a) straight and (b) serpentine interconnects. (c) The schematic diagram of the process for fabricating the island–bridge design. Reprinted with permission from Ref. [38], Copyright 2009, AIP Publishing.

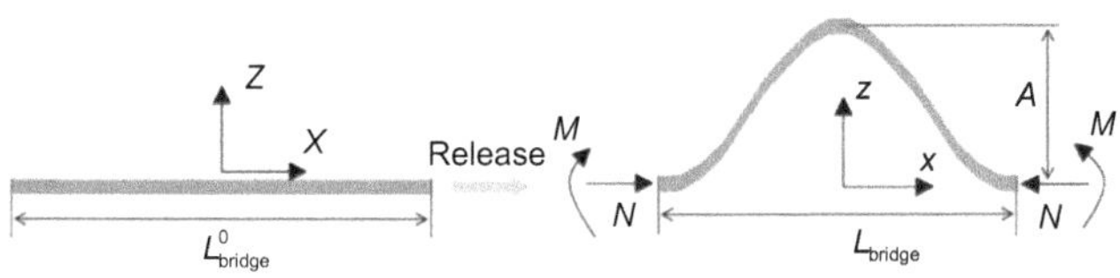

Figure 10.6 Schematic of the mechanics model for straight interconnects. Reprinted with permission from Ref. [21], Copyright 2009, AIP Publishing.

in the reference configuration X of the beam and the buckled configuration x in Fig. 10.6. L^0_{bridge} and L_{bridge} are the distances between the islands before and after relaxation of the substrate, respectively. The amplitude of the interconnection can be solved by minimizing the total energy of the system. Then, amplitude A is equal to

$$A = \frac{2L^0_{\text{bridge}}}{\pi}\sqrt{\frac{L^0_{\text{bridge}} - L_{\text{bridge}}}{L^0_{\text{bridge}}} - \varepsilon_c}\ , \qquad (10.19)$$

where $\varepsilon_c = \pi^2 h^2_{\text{bridge}} / \left[3(L^0_{\text{bridge}})^2\right]$ is the critical buckling strain for Euler buckling of a double clamped beam. The maximum strain in the interconnect is the sum of the bending and the membrane strain and can be expressed as

$$\begin{aligned}\varepsilon_{\text{bridge}}^{\max} &= 2\pi \frac{h_{\text{bridge}}}{L_{\text{bridge}}^0}\sqrt{\frac{L_{\text{bridge}}^0 - L_{\text{bridge}}}{L_{\text{bridge}}^0} - \varepsilon_{\text{c}}} \\ &\approx 2\pi \frac{h_{\text{bridge}}}{L_{\text{bridge}}^0}\sqrt{\frac{L_{\text{bridge}}^0 - L_{\text{bridge}}}{L_{\text{bridge}}^0}}\end{aligned}. \quad (10.20)$$

The approximation of $\varepsilon_{\text{bridge}}^{\max}$ is based on

$$(L_{\text{bridge}}^0 - L_{\text{bridge}})/L_{\text{bridge}}^0 >> h_{\text{bridge}}^2/(L_{\text{bridge}}^0)^2.$$

Considering that, for the interconnect, $\varepsilon_{\text{pre}} = (L_{\text{bridge}}^0 - L_{\text{bridge}})/L_{\text{bridge}}$, its maximum strain is

$$\varepsilon_{\text{bridge}}^{\max} = 2\pi \frac{h_{\text{bridge}}}{L_{\text{bridge}}^0}\sqrt{\frac{\varepsilon_{\text{pre}}}{1+\varepsilon_{\text{pre}}}}. \quad (10.21)$$

So for Eq. (10.21), the maximum strain in the interconnect, $\varepsilon_{\text{bridge}}^{\max}$, is proportional to the ratio of the interconnect's thickness to its length, $h_{\text{bridge}}/L_{\text{bridge}}^0$. Equation (10.21) reveals that thinner, longer interconnects improve the stretchability of the island–bridge design, which should guide its further development.

The rigid island ($L_{\text{island}}^0 \times L_{\text{island}}^0$) is fully boned on the substrate and is subjected to axial force and the bending moment from the interconnects:

$$N = E_{\text{bridge}} h_{\text{bridge}} \varepsilon_{\text{c}} \quad (10.22)$$

$$\begin{aligned}M &= (E_{\text{bridge}} h_{\text{bridge}}^3/12) \times \left[2\pi^2 A/(L_{\text{bridge}}^0)^2\right] \\ &\approx \pi E_{\text{bridge}} h_{\text{bridge}}^3/(3L_{\text{bridge}}^0)\sqrt{\varepsilon_{\text{pre}}/(1+\varepsilon_{\text{pre}})}\end{aligned}. \quad (10.23)$$

Simulation results show that the axial force is negligible compared to the effect of the bending moment. Dimensionless analysis shows that the maximum strain in the island is $\varepsilon_{\text{island}}^{\max} = \alpha 6(1-v_{\text{island}}^2)M/(E_{\text{island}} h_{\text{island}}^2)$, where α is a design-dependent constant and is essential to the stretchability of the system. The maximum strain of the island is

$$\begin{aligned}\varepsilon_{\text{island}}^{\max} &\approx \frac{6(1-v_{\text{island}}^2)M}{E_{\text{island}} h_{\text{island}}^2} \\ &= 2\pi \frac{(1-v_{\text{island}}2)E_{\text{bridge}} h_{\text{bridge}}^3}{E_{\text{island}} h_{\text{island}}^2 L_{\text{bridge}}^0}\sqrt{\frac{\varepsilon_{\text{pre}}}{1+\varepsilon_{\text{pre}}}}\end{aligned}. \quad (10.24)$$

Considering the maximum strain of the interconnect, Eq. (10.24) can be rewritten as

$$\varepsilon_{\text{island}}^{\max} = \frac{(1-v_{\text{island}}^2)E_{\text{bridge}}h_{\text{bridge}}^2}{E_{\text{island}}h_{\text{island}}^2}\varepsilon_{\text{bridge}}^{\max}. \tag{10.25}$$

Hence, stiff and thick islands can reduce the strain level in the island.

The maximum prestrain $\varepsilon_{\text{pre}}^{\max}$ should not induce bridge failure, and the maximum strains of the island and bridge should be smaller than those of the materials of the island and bridge, respectively. When $a < 1$,

$$\varepsilon_{\text{pre}}^{\max} < \frac{\alpha^2}{1-\alpha^2}, \tag{10.26}$$

where

$$a = \frac{L_{\text{bridge}}^0}{2\pi h_{\text{bridge}}}\min\left[\varepsilon_{\text{bridge}}^{\text{failure}}, \frac{E_{\text{island}}h_{\text{island}}^2}{(1-v_{\text{island}}^2)E_{\text{bridge}}h_{\text{bridge}}^2}\varepsilon_{\text{island}}^{\text{failure}}\right]. \tag{10.27}$$

For $a \geq 1$, the maximum prestrain is governed by the failure of the PDMS substrate.

Maximum stretchability is reached when the interconnect flattens and cannot sustain additional stretching,

$$\varepsilon_{\text{stretchability}} = \frac{L_{\text{bridge}}^0 - L_{\text{bridge}}}{L_{\text{bridge}} + L_{\text{island}}^0} = \frac{\varepsilon_{\text{pre}}}{1+(1+\varepsilon_{\text{pre}})\dfrac{L_{\text{island}}^0}{L_{\text{bridge}}^0}}. \tag{10.28}$$

This result shows that long interconnects, short islands, and large prestrains can increase the stretchability of an island–bridge design. For long interconnects ($L_{\text{bridge}}^0 >> L_{\text{island}}^0$), the stretchability is prestrain ε_{pre}; for short interconnects ($L_{\text{bridge}}^0 << L_{\text{island}}^0$), the stretchability is $L_{\text{bridge}}^0/L_{\text{island}}^0[\varepsilon_{\text{pre}}/(1+\varepsilon_{\text{pre}})]$.

The maximum compressibility is reached when the system reaches the corresponding failure strains of the island and bridge, or when neighboring islands come into contact. This gives

$$\varepsilon_{\text{compressibility}} = \min\left[\frac{(1+\varepsilon_{\text{pre}})\alpha^2 - \varepsilon_{\text{pre}}}{1+\left(1+\varepsilon_{\text{pre}}\dfrac{L_{\text{island}}^0}{L_{\text{bridge}}^0}\right)}, \frac{1}{1+(1+\varepsilon_{\text{pre}})\dfrac{L_{\text{island}}^0}{L_{\text{bridge}}^0}}\right], \tag{10.29}$$

where a is expressed as Eq. (10.27). For long interconnects, contact between neighboring islands limits compressibility ($\varepsilon_{\text{compressibility}} = 1/[1+(1+\varepsilon_{\text{pre}})\times(L^0_{\text{island}}/L^0_{\text{bridge}})]$).

For short interconnects, materials failure limits compressibility ($\varepsilon_{\text{compressibility}} = [(1+\varepsilon_{\text{pre}})a^2 - \varepsilon_{\text{pre}}]/[1+(1+\varepsilon_{\text{pre}})\times(L^0_{\text{bridge}}/L^0_{\text{bridge}})]$).

From the above discussion, a simple way to increase stretchability and compressibility is to increase the length of the interconnects.

10.3.2 Serpentine Interconnects

As mentioned above, a simple way to increase the stretchability of the island–bridge design is to increase the length of interconnects. Serpentine interconnects, shown in Fig. 10.5b, were developed and can achieve high stretchability, up to ~140% [18]. The fabrication of serpentine interconnects is shown in Fig. 10.7. A representative unit cell of an interconnect is shown in Fig. 10.7d, and its geometry can be described by space l_1, length l_2, width w, and thickness t. The critical buckling strain for the entire structure can be described by

$$\varepsilon_{\text{cr}} = \frac{\alpha(\alpha^2+6)+3\pi/4(2\alpha^2+1)-\dfrac{9(\alpha^2+\pi\alpha+2)^2}{8\alpha+\pi+8(2\alpha+\pi)m^2}}{f_1(m)+f_2(m)\alpha+f_3(m)\alpha^2}\sqrt{\frac{G}{E}}\frac{t^2}{w^2}, \tag{10.30}$$

where $\alpha = l_2/l_1$ and f_i, $i = 1, 2, 3$, are the contributions of each second-order term of length scaled to the lateral buckling [15]; G and E are the shear and Young's moduli of the serpentine interconnect; and m is the number of units in an interconnect, i.e., $m = 3$ in Fig. 10.7d. Equation (10.30) shows that the critical strain of the interconnect is proportional to t^2/w^2.

When the applied strain is large, the interconnects undergo lateral buckling to reduce the strain energy and the strain in the interconnects. The out-of-plane and in-plane bending strains of straight interconnects during lateral buckling are proportional to the thickness and width, respectively [19]. Here, for serpentine structures, the maximum principal strain during reversible deformation (without plastic deformation) is

$$\varepsilon_{\max} = g_1(m,\alpha)\sqrt{\varepsilon_{\text{appl}}}\frac{t}{l_1} + g_2(m,\alpha)\varepsilon^2_{\text{appl}}\frac{w}{l_1}, \tag{10.31}$$

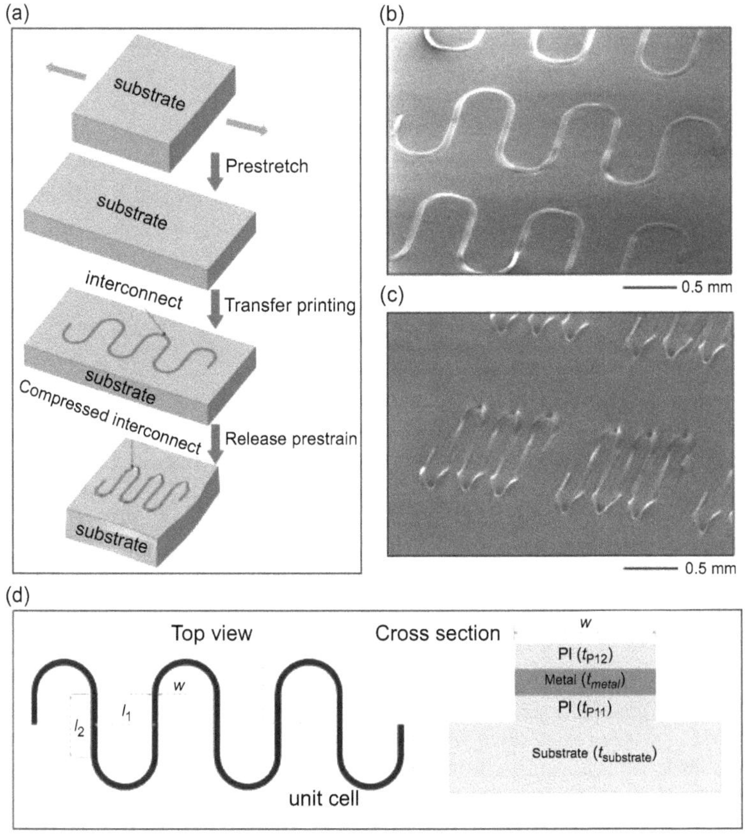

Figure 10.7 (a) Schematic illustration of the prestrain strategy for fabricating serpentine interconnects with enhanced stretchability. (b) SEM image of the serpentine structures before releasing the prestrain on the substrate; (c) SEM image of serpentine interconnects after releasing the prestrain on the substrate; (d) geometric parameters for serpentine interconnects with top and cross-sectional views. Reprinted with permission from Ref. [16], Copyright 2013, John Wiley & Sons.

Where g_1 and g_2 are functions depending on the number of units cells m and length/spacing ratio α [19]. The out-of-plane bending strain (the first right-hand term of Eq. (10.31)) dominates in the initial stage of post-buckling, where the applied strain is small ($\sqrt{\varepsilon_{\text{appl}}} >> \varepsilon_{\text{appl}}^2$). The in-plane bending strain (the second term on the right-hand side of Eq. (10.31)) takes effect only when the applied strain becomes sufficiently large ($\sqrt{\varepsilon_{\text{appl}}} \sim \varepsilon_{\text{appl}}^2$). Considering the yield criterion of the interconnect—that the maximum strain in Eq. (10.31) is equal to that of the yield strain of the interconnect material—$\varepsilon_{max} = \varepsilon_{yield}$. To optimize the structural design, the stretchability of the interconnect is set to have the relation $\varepsilon_{elasti\text{-}stretchhability} = \lambda^2$, where $\lambda > 0$ is the solution to the following fourth-order algebraic equation:

$$g_2(m,\alpha)\lambda^4 + g_1(m,\alpha)\frac{t}{w}\lambda - \frac{\varepsilon_{yield}}{w}l_1 = 0. \qquad (10.32)$$

Equation (10.32) indicates that elastic stretchability increases with decreasing thickness or width, or with increasing $e_{yield}\, l_1/w$, number of the cells m, or length/space ratio α.

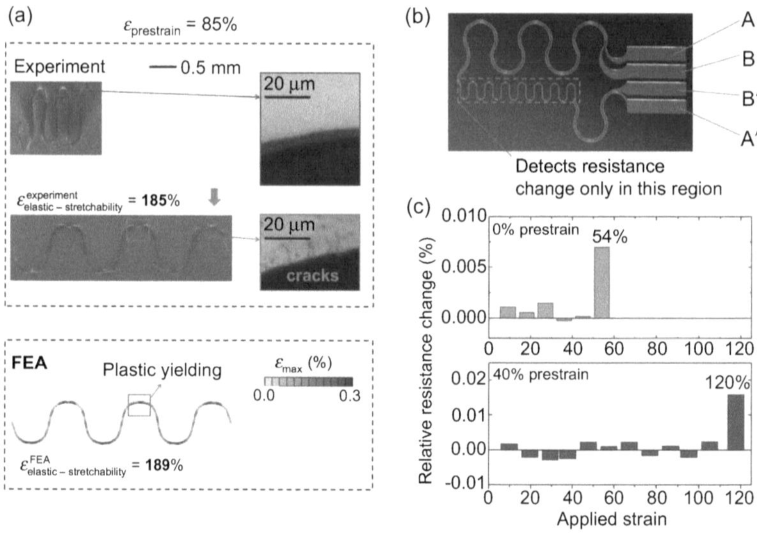

Figure 10.8 (a) Experimental image of serpentine interconnect and fracture sites due to cyclic stretching (with an amplitude of 185%), and finite element analysis results for the strain distribution when stretched to the predicted elastic stretchability (189%), for an 85% prestrain. (b) Four-point probe samples, with terminals marked as A, B, B′, and A′. Current was supplied via A and A′ and the generated voltage was measured via B and B′. Separating the current (A and A′) and voltage electrodes (B and B′) eliminates the resistance contribution of the wiring and contact resistances, giving more accurate resistance measurements. (c) Relative resistance changes versus applied strain. The interconnect consists of 0.3 μm thick copper sandwiched by two 1.2 μm thick PI (polyimide) layers. Reprinted with permission from Ref. [16], Copyright 2013, John Wiley & Sons.

During plastic deformation without material failure, total stretchability falls in the range $\varepsilon_{\text{total-stretchability}} = 81\%$. For comparison, elastic stretchability is about $\varepsilon_{\text{elastic-stretchability}} = 43\%$. Theoretical predictions match experimental observations well: $85\% < \varepsilon_{\text{total-stretchability}} < 90\%$ and $40\% < \varepsilon_{\text{elastic-stretchability}} < 50\%$. Note that hardening of the plastic work is not considered in theoretical predictions, which underestimates the total stretchability. From the above discussion, stretchability can be increased by decreasing the thickness of the interconnect.

Serpentine interconnects fracture, with cracks visible on their surface (Fig. 10.8a), when subjected to cyclic load with high strain. Plastic deformation of serpentine interconnects has also been studied; results show that finite element model (FEM) analysis predicts experimental results well. Interconnect resistance was also studied experimentally (Fig. 10.8b), confirming that the resistance increases considerably when the interconnect is reading a plastic deformation region (Fig. 10.8c). These results indicate the importance of the prestrain on the substrate and the maximum stretchability analysis of serpentine interconnects [16].

10.3.3 Fractal Interconnects

The geometric construction of fractal interconnects is illustrated in Fig. 10.9. First-order interconnects consist of straight wires and half circles. Second-order interconnects are created using the structure of the first-order interconnects and reproduce the original first-order layout. Likewise, third-order interconnects are created with replacing second-order interconnects with the layout of first-order interconnects. The scale of the interconnects decreases with increasing order.

The geometry of fractal interconnects depends on their order, where η denotes the height/spacing aspect ratio at each order. The height $h^{(i)}$ and spacing $l^{(i)}$ of the ith ($i = 1 \ldots n$) have the relationship $h^{(i)} = nl^{(i)}$ (Fig. 10.10). Height $h^{(i)}$ also has the relationship $h^{(i)} = 2ml^{(i-1)}$ ($i = 2 \ldots n$), where m is the number of unit cells. With the above relationship, we can obtain the spacing and height at any order i, which can be scaled with the spacing of the highest order $l^{(n)}$ by

$$l^{(i)} = \left(\frac{\eta}{2m}\right)^{n-i} l^{(n)}, h(i) = \eta \left(\frac{\eta}{2m}\right)^{n-i} l^{(n)} (i = 1 \ldots n). \qquad (10.33)$$

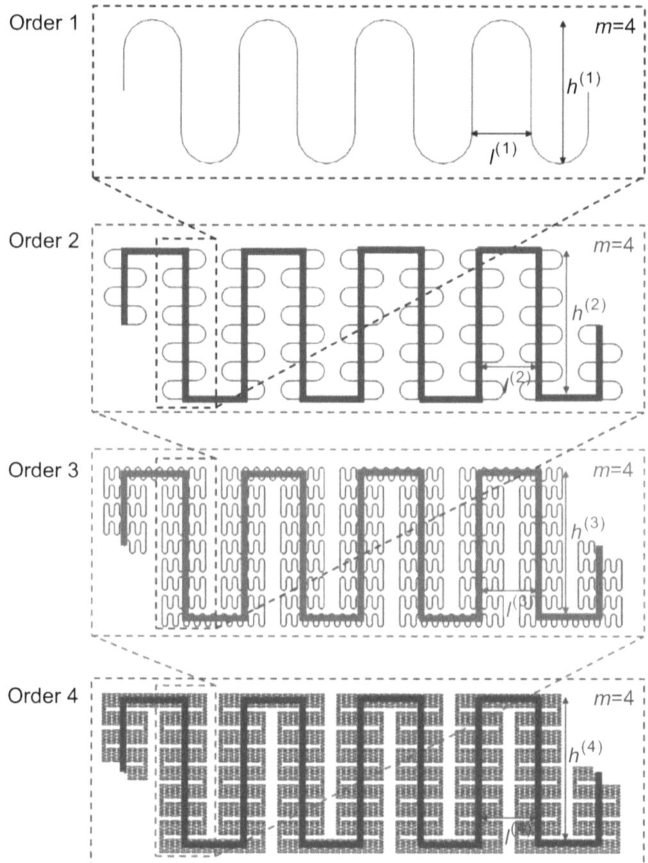

Figure 10.9 Schematic of the geometry of a fractal-inspired interconnect. Reprinted with permission from Ref. [24], Copyright 2014, Elsevier.

With fractal interconnects, elastic stretchability can be enhanced ~200 times, which is high enough for most biomedical applications, such as inflatable catheters, where deformation is large (>20%), such as in the heart or elbow.

The maximum strain of fractal interconnects is difficult to calculate or model because of their complex deformation and the many elements required for a conventional finite element analysis (FEA) approach (i.e., >1 million for a fractal order of 4).

Zhang et al. developed an effective and robust hierarchical computational model (HCM) to calculate the deformation of the fractal interconnect [24]. The main two stages are illustrated in

Fig. 10.11. The first stage is using a straight line to replace the first-order interconnect. FEA is used for the equivalent structure of straight beams to determine the deformation. Stage I is complete when the equivalent structure of straight beams is fully unraveled. Stage II is shown in Fig. 10.11b, in which the first-order fractal interconnect is unraveled. The second stage is mainly about calculating the horizontally aligned first-order fractal interconnects (blue in Fig. 10.11a) because their higher flexibility compared to the straight beams (brown). Hence, the deformation of the beams is negligible, and only a single first-order fractal interconnect (Fig. 10.11b) is analyzed by FEA. Stage II is complete when each first-order interconnect is fully unraveled.

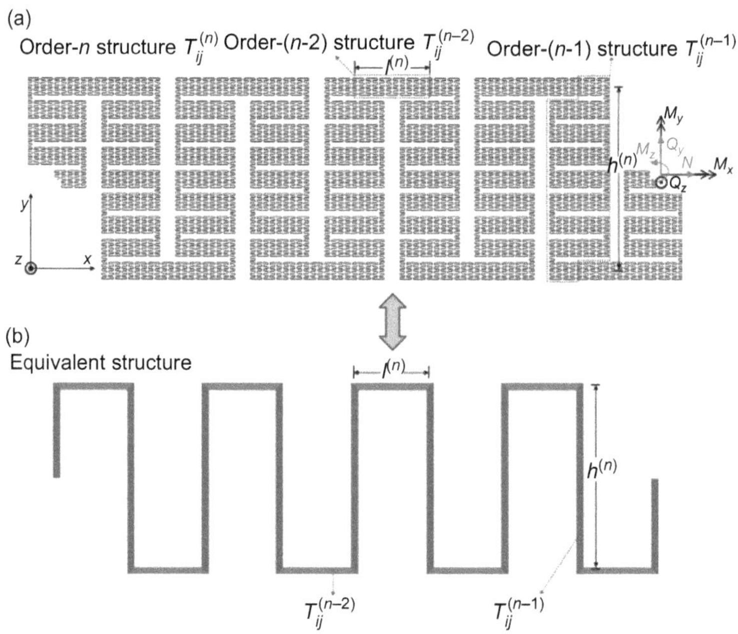

Figure 10.10 Schematic of the equivalent structure for a fractal-inspired interconnect: (a) An order-*n* fractal interconnect composed of vertically aligned order-(*n*-1) interconnects and horizontally aligned order-(*n*-2) interconnects and (b) an equivalent structure consisting of only straight beams. Reprinted with permission from Ref. [24], Copyright 2014, Elsevier.

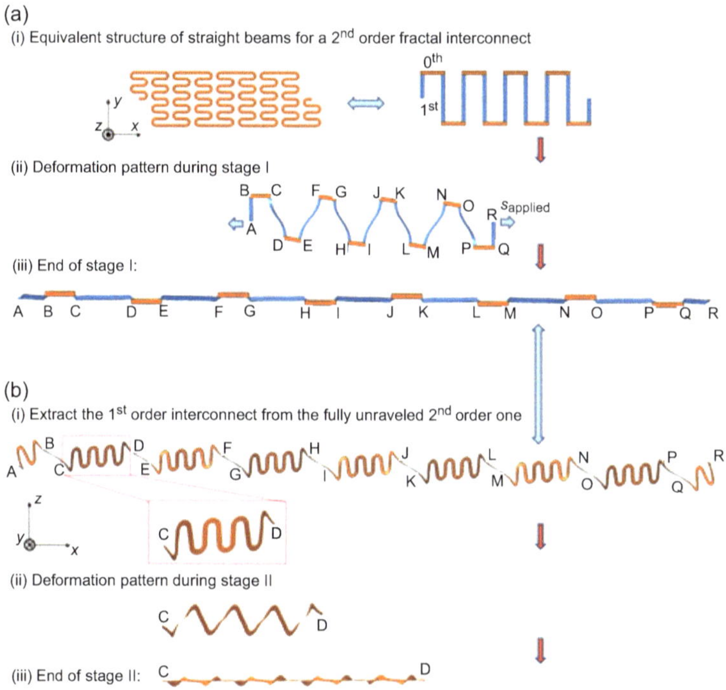

Figure 10.11 Illustration of the hierarchal computational model (HCM) for a second-order fractal interconnect: (a) Stage I—unraveling the second-order structure, in which the interconnect is modeled by an equivalent structure of straight beams; and (b) Stage II—unraveling each first-order structure, studied using the structure's original geometry. Reprinted with permission from Ref. [24], Copyright 2014, Elsevier.

Results show that the HCM is applicable for most combinations of geometric parameters if $n^2 < 2m$. Calculation indicates that such designs can achieve elastic stretchabilities of 747% and 2140% with third- and fourth-order fractal interconnects, respectively.

The unit cell of a fractal interconnect need not be only serpentine; it can also be zigzag or sinusoidal (Fig. 10.12) [23]. Su et al. studied the elastic stiffnesses of fractal interconnects, including these three shapes.

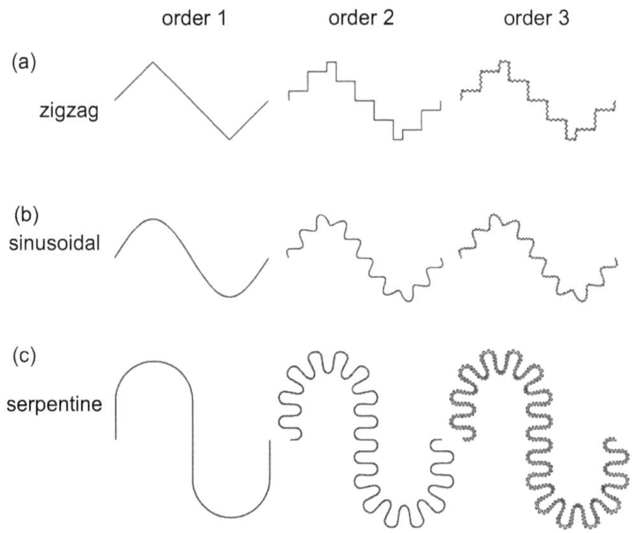

Figure 10.12 Schematic of order 1, 2, and 3 fractal interconnects with (a) zigzag, (b) sinusoidal, and (c) serpentine shapes. Reprinted with permission from Ref. [23], Copyright 2014, John Wiley & Sons.

Fractal interconnects are much longer than straight ones, inducing a large electrical resistance that needs to be considered in the design. Increasing the cross-sectional area can both overcome this effect and enhance elastic stiffness. The high elastic stiffness causes the deformation that occurs to tend to be in-plane bending instead of buckling [23], which is also helpful for avoiding interfacial peeling under stretching [25, 26].

10.4 Origami/Kirigami Designs

In addition to ribbon (film)/substrate system [14–24, 40–45, 49, 52], origami/kirigami-based approaches provide another method for achieving deformable electronics [53–59]. For example, Song et al. developed an integrated, deformable LIB [54], the manufacture of which is illustrated in Fig. 10.13. Figure 10.14 shows a photograph of the battery; observe that its width can be compressed from 9.467 cm to 0.657 cm without any performance degradation. Simulation results (Fig. 10.14c,d) show a small strain (on the order of 0.1%) of the battery with twisting and bending [51].

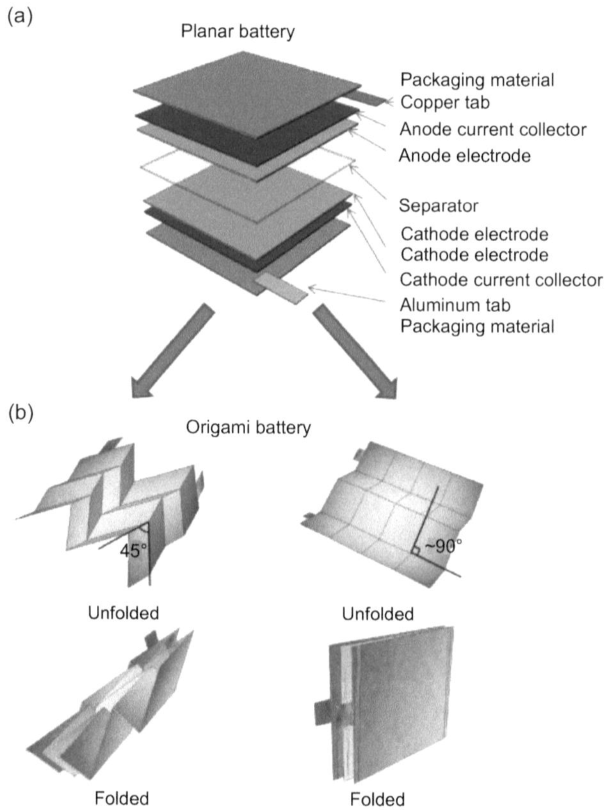

Figure 10.13 Origami lithium-ion batteries (LIBs). (a) Exploded view of the multilayer structure of conventional LIBs in a planar state. (b) Two examples of origami LIBs using Miura folding. Reprinted with permission from Ref. [51], Copyright 2014, Springer Nature.

The mechanics of the Miura-ori were described by Lv et al. [60]. Here, we briefly introduce its results with the schematic shown in Fig. 10.15. The geometry of a Miura-ori is defined by many identical rigid parallelogram faces linked by edges that can be folded into "mountain" and "valley" creases (Fig. 10.15b) [60].

The Miura-ori is a periodic structure; the unit cell is shown in Fig. 10.15c, in which the four parallelograms are identical, with dimensions α, b, and $\beta \in [0°, 90°]$. The rigid origami has $n - 3$ degrees of freedom, where n is the number of edges at the vertex (Fig. 10.15d). For Miura-ori, $n = 4$ and it only has one degree of freedom. When the shape of a parallelogram face is prescribed (i.e., α, b, and β), $\phi \in [0°, 2\beta]$ can be used to characterize the folding of the Miura-ori unit cell. When $\phi = 2\beta$, the Miura-ori is in its planar state; when $\phi = 0°$, it is in the collapsed state. The dimensions of the unit cell are

$$l = 2b\sin(\phi/2), w = 2a\frac{\cos\beta}{\cos(\phi/2)}, h = \frac{a\sqrt{\sin^2\beta - \sin^2(\phi/2)}}{\cos(\phi/2)} \tag{10.34}$$

and α_1 and α_2 are

$$\alpha_1 = \cos^{-1}\left[1 - 2\frac{\sin^2(\phi/2)}{\sin^2\beta}\right], \quad \alpha_2 = \cos^{-1}[1 - 2\cot^2\beta\tan^2(\phi/2)]. \tag{10.35}$$

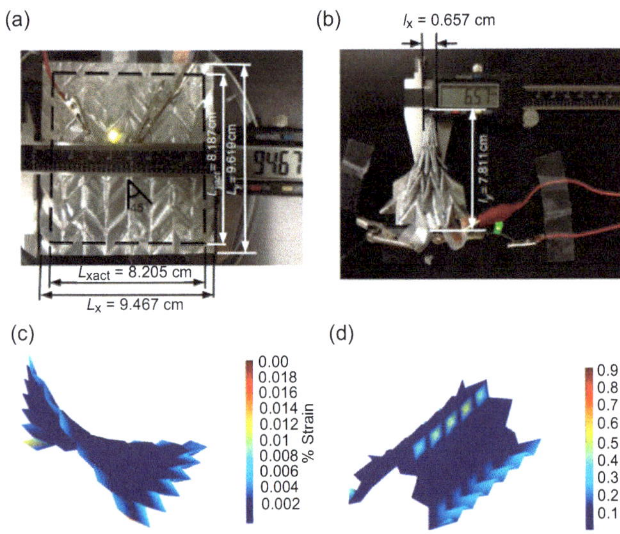

Figure 10.14 (a) Photograph of the origami battery in its completely unfolded state; the battery was used to power a light-emitting diode (LED). (b) Photograph of the origami battery operating an LED in a compressed state. (c) Finite element results of the strain contour of a 45° Miura pattern subjected to twisting, with a twisting angle of 90° per unit cell. (d) Finite element results of the strain contour of a 45° Miura pattern subjected to bending with a radius of two unit cells. Reprinted with permission from Ref. [51], Copyright 2014, Springer Nature.

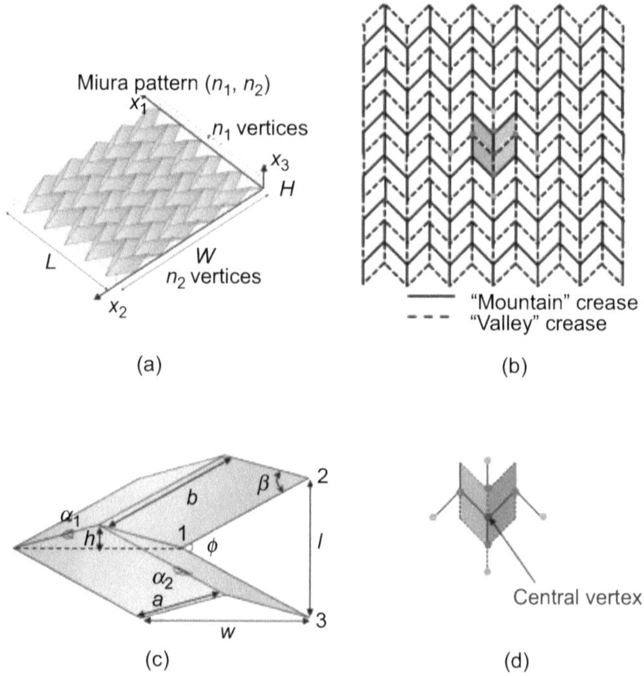

Figure 10.15 (a) A Miura-ori (n_1, n_2) in its folded state with n_1 vertices in the x_1 direction and n_2 vertices in the x_2 direction; x_3 is the out-of-plane direction. Specifically, for this illustration, n_1 = 11, n_2 = 11, β = 45°, and $a/b = 1/\sqrt{2}$. (b) A Miura-ori in its planar state, corresponding to (a). The solid lines represent "mountain" creases that remain on the top after folding. The dashed lines represent "valley" creases, which remain on the bottom after folding. (c) A unit cell of a Miura-ori: α_1 and α_2 are two dihedral angles. In each parallelogram, the length of the short side is a and that of the long side is b, with an acute angle of β. The projected angle between the two ridges is ϕ. The size of unit cell is l, w, and h in the x_1, x_2, and x_3 directions, respectively. (d) A nonlocal element for Miura-ori that focuses on the central vertex. Reprinted with permission from Ref. [60], Copyright 2014, Springer Nature.

When α_1 and α_2 equal nearly 180°, the Miura-ori is in the planar state: when they are equal to 0°, the Miura-ori is completely collapsed. The boundaries of a Miura-ori (Fig. 10.15d) are

$$L=(n_1-1)\,b\sin(\phi/2),$$

$$w=(n_2-1)\,a\frac{\cos\beta}{\cos(\phi/2)}+b\cos(\phi/2). \tag{10.36}$$

$$H=\frac{a\sqrt{\sin^2\beta-\sin^2(\phi/2)}}{\cos(\phi/2)}$$

The space the Miura-ori occupies equals $V = L \times H \times W$. With calculation, the in-plane Poisson's ratio of a Miura-ori can be expressed as

$$v_{12} = -\cot^2(\phi/2)\frac{(n_2-1)\,\eta\cos\beta + \cos^2(\phi/2)}{(n_2-1)\,\eta\cos\beta - \cos^2(\phi/2)}, \tag{10.37}$$

where $\eta = a/b$. The other in-plane Poisson's ratio, v_{12}, is the reciprocal of v_{21}. According to Eq. (10.37), the ranges of Poisson's ratios are, specifically, $v_{12} \in [-\infty, +\infty]$, $v_{13} \in [0,]$, $v_{23} \in [-\infty, +\infty]$ when Poisson's ratio of the common material is in the range $v \in [-1, 0.5]$ [60].

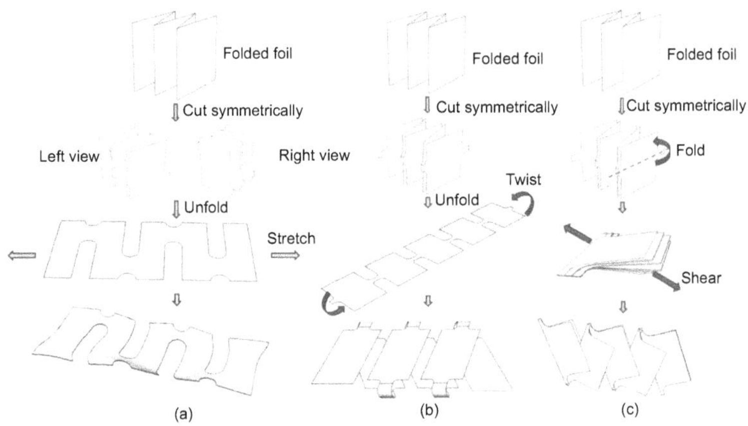

Figure 10.16 Illustrations of three kirigami patterns. (a) A zigzag-cut pattern, where out-of-plane deformation occurs to accommodate stretching. (b) A cut-and-twist pattern, where rotation is involved to accommodate stretching and prevent out-of-plane deformation. (c) A cut-and-shear pattern, where the packing density doubles compared with that of the cut-and-twist pattern. Rotation is involved to accommodate stretching and prevent out-of-plane deformation. Reprinted from Ref. [54] under Creative Commons license.

The concept of kirigami is similar to that of origami, but kirigami involves cutting of the structures (Fig. 10.16). Song et al. used the concept of kirigami to develop an LIB that achieved great stretchability (over 150%) [54]. Three kirigami patterns were utilized (as illustrated in Fig. 10.16), including the zigzag-cut, cut-and-twist, and cut-and-shear patterns. One advantage of using kirigami is that little or no out-of-plane deformation occurs. Results show that the packing density of the cut-and-shear pattern is twice that of the cut-and-twist pattern. The fracture of the kirigami structure due to cutting and folding is suppressed by plastic rolling, and the amplitude and direction of folding must be calculated and optimized carefully.

Lamoureux et al. utilized kirigami structures to build the integrated solar-tracking system shown in Figs. 10.17 and 10.18. With this design, thin-film GaAs solar cells can be stretched to produce an array of tilted surface elements that can be controlled to within ±1°. They are also mechanically robust, with no appreciable decrease in performance after >300 cycles. From the geometrical relationship shown in Fig. 10.18b, the change in feature angle θ and strain ε_T are

$$\theta = \cos^{-1}\left(1/\left(\varepsilon_A + 1\right)\right) \tag{10.38}$$

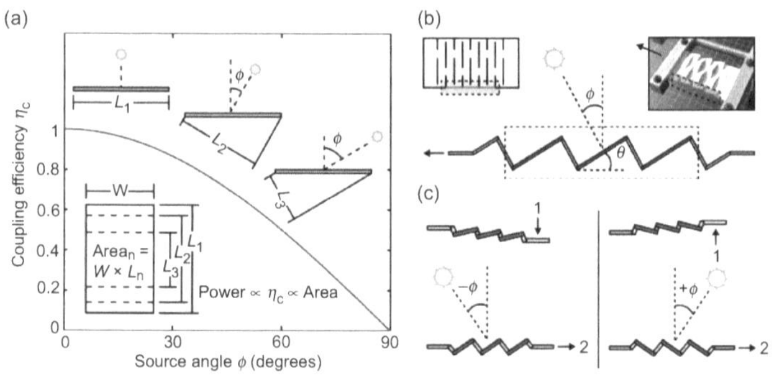

Figure 10.17 (a) Coupling efficiency n_c versus source angle ϕ for a planar solar panel. The panel's projected area decreases with cos (ϕ). (b) A kirigami tracking structure that, upon stretching, simultaneously changes the angle of the elements comprising the sheet. By incorporating thin-film solar cells into this structure, it may be used as a low-profile alternative to conventional single-axis solar tracking. (c) The direction of feature tilt (that is, clockwise or counterclockwise with respect to the original plane) is controlled by lifting or lowering one end of the sheet (Step 1) before straining (Step 2). Reprinted from Ref. [53] under Creative Commons license.

$$\varepsilon_T = \frac{R_1 - 1}{R_1 + 1}\left[\cos\left(\sin^{-1}\left(\frac{2R_1 \tan\theta}{R_1 R_2 - R_2}\right)\right) - 1\right], \tag{10.39}$$

where the axial and transverse strains are $\varepsilon_A = (L - L_0)/L_0$ and $\varepsilon_T = (W - W_0)/W_0$, respectively. $R_1 = L_c/x$ and $R_2 = L_c/y$ are dimensionless parameters. The main failure of kirigami structures results from plastic deformation of their material. Considering the pseudo-plastic limits for different superstructures, the maximum feature angle depends only on the cut geometry and can be expressed as

$$\theta_{max} = \tan^{-1}\left(\frac{R_1R_2 - R_2}{2R_1}\right). \tag{10.40}$$

For the Kapton/GaAs tracker ($R_1 = R_2 = 5$), energy stored in a system undergoing deformation was shown to decrease by 36% after 350 cycles. However, by optimizing the cut geometry and minimizing stress at the cuts, the strain fade can be decreased to 3% over 1000 cycles with $R_1 = R_2 = 10$.

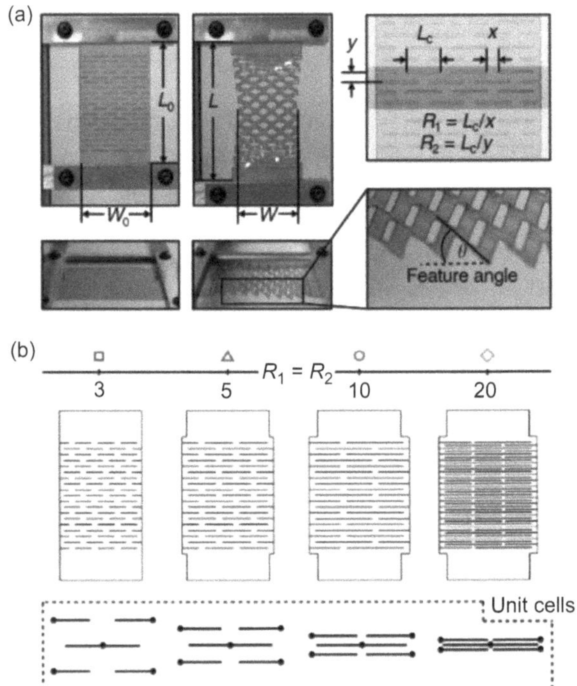

Figure 10.18 Kirigami cut geometry and geometric system response. (a) Response of a Kapton kirigami structure to stretching in the axial direction (ε_A) is accompanied by a decrease in sample width ε_T and a change in feature angle θ. Also shown are the geometric parameters that define the kirigami structure: cut length L_c and the spacing between cuts in the transverse (x) and axial (y) directions, which can be expressed as dimensionless parameters R_1 and R_2. (b) Schematics of four kirigami structures, where $R_1 = R_2 = 3$, 5, 10, and 20, along with their corresponding unit cells. Reprinted from Ref. [53] under Creative Commons license.

Recently mechanically guided assembly with kirigami/origami has shown new potential for the development of stretchable electronics [56]. Figure 10.19 shows the fabrication of 3D mesostructures from 2D precursors. First, a thin 2D precursor is formed on a source wafer or other planar substrate. Second, the 2D precursor is precisely patterned on the prestretched substrate, of which selective sites bond strongly with the precursor. Third, releasing the prestretch induces compressive stress at the boned sites and buckling of the 2D precursor. Guided by mechanics modeling, a wide range of filamentary mesostructures are possible, such as helices, toroid and conical spirals, and spherical baskets [31, 33, 56, 61–63].

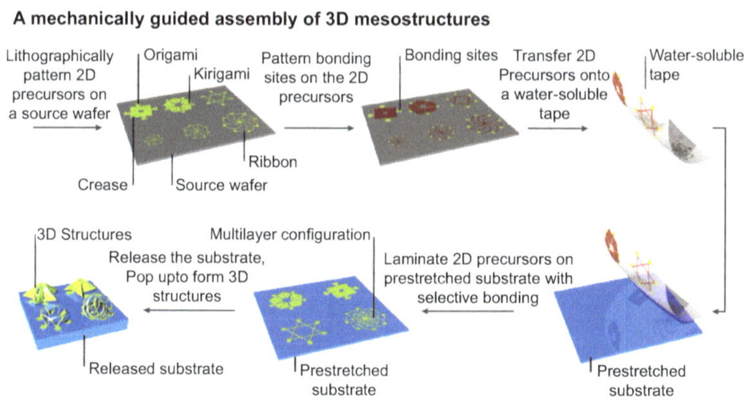

Figure 10.19 Schematic of the steps for fabricating 3D mesostructures by controlled mechanical buckling of 2D precursors formed using lithographic techniques. . Reprinted with permission from Ref. [56], Copyright 2017, Springer Nature.

There are many routines for tuning the deformation of 2D precursor-formed structures. For example, controlling the distribution of the thicknesses of the 2D precursors can guide folding at specific locations due to variations in bending stiffness. Combining structures (membranes, plates, ribbons, and wires) creates more possibilities for 3D mesostructures [56], and releasable, multilayer 2D precursors allow for dense 3D architectures with nested layouts, controlled points of entanglement, and other features that cannot be obtained with single-layer 2D precursors [61].

One advantage of this method is that different materials—such as semiconductors, metals, polymers, and heterogeneous combinations of these—can be used. Moreover, the dimensions of the structure can range from sub-micrometer to meter sizes [64].

There are two representative applications of a mechanically guided assembly. The first is an NFC device (Fig. 10.20a) that offers enhanced quality, an expanded working angle, and higher induced voltage in comparison with similar conventional 2D counterparts [61]. Note that the induced voltage of the new NFC device is about twice that of its conventional 2D counterparts (Fig. 10.20b). The other application is mechanically tunable spiral micro-inductors, whose quality and resonance frequency can be adjusted by controlling the deformation of the elastomer substrate [62].

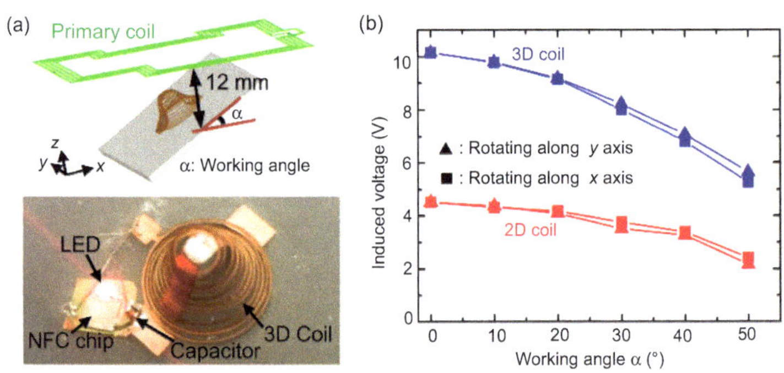

Figure 10.20 (a) Image demonstrating the use of the 3D near-field communication (NFC) device for lighting a commercial red LED. (b) Computed results of the induced voltage as a function of the working angle and applied strain when the devices are coupled with a commercial primary coil. Reprinted from Ref. [61] under Creative Commons license.

The future of mechanically guided assemblies may lead to the development of freestanding, isolated 3D mesostructures without the support of a substrate and the extension of the range of dimensions from mesoscale to nanoscale with 2D materials [56].

10.5 Conclusion

There are two main routes for developing stretchable electronics. One is to develop new materials, and the other is to develop new electronic structures. Structural designs may be the simplest way to use current materials and conventional layout to achieve satisfactory performance. In this chapter, we introduced the mechanical control and analysis of the primary structural design of stretchable electronics—including wavy deigns, island–bridge designs, and origami/kirigami designs—focusing on predicting the shape and stretchability of electronics structures. Results show that current technologies enable electronics to be stretched to 200 times their original length with fractal interconnects and to form complex 3D structures from 2D structures using origami/kirigami designs [24, 56]. These structural designs have been used in wearable devices, biomedical sensors, stretchable energy packages, electronic eye cameras, and many other applications [1, 6, 37, 65, 66]. However, there are still many challenges for stretchable electronics, such as decreasing their cost, increasing their lifetime, and decreasing the complexity of their fabrication. For further development, multidisciplinary collaboration will be crucial, and mechanics can be enhanced with improvements in materials science and electronics.

References

1. Rogers, J. A., Someya, T., and Huang, Y. 2010. Materials and mechanics for stretchable electronics, *Science*, **327**(5973), pp. 1603–1607.
2. Kim, D. H., Wang, S., Keum, H., Ghaffari, R., Kim, Y. S., Tao, H., Panilaitis, B., Li, M., Kang, Z., and Omenetto, F. 2012. Thin, flexible sensors and actuators as "instrumented" surgical sutures for targeted wound monitoring and therapy, *Small*, **8**(21), pp. 3263–3268.
3. Lacour, S. P., Jones, J., Suo, Z., and Wagner, S. 2004. Design and performance of thin metal film interconnects for skin-like electronic circuits, *IEEE Electron. Device Lett.*, **25**(4), pp. 179–181.
4. Kim, D.-H., Lu, N., Ma, R., Kim, Y.-S., Kim, R.-H., Wang, S., Wu, J., Won, S. M., Tao, H., Islam, A., Yu, K. J., Kim, T.-I., Chowdhury, R., Ying, M., Xu, L., Li, M., Chung, H.-J., Keum, H., McCormick, M., Liu, P., Zhang, Y.-W., Omenetto, F. G., Huang, Y., Coleman, T., and Rogers, J. A. 2011. Epidermal electronics, *Science*, **333**(6044), pp. 838–843.
5. Fu, H., Nan, K., Bai, W., Huang, W., Bai, K., Lu, L., Zhou, C., Liu, Y., Liu, F., Wang, J., Han, M., Yan, Z., Luan, H., Zhang, Y., Zhang, Y., Zhao, J., Cheng, X., Li, M., Lee, J. W., Liu, Y., Fang, D., Li, X., Huang, Y., Zhang, Y., and Rogers, J. A. 2018. Morphable 3D mesostructures and microelectronic devices by multistable buckling mechanics, *Nat. Mater.*, **17**(3), pp. 268–276.

6. Ko, H. C., Stoykovich, M. P., Song, J., Malyarchuk, V., Choi, W. M., Yu, C.-J., Geddes III, J. B., Xiao, J., Wang, S., and Huang, Y. 2008. A hemispherical electronic eye camera based on compressible silicon optoelectronics, *Nature*, **454**(7205), p. 748.
7. Garnier, F., Hajlaoui, R., Yassar, A., and Srivastava, P. 1994. All-polymer field-effect transistor realized by printing techniques, *Science*, **265**(5179), pp. 1684–1686.
8. Bao, Z., Feng, Y., Dodabalapur, A., Raju, V., and Lovinger, A. J. 1997. High-performance plastic transistors fabricated by printing techniques, *Chem. Mater.*, **9**(6), pp. 1299–1301.
9. Rogers, J. A., Bao, Z., Baldwin, K., Dodabalapur, A., Crone, B., Raju, V., Kuck, V., Katz, H., Amundson, K., and Ewing, J. 2001. Paper-like electronic displays: Large-area rubber-stamped plastic sheets of electronics and microencapsulated electrophoretic inks, *Proc. Nat. Acad. Sci.*, **98**(9), pp. 4835–4840.
10. Gelinck, G. H., Huitema, H. E. A., van Veenendaal, E., Cantatore, E., Schrijnemakers, L., van der Putten, J. B., Geuns, T. C., Beenhakkers, M., Giesbers, J. B., and Huisman, B.-H. 2004. Flexible active-matrix displays and shift registers based on solution-processed organic transistors, *Nat. Mater.*, **3**(2), p. 106.
11. Park, S. I., Ahn, J. H., Feng, X., Wang, S., Huang, Y., and Rogers, J. A. 2008. Theoretical and experimental studies of bending of inorganic electronic materials on plastic substrates, *Adv. Funct. Mater.*, **18**(18), pp. 2673–2684.
12. Kim, D.-H., Ahn, J.-H., Choi, W. M., Kim, H.-S., Kim, T.-H., Song, J., Huang, Y. Y., Liu, Z., Lu, C., and Rogers, J. A. 2008. Stretchable and foldable silicon integrated circuits, *Science*, **320**(5875), pp. 507–511.
13. Khang, D.-Y., Jiang, H., Huang, Y., and Rogers, J. A. 2006. A stretchable form of single-crystal silicon for high-performance electronics on rubber substrates, *Science*, **311**(5758), pp. 208–212.
14. Jiang, H., Khang, D.-Y., Song, J., Sun, Y., Huang, Y., and Rogers, J. A. 2007. Finite deformation mechanics in buckled thin films on compliant supports, *Proc. Nat. Acad. Sci.*, **104**(40), pp. 15607–15612.
15. Zhang, Y., Xu, S., Fu, H., Lee, J., Su, J., Hwang, K. C., Rogers, J. A., and Huang, Y. 2013. Buckling in serpentine microstructures and applications in elastomer-supported ultra-stretchable electronics with high areal coverage, *Soft Matter*, **9**(33), p. 8062.
16. Zhang, Y., Wang, S., Li, X., Fan, J. A., Xu, S., Song, Y. M., Choi, K.-J., Yeo, W.-H., Lee, W., Nazaar, S. N., Lu, B., Yin, L., Hwang, K.-C., Rogers, J. A., and

Huang, Y. 2014. Experimental and theoretical studies of serpentine microstructures bonded to prestrained elastomers for stretchable electronics, *Adv. Funct. Mater.*, **24**(14), pp. 2028–2037.

17. Fan, Z., Zhang, Y., Ma, Q., Zhang, F., Fu, H., Hwang, K.-C., and Huang, Y. 2016. A finite deformation model of planar serpentine interconnects for stretchable electronics, *Int. J. Solids Struct.*, **91**, pp. 46–54.

18. Kim, D.-H., Song, J., Choi, W. M., Kim, H.-S., Kim, R.-H., Liu, Z., Huang, Y. Y., Hwang, K.-C., Zhang, Y.-W., and Rogers, J. A. 2008. Materials and noncoplanar mesh designs for integrated circuits with linear elastic responses to extreme mechanical deformations, *Proc. Nat. Acad. Sci.*, **105**(48), pp. 18675–18680.

19. Su, Y., Wu, J., Fan, Z., Hwang, K.-C., Song, J., Huang, Y., and Rogers, J. A. 2012. Postbuckling analysis and its application to stretchable electronics, *J. Mech. Phys. Solids*, **60**(3), pp. 487–508.

20. Kim, D. H., Choi, W. M., Ahn, J. H., Kim, H. S., Song, J., Huang, Y., Liu, Z., Lu, C., Chan, G. K., and Rogers, J. A. 2008. Complementary metal oxide silicon integrated circuits incorporating monolithically integrated stretchable wavy interconnects, *Appl. Phys. Lett.*, **93**(4), p. 1459.

21. Song, J., Huang, Y., Xiao, J., Wang, S., Hwang, K., Ko, H., Kim, D.-H., Stoykovich, M., and Rogers, J. 2009. Mechanics of noncoplanar mesh design for stretchable electronic circuits, *J. Appl. Phys.*, **105**(12), p. 123516.

22. Lacour, S. P., Jones, J., Wagner, S., Li, T., and Suo, Z. 2005. Stretchable interconnects for elastic electronic surfaces, *Proc. IEEE*, **93**(8), pp. 1459–1467.

23. Su, Y., Wang, S., Huang, Y., Luan, H., Dong, W., Fan, J. A., Yang, Q., Rogers, J. A., and Huang, Y. 2015. Elasticity of fractal inspired interconnects, *Small*, **11**(3), pp. 367–373.

24. Zhang, Y., Fu, H., Xu, S., Fan, J. A., Hwang, K.-C., Jiang, J., Rogers, J. A., and Huang, Y. 2014. A hierarchical computational model for stretchable interconnects with fractal-inspired designs, *J. Mech. Phys. Solids*, **72**, pp. 115–130.

25. Hsu, Y.-Y., Gonzalez, M., Bossuyt, F., Axisa, F., Vanfleteren, J., and De Wolf, I. 2009. In situ observations on deformation behavior and stretching-induced failure of fine pitch stretchable interconnect, *J. Mater. Res.*, **24**(12), pp. 3573–3582.

26. Duan, Y., Huang, Y., Yin, Z., Bu, N., and Dong, W. 2014. Non-wrinkled, highly stretchable piezoelectric devices by electrohydrodynamic direct-writing, *Nanoscale*, **6**(6), pp. 3289–3295.

27. Xu, S., Zhang, Y., Cho, J., Lee, J., Huang, X., Jia, L., Fan, J. A., Su, Y., Su, J., Zhang, H., Cheng, H., Lu, B., Yu, C., Chuang, C., Kim, T.-I., Song, T., Shigeta, K., Kang, S., Dagdeviren, C., Petrov, I., Braun, P. V., Huang, Y., Paik, U., and Rogers, J. A. 2013. Stretchable batteries with self-similar serpentine interconnects and integrated wireless recharging systems, *Nat. Commun.*, **4**, p. 1543.

28. Choi, W. M., Song, J., Khang, D., Jiang, H., Huang, Y., and Rogers, J. A. 2007. Biaxially stretchable "Wavy" silicon nanomembranes, *Nano Lett.*, **7**(6), pp. 1655–1663.

29. Wang, X., Zeng, W., Hong, L., Xu, W., Yang, H., Wang, F., Duan, H., Tang, M., and Jiang, H. 2018. Stress-driven lithium dendrite growth mechanism and dendrite mitigation by electroplating on soft substrates, *Nat. Energy*, **3**(3), pp. 227–235.

30. Xu, S., Yan, Z., Jang, K.-I., Huang, W., Fu, H., Kim, J., Wei, Z., Flavin, M., McCracken, J., Wang, R., Badea, A., Liu, Y., Xiao, D., Zhou, G., Lee, J., Chung, H. U., Cheng, H., Ren, W., Banks, A., Li, X., Paik, U., Nuzzo, R. G., Huang, Y., Zhang, Y., and Rogers, J. A. 2015. Assembly of micro/nanomaterials into complex, three-dimensional architectures by compressive buckling, *Science*, **347**(6218), pp. 154–159.

31. Yan, Z., Zhang, F., Wang, J., Liu, F., Guo, X., Nan, K., Lin, Q., Gao, M., Xiao, D., Shi, Y., Qiu, Y., Luan, H., Kim, J. H., Wang, Y., Luo, H., Han, M., Huang, Y., Zhang, Y., and Rogers, J. A. 2016. Controlled mechanical buckling for origami-inspired construction of 3D microstructures in advanced materials, *Adv. Funct. Mater.*, **26**(16), pp. 2629–2639.

32. Yan, Z., Zhang, F., Liu, F., Han, M., Ou, D., Liu, Y., Lin, Q., Guo, X., Fu, H., Xie, Z., Gao, M., Huang, Y., Kim, J., Qiu, Y., Nan, K., Kim, J., Gutruf, P., Luo, H., Zhao, A., Hwang, K.-C., Huang, Y., Zhang, Y., and Rogers, J. A. 2016. Mechanical assembly of complex, 3D mesostructures from releasable multilayers of advanced materials, *Sci. Adv.*, **2**(9), p. e1601014.

33. Zhang, Y., Yan, Z., Nan, K., Xiao, D., Liu, Y., Luan, H., Fu, H., Wang, X., Yang, Q., Wang, J., Ren, W., Si, H., Liu, F., Yang, L., Li, H., Wang, J., Guo, X., Luo, H., Wang, L., Huang, Y., and Rogers, J. A. 2015. A mechanically driven form of Kirigami as a route to 3D mesostructures in micro/nanomembranes, *Proc. Nat. Acad. Sci.*, **112**(38), pp. 11757–11764.

34. Hung, P. J., Jeong, K., Liu, G. L., and Lee, L. P. 2004. Microfabricated suspensions for electrical connections on the tunable elastomer membrane, *Appl. Phys. Lett.*, **85**(24), pp. 6051–6053.

35. Dinyari, R., Rim, S.-B., Huang, K., Catrysse, P. B., and Peumans, P. 2008. Curving monolithic silicon for nonplanar focal plane array applications, *Appl. Phys. Lett.*, **92**(9), p. 091114.

36. Huang, K. and Peumans, P. 2006. Stretchable silicon sensor networks for structural health monitoring, *Proc. Smart Structures and Materials 2006: Sensors and Smart Structures Technologies for Civil, Mechanical, and Aerospace Systems*, International Society for Optics and Photonics, p. 617412.

37. Someya, T., Kato, Y., Sekitani, T., Iba, S., Noguchi, Y., Murase, Y., Kawaguchi, H., and Sakurai, T. 2005. Conformable, flexible, large-area networks of pressure and thermal sensors with organic transistor active matrixes, *Proc. Natl. Acad. Sci. U.S.A.*, **102**(35), pp. 12321–12325.

38. Song, J., Jiang, H., Huang, Y., and Rogers, J. 2009. Mechanics of stretchable inorganic electronic materials, *J. Vac. Sci. Technol. A*, **27**(5), pp. 1107–1125.

39. Song, J. 2015. Mechanics of stretchable electronics, *Curr. Opin. Solid State Mat. Sci.*, **19**(3), pp. 160–170.

40. Song, J., Jiang, H., Liu, Z. J., Khang, D. Y., Huang, Y., Rogers, J. A., Lu, C., and Koh, C. G. 2008. Buckling of a stiff thin film on a compliant substrate in large deformation, *Int. J. Solids Struct.*, **45**(10), pp. 3107–3121.

41. Cheng, H. and Song, J. 2013. A simply analytic study of buckled thin films on compliant substrates, *J. Appl. Mech.*, **81**(2), pp. 024501-024501-024503.

42. Sun, Y., Choi, W. M., Jiang, H., Huang, Y. Y., and Rogers, J. A. 2006. Controlled buckling of semiconductor nanoribbons for stretchable electronics, *Nat. Nanotechnol.*, **1**(3), p. 201.

43. Jiang, H., Sun, Y., Rogers, J. A., and Huang, Y. 2007. Mechanics of precisely controlled thin film buckling on elastomeric substrate, *Appl. Phys. Lett.*, **90**(13), p. 133119.

44. Song, J., Jiang, H., Choi, W., Khang, D., Huang, Y., and Rogers, J. 2008. An analytical study of two-dimensional buckling of thin films on compliant substrates, *J. Appl. Phys.*, **103**(1), p. 014303.

45. Choi, W. M., Song, J., Khang, D.-Y., Jiang, H., Huang, Y. Y., and Rogers, J. A. 2007. Biaxially stretchable "wavy" silicon nanomembranes, *Nano Lett.*, **7**(6), pp. 1655–1663.

46. Huang, Z., Hong, W., and Suo, Z. 2005. Nonlinear analyses of wrinkles in a film bonded to a compliant substrate, *J. Mech. Phys. Solids*, **53**(9), pp. 2101–2118.

47. Huang, Z., Hong, W., and Suo, Z. 2004. Evolution of wrinkles in hard films on soft substrates, *Phys. Rev. E*, **70**(3), p. 030601.

48. Chen, X. and Hutchinson, J. W. 2004. A family of herringbone patterns in thin films, *Scripta Mater.*, **50**(6), pp. 797–801.

49. Chen, X. and Hutchinson, J. W. 2004. Herringbone buckling patterns of compressed thin films on compliant substrates, *J. Appl. Mech. Trans. ASME*, **71**(5), pp. 597–603.

50. Huang, R. and Im, S. H. 2006. Dynamics of wrinkle growth and coarsening in stressed thin films, *Phys. Rev. E*, **74**(2), p. 026214.

51. Song, Z., Ma, T., Tang, R., Cheng, Q., Wang, X., Krishnaraju, D., Panat, R., Chan, C. K., Yu, H., and Jiang, H. 2014. Origami lithium-ion batteries, *Nat. Commun.*, **5**, p. 3140.

52. Stafford, C. M., Vogt, B. D., Harrison, C., Julthongpiput, D., and Huang, R. 2006. Elastic moduli of ultrathin amorphous polymer films, *Macromolecules*, **39**(15), pp. 5095–5099.

53. Lamoureux, A., Lee, K., Shlian, M., Forrest, S. R., and Shtein, M. 2015. Dynamic kirigami structures for integrated solar tracking, *Nat. Commun.*, **6**, p. 8092.

54. Song, Z., Wang, X., Lv, C., An, Y., Liang, M., Ma, T., He, D., Zheng, Y. J., Huang, S. Q., and Yu, H. 2015. Kirigami-based stretchable lithium-ion batteries, *Sci. Rep.*, **5**, p. 10988.

55. Wu, C., Wang, X., Lin, L., Guo, H., and Wang, Z. L. 2016. Paper-based triboelectric nanogenerators made of stretchable interlocking kirigami patterns, *ACS Nano*, **10**(4), pp. 4652–4659.

56. Zhang, Y., Zhang, F., Yan, Z., Ma, Q., Li, X., Huang, Y., and Rogers, J. A. 2017. Printing, folding and assembly methods for forming 3D mesostructures in advanced materials, *Nat. Rev. Mater.*, **2**, p. 17019.

57. Schroeder, T. B., Guha, A., Lamoureux, A., VanRenterghem, G., Sept, D., Shtein, M., Yang, J., and Mayer, M. 2017. An electric-eel-inspired soft power source from stacked hydrogels, *Nature*, **552**(7684), p. 214.

58. Rafsanjani, A. and Bertoldi, K. 2017. Buckling-induced kirigami, *Phys. Rev. Lett.*, **118**(8), p. 084301.

59. Bertoldi, K., Vitelli, V., Christensen, J., and van Hecke, M. 2017. Flexible mechanical metamaterials, *Nat. Rev. Mater.*, **2**(11), p. 17066.

60. Lv, C., Krishnaraju, D., Konjevod, G., Yu, H., and Jiang, H. 2014. Origami based mechanical metamaterials, *Sci. Rep.*, **4**, p. 5979.

61. Yan, Z., Zhang, F., Liu, F., Han, M., Ou, D., Liu, Y., Lin, Q., Guo, X., Fu, H., and Xie, Z. 2016. Mechanical assembly of complex, 3D mesostructures from releasable multilayers of advanced materials, *Sci. Adv.*, **2**(9), pp. e1601014–e1601014.

62. Xu, S., Yan, Z., Jang, K. I., Huang, W., Fu, H., Kim, J., Wei, Z., Flavin, M., Mccracken, J. M., and Wang, R. 2015. Assembly of micro/nanomaterials into complex, three-dimensional architectures by compressive buckling, *Science*, **347**(6218), pp. 154–159.

63. Liu, Y., Yan, Z., Lin, Q., Guo, X., Han, M., Nan, K., Hwang, K. C., Huang, Y., Zhang, Y., and Rogers, J. A. 2016. Guided formation of 3D helical mesostructures by mechanical buckling: Analytical modeling and experimental validation, *Adv. Funct. Mater.*, **26**(17), p. 2909.

64. Yan, Z., Han, M. D., Yang, Y. Y., Nan, K. W., Luan, H. W., Luo, Y. Y., Zhang, Y. H., Huang, Y. G., and Rogers, J. A. 2017. Deterministic assembly of 3D mesostructures in advanced materials via compressive buckling: A short review of recent progress, *Extreme Mech. Lett.*, **11**, pp. 96–104.

65. Gong, S. and Cheng, W. 2017. Toward soft skin-like wearable and implantable energy devices, *Adv. Energy Mater.*, **7**(23), p. 1700648.

66. Liu, W., Song, M.-S., Kong, B., and Cui, Y. 2017. Flexible and stretchable energy storage: Recent advances and future perspectives, *Adv. Mater.*, **29**(1), p. 1603436.

Index